Ionic Liquids: Modern Concepts

Ionic Liquids: Modern Concepts

Edited by **Pablo Rickman**

NY RESEARCH
P R E S S
New York

Published by NY Research Press,
23 West, 55th Street, Suite 816,
New York, NY 10019, USA
www.nyresearchpress.com

Ionic Liquids: Modern Concepts
Edited by Pablo Rickman

International Standard Book Number: 978-1-63238-305-1 (Hardback)

The publisher's policy is to use permanent paper from mills that operate a sustainable forestry policy. Furthermore, the publisher ensures that the text paper and cover boards used have met acceptable environmental accreditation standards.

Trademark Notice: Registered trademark of products or corporate names are used only for explanation and identification without intent to infringe.

Printed in the United States of America.

Contents

Preface

This book presents updated scientific developments in theoretical, specific and applied domains of ionic liquids. Ionic liquids studies are a rapidly evolving field in physical chemistry, material science, technology and engineering. Use of ionic liquids for research in biology and natural resource domain has received significant attention. This book encompasses the latest developments in ionic liquid research on organic reactions and biological applications, and materials and processing as its relevant fields. It is a valuable source of information for scientists, engineers and academicians engaged in the research related to ionic liquids.

The researches compiled throughout the book are authentic and of high quality, combining several disciplines and from very diverse regions from around the world. Drawing on the contributions of many researchers from diverse countries, the book's objective is to provide the readers with the latest achievements in the area of research. This book will surely be a source of knowledge to all interested and researching the field.

In the end, I would like to express my deep sense of gratitude to all the authors for meeting the set deadlines in completing and submitting their research chapters. I would also like to thank the publisher for the support offered to us throughout the course of the book. Finally, I extend my sincere thanks to my family for being a constant source of inspiration and encouragement.

Editor

Organic Reactions and Biological Applications

New Brønsted Ionic Liquids: Synthesis, Thermodynamics and Catalytic Activity in Aldol Condensation Reactions

I. Cota, R. Gonzalez-Olmos, M. Iglesias and
F. Medina

Additional information is available at the end of the chapter

1. Introduction

It is a continuous challenge to find new catalysts able to perform with good activities and selectivity condensation reactions for the synthesis of pharmaceutical and fine chemicals. In the last years room temperature ionic liquids (ILs) have received a lot of interest as environmental friendly or "green" alternatives to conventional molecular solvents. They differ from molecular solvents by their unique ionic character and their "structure and organization" which can lead to specific effects [1].

Room-temperature ILs have been used as clean solvents and catalysts for green chemistry, stabilizing agents for the catalysts or intermediates, electrolytes for batteries, in photochemistry and electrosynthesis etc [2-6]. Their success as environmental benign solvents or catalysts is described in numerous reactions [7-11], such as Diels-Alder reactions [12, 13], the Friedel-Crafts reaction [14-17], esterification [18-20], cracking rections [21], and so on. The link between ionic ILs and green chemistry is related to the solvent properties of ILs. Some of the properties that make ILs attractive media for catalysis are: they have no significant vapour pressure and thus create no volatile organic pollution during manipulation; ILs have good chemical and thermal stability, most ILs having liquid ranges for more than 300⁰C; they are immiscible with some organic solvents and therefore can be used in two-phase systems; ILs polarity can be adjusted by a suitable choice of cation/anion; they are able to dissolve a wide range of organic, inorganic and organometallic compounds; ILs are often composed of weakly coordinating anions and therefore have the potential to be highly polar.

The number of ILs has increased exponentially in the recent years. Many of them are based on the imidazolium cation and in a lesser proportion, alkyl pyridiniums and trialkylamines (Scheme 1). By changing the anion or the alkyl chain of the cation, a wide variety of ILs may be designed for specific applications. They can be of hydrophobic or hydrophilic nature depending on the chemical structures involved.

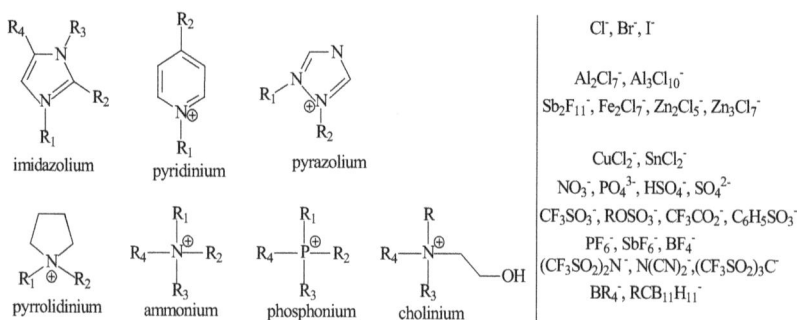

Structures of cations: imidazolium, pyridinium, pyrazolium, pyrrolidinium, ammonium, phosphonium, cholinium.

Anions:
Cl^-, Br^-, I^-

$Al_2Cl_7^-, Al_3Cl_{10}^-$
$Sb_2F_{11}^-, Fe_2Cl_7^-, Zn_2Cl_5^-, Zn_3Cl_7^-$

$CuCl_2^-, SnCl_2^-$
$NO_3^-, PO_4^{3-}, HSO_4^-, SO_4^{2-}$
$CF_3SO_3^-, ROSO_3^-, CF_3CO_2^-, C_6H_5SO_3^-$
PF_6^-, SbF_6^-, BF_4^-
$(CF_3SO_2)_2N^-, N(CN)_2^-, (CF_3SO_2)_3C^-$
$BR_4^-, RCB_{11}H_{11}^-$

Scheme 1. Main cations and anions described in literature [1].

ILs can be divided into two broad categories: aprotic ionic liquids (AILs) and protic ionic liquids (PILs).

AILs largely dominate the open literature due to their relative inertness to organometallic compounds and their potential of applications, particularly in catalysis. They are synthesized by transferring an alkyl group to the basic nitrogen site through S_N2 reactions [1].

PILs are formed through proton transfer from a Brønsted acid to a Brønsted base. Recently there has been an increasing interest in PILs due to their greater potential as environmental friendly solvents and promising applications. Moreover, they present the advantage of being cost-effective and easily prepared as their formation does not involve the formation of residual by-products. A specific feature of the PILs is that they are capable of developing a certain hydrogen bonding potency, including proton acceptance and proton donation and they are highly tolerant to hydroxylic media [22-23].

The application of new policies on terms of environment, health and safety deals towards minimizing or substituting organic volatile solvents by green alternatives, placing a renewed emphasis on research and development of lesser harmful compounds as ILs. On the other hand, recently the interest in the use of PILs to tailor the water properties for cleaning applications in processes of minimization of CO_2/SO_2 emissions has increased [24-26].

In the last years numerous studies report the use of ILs as selective catalysts for different reactions, like aldol condensation reactions where several ILs have been successfully applied as homogeneous and heterogeneous catalysts [27-30]. Abelló et al. [28] described the use of choline hydroxide as basic catalyst for aldol condensation reactions between several ketones and aldehydes. Better conversions and selectivities were obtained when compared to other well-known catalysts, such as rehydrated hydrotalcites, MgO and NaOH. In addition, higher performance was obtained when choline was immobilized on MgO.

Zhu et al. [27] described the use of 1,1,3,3-tetramethylguanidine lactate ([TMG] [Lac]) as recyclable catalyst for direct aldol condensation reactions at room temperature without any solvent. It was demonstrated that for each reaction only the aldol adduct was produced when the molar ratio of the IL and substrate was smaller than 1. Moreover, after the reaction the IL was easily recovered and recycled without considerably decrease of activity.

Kryshtal et al. [29] described the application of tetraalkylammonium and 1,3-dialkylimidazolium perfluoro-borates and perfluoro-phosphates as recoverable phase-transfer catalysts in multiphase reactions of CH-acids, in particular in solid base-promoted cross-aldol condensations. The catalysts retained their catalytic activity over several reaction cycles.

In the study of Lombardo et al. [30] two onium ion-tagged prolines, imidazolium bis (trifluoromethylsulfonyl)imide-substituted proline and butyldimethylammonium bis (trifluoromethylsulfonyl) imide-substituted proline, were synthesized and their catalytic activity in the direct asymmetric aldol condensation was studied. The catalytic protocol developed by this group makes use of a 6-fold lower amount of catalyst with respect to the preceding reports [31, 32] and affords greater chemical yields and higher enantioselectivity.

The main objective of this chapter is to develop and study the applications of a new family of ILs based on substituted amine cations of the form RNH_3^+ combined with organic anions of the form $R'COO^-$ (being of different nature R and R'). The variations in the anion alkyl chain, in conjunction with the cations, lead to a large matrix of materials.

This kind of compounds show interesting properties for industrial use of ILs: low cost of preparation, simple synthesis and purification methods. Moreover, the very low toxicity and the degradability of this kind of ILs have been verified. Thus, sustainable processes can be originated from their use.

Recently, many studies dealing with the application of ILs in organic synthesis and catalysis have been published, pointing out the vast interest in this type of compounds [33-36]. With these facts in mind, we studied their catalytic potential for two condensation reactions of carbonyl compounds. The products obtained from these reactions are applied in pharmacological, flavor and fragrance industry.

2. Experimental

2.1. Preparation of ILs and supported ILs

The ILs synthesized in this work are: 2-hydroxy ethylammonium formate (2-HEAF), 2-hydroxy ethylammonium acetate (2-HEAA), 2-hydroxy ethylammonium propionate (2-HEAP), 2-hydroxi ethylammonium butanoate (2-HEAB), 2-hydroxy ethylammonium isobutanoate (2-HEAiB) and 2-hydroxi ethylammonium pentanoate (2-HEAPE).

The amine (Merck Synthesis, better than 99%) was placed in a three necked flask all-made-in-glass equipped with a reflux condenser, a PT-100 temperature sensor for controlling temperature and a dropping funnel. The flask was mounted in a thermal bath. A slight heating is necessary for increasing miscibility between reactants and then allow reaction. The organic acid (Merck Synthesis, better than 99%) was added drop wise to the flask under stirring with a magnetic bar. Stirring was continued for 24 h at laboratory temperature, in order to obtain a final viscous liquid. Lower viscosity was observed in the final product by decreasing molecular weight of reactants. No solid crystals or precipitation was noticed when the liquid sample was purified or stored at freeze temperature for a few months after synthesis. The reaction is a simple acid–base neutralization creating the formiate, acetate, propionate, butanoate, isobutanoate or pentanoate salt of ethanolamine that in a general form should be expressed as follows:

$$(HOCH_2CH_2)NH_2 + HOOCR \rightarrow (HOCH_2CH_2)NH_3^+(^-OOCR) \tag{1}$$

For example, when formic acid is used this equation shows the chemical reaction for the reactants ethanolamine + formic acid, with 2-HEAF as neutralization product.

Because these chemical reactions are highly exothermic, an adequate control of temperature is essential throughout the chemical reaction; otherwise heat evolution may produce the dehydration of the salt to the corresponding amide, as in the case for nylon salts (salts of diamines with dicarboxy acids).

As observed in our laboratory during IL synthesis, dehydration begins around 423.15 K for the lightest ILs. The color varied in each case from transparent to dark yellow when the reaction process and purification (strong agitation and slight heating for the vaporization of residual non-reacted acid for at least for 24 h) were completed.

There was no detectable decomposition for the ILs studied here when left for over 12 months at laboratory temperature. Less than 1% amide was detected after this period of time. On the basis of these results it appears obvious that the probability of amide formation is low for this kind of structures.

In order to obtain the supported ILs, 1 g of IL was dissolved in 7 ml of ethanol and after stirring at room temperature for 30 min, 1 g of alanine (Fluka, better than 99%) was added. The mixture was stirred for 2 h and then heated at 348 K under vacuum to remove ethanol. The supported ILs thus obtained were labelled hereafter as a-ILs.

2.2. Spectroscopy test

FT-IR spectrum was taken by a Jasco FT/IR 680 plus model IR spectrometer, using a NaCl disk.

2.3. Physical properties equipment

During the course of the experiments, the purity of ILs was monitored by different physical properties measurements. The pure ILs were stored in sun light protected form, constant humidity and low temperature. Usual manipulation and purification in our experimental work was applied [22].

The densities and ultrasonic velocities of pure components were measured with an Anton Paar DSA-5000 vibrational tube densimeter and sound analyzer, with a resolution of 10^{-5} g cm^{-3} and 1 m s^{-1}. Apparatus calibration was performed periodically in accordance with provider's instructions using a double reference (millipore quality water and ambient air at each temperature). Accuracy in the temperature of measurement was better than $\pm 10^{-2}$ K by means of a temperature control device that apply the Peltier principle to maintain isothermal conditions during the measurements.

The ion conductivity was measured by a Jenway Model 4150 Conductivity/TDS Meter with resolution of 0.01μS to 1 mS and accuracy of ±0.5% at the range temperature. The accuracy of temperature into the measurement cell was ±0.5 °C.

2.4. Catalytic studies

The studied reactions were the condensation between citral and acetone and between benzaldehyde and acetone. The reactions were performed in liquid phase using a 100 mL batch reactor equipped with a condenser system. To a stirred solution of substrate and ketone (molar ratio ketone/substrate = 4.4) was added 1 g of IL, and the flask was maintained at 333 K using an oil bath. Samples were taken at regular time periods and analyzed by gas chromatography using a flame ionization detector and an AG Ultra 2 column (15 m x 0.32 mm x 0.25 μm). Tetradecane was used as the internal standard. Reagents were purchase from Aldrich and used without further purification.

In order to separate the ILs from the reaction mixture, at the end of the reaction 6 mL of H_2O were added. The mixture was stirred for 2 h and then left 15 h to repose. Two phases were separated: the organic phase which contains the reaction products and the aqueous phase which contains the IL. In order to separate the IL, the aqueous phase was heated up to 393 K under vacuum.

3. Results and discussion

As Figure 1 shows, the broad band in the 3500-2400 cm^{-1} range exhibits characteristic ammonium structure for all the neutralization products. The OH stretching vibration is embedded in this band. The broad band centered at 1600 cm^{-1} is a combined band of the carbonyl stretching and N-H plane bending vibrations. FT-IR results clearly demonstrate the IL characteristics of compounds synthesized in this work.

Due to space considerations, we will present the thermodynamic properties only for two of the studied ILs: 2-HEAF and 2-HEAPE.

The molar mass and experimental results at standard condition for 2-HEAF and 2-HEAPE are shown in Table 1.

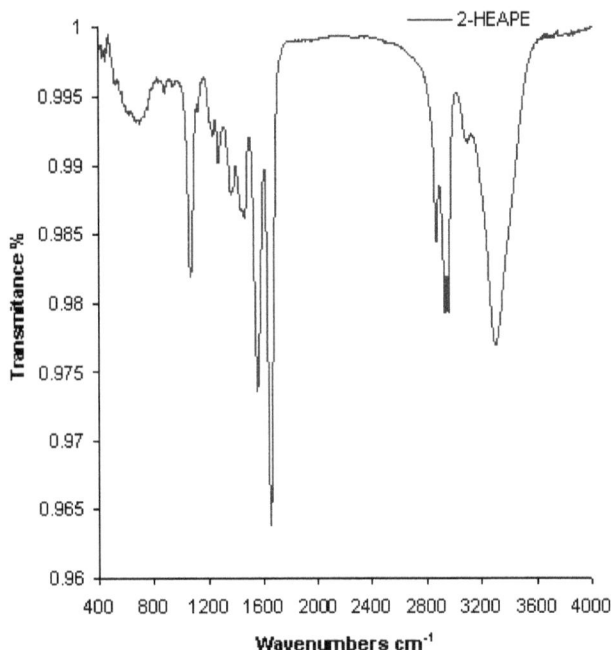

Figure 1. FT-IR spectrum for 2-HEAPE.

IL	Molecular Weight (g•mol⁻¹)	Exp. Density (g•cm⁻³)	Exp. Ultrasonic Velocity (ms⁻¹)	Exp. Conductivity (µS•cm⁻¹)
2-HEAF	107.11	1.176489	1709.00	4197.6
2-HEAPE	163.21	1.045479	1591.59	239.6

ªOther experimental data for comparison are not available from the literature.

Table 1. Experimental data for pure ionic liquids at 298.15 K and other relevant information[a]

The densities, ultrasonic velocities and isobaric expansibility of 2-HEAF and 2-HEAPE are given in Table 2, and the ionic conductivities are given in Table 3. From the results obtained it can be observed that an increase in temperature diminishes the interaction among ions, lower values of density and ultrasonic velocity being gathered for rising temperatures in each case.

2-hydroxy ethylammonium formate (2-HEAF)									
T	ρ	u	κ_S	$10^3 \cdot \alpha$	T	ρ	u	κ_S	$10^3 \cdot \alpha$
(K)	(gcm^{-3})	(ms^{-1})	(TPa^{-1})	(K^{-1})	(K)	(gcm^{-3})	(ms^{-1})	(TPa^{-1})	(K^{-1})
338.15	1.148091	1613.59	334.53	0.6188	327.16	1.155890	1639.38	321.90	0.6148
337.90	1.148254	1614.14	334.26	0.6187	326.91	1.156069	1639.97	321.62	0.6147
337.66	1.148433	1614.71	333.97	0.6186	326.66	1.156247	1640.57	321.34	0.6146
337.40	1.148608	1615.30	333.67	0.6185	326.41	1.156426	1641.16	321.06	0.6145
337.15	1.148785	1615.87	333.39	0.6184	326.16	1.156603	1641.75	320.78	0.6144
336.91	1.148963	1616.46	333.09	0.6183	325.91	1.156780	1642.34	320.50	0.6143
336.66	1.149139	1617.04	332.80	0.6182	325.65	1.156957	1642.94	320.21	0.6142
336.41	1.149316	1617.63	332.51	0.6182	325.40	1.157136	1643.53	319.93	0.6141
336.16	1.149494	1618.22	332.21	0.6181	325.16	1.157314	1644.12	319.66	0.6140
335.90	1.149669	1618.81	331.92	0.6180	324.90	1.157490	1644.72	319.37	0.6139
335.65	1.149848	1619.38	331.64	0.6179	324.65	1.157669	1645.32	319.09	0.6138
335.40	1.150027	1619.96	331.35	0.6178	324.40	1.157846	1645.91	318.81	0.6137
335.16	1.150205	1620.55	331.05	0.6177	324.15	1.158023	1646.50	318.54	0.6136
334.90	1.150384	1621.13	330.77	0.6176	323.90	1.158201	1647.09	318.26	0.6135
334.66	1.150560	1621.71	330.48	0.6175	323.65	1.158378	1647.68	317.98	0.6134
334.40	1.150740	1622.30	330.19	0.6174	323.40	1.158556	1648.28	317.70	0.6133
334.16	1.150916	1622.89	329.90	0.6173	323.15	1.158734	1648.90	317.42	0.6132
333.90	1.151094	1623.48	329.61	0.6173	322.90	1.158910	1649.47	317.15	0.6131
333.65	1.151271	1624.06	329.32	0.6172	322.66	1.159088	1650.06	316.87	0.6130
333.41	1.151449	1624.64	329.03	0.6171	322.41	1.159265	1650.66	316.59	0.6129
333.16	1.151625	1625.23	328.75	0.6170	322.16	1.159442	1651.25	316.32	0.6128
332.90	1.151804	1625.82	328.46	0.6169	321.91	1.159620	1651.85	316.04	0.6127
332.65	1.151981	1626.41	328.17	0.6168	321.65	1.159797	1652.43	315.77	0.6126
332.41	1.152159	1626.99	327.88	0.6167	321.40	1.159976	1653.03	315.49	0.6125
332.15	1.152338	1627.58	327.59	0.6166	321.15	1.160154	1653.63	315.22	0.6124
331.90	1.152514	1628.16	327.31	0.6165	320.91	1.160330	1654.22	314.94	0.6124
331.65	1.152694	1628.75	327.02	0.6164	320.66	1.160509	1654.81	314.67	0.6123
331.40	1.152871	1629.34	326.74	0.6163	320.40	1.160688	1655.41	314.39	0.6122
331.16	1.153048	1629.93	326.45	0.6162	320.15	1.160863	1656.01	314.12	0.6121
330.90	1.153225	1630.52	326.16	0.6162	319.90	1.161042	1656.60	313.85	0.6120
330.65	1.153405	1631.11	325.88	0.6161	319.65	1.161218	1657.19	313.58	0.6119
330.41	1.153582	1631.69	325.59	0.6160	319.40	1.161398	1657.79	313.30	0.6118

330.15	1.153761	1632.29	325.30	0.6159	319.15	1.161574	1658.39	313.03	0.6117
329.90	1.153939	1632.88	325.02	0.6158	318.91	1.161750	1658.98	312.76	0.6116
329.65	1.154114	1633.47	324.73	0.6157	318.65	1.161930	1659.58	312.48	0.6115
329.41	1.154294	1634.06	324.45	0.6156	318.40	1.162110	1660.18	312.21	0.6114
329.15	1.154469	1634.65	324.17	0.6155	318.16	1.162286	1660.78	311.93	0.6113
328.91	1.154648	1635.24	323.88	0.6154	317.90	1.162462	1661.37	311.67	0.6112
328.65	1.154826	1635.84	323.59	0.6153	317.65	1.162643	1661.97	311.39	0.6111
328.40	1.155003	1636.43	323.31	0.6152	317.41	1.162820	1662.56	311.12	0.6110
328.15	1.155181	1637.02	323.03	0.6151	317.15	1.162998	1663.16	310.85	0.6109
327.90	1.155360	1637.61	322.75	0.6150	316.91	1.163174	1663.75	310.58	0.6108
327.66	1.155535	1638.20	322.47	0.6149	316.65	1.163352	1664.35	310.31	0.6107
316.15	1.163706	1665.55	309.77	0.6105	303.90	1.172408	1695.01	296.88	0.6054
315.90	1.163885	1666.15	309.50	0.6104	303.65	1.172587	1695.62	296.62	0.6053
315.65	1.164062	1666.74	309.23	0.6103	303.40	1.172764	1696.23	296.36	0.6052
315.40	1.164240	1667.34	308.96	0.6102	303.15	1.172937	1696.81	296.11	0.6051
315.15	1.164417	1667.94	308.70	0.6101	302.90	1.173120	1697.43	295.85	0.6050
314.90	1.164597	1668.54	308.43	0.6100	302.65	1.173295	1698.04	295.59	0.6049
314.65	1.164774	1669.14	308.16	0.6099	302.40	1.173473	1698.64	295.34	0.6048
314.40	1.164951	1669.73	307.89	0.6098	302.15	1.173648	1699.25	295.09	0.6047
314.15	1.165128	1670.33	307.63	0.6097	301.90	1.173826	1699.86	294.83	0.6046
313.90	1.165305	1670.94	307.35	0.6096	301.65	1.174003	1700.47	294.57	0.6045
313.65	1.165485	1671.54	307.09	0.6095	301.40	1.174180	1701.07	294.32	0.6043
313.40	1.165661	1672.13	306.82	0.6094	301.15	1.174361	1701.68	294.06	0.6042
313.15	1.165839	1672.72	306.56	0.6093	300.90	1.174535	1702.29	293.81	0.6041
312.90	1.166018	1673.34	306.29	0.6092	300.65	1.174714	1702.90	293.56	0.6040
312.65	1.166194	1673.94	306.02	0.6091	300.40	1.174891	1703.50	293.30	0.6039
312.40	1.166372	1674.54	305.75	0.6090	300.15	1.175070	1704.12	293.05	0.6038
312.15	1.166549	1675.14	305.49	0.6089	299.90	1.175247	1704.73	292.79	0.6037
311.90	1.166726	1675.74	305.22	0.6088	299.65	1.175425	1705.33	292.54	0.6036
311.65	1.166903	1676.34	304.96	0.6086	299.40	1.175602	1705.95	292.29	0.6035
311.40	1.167085	1676.95	304.69	0.6085	299.15	1.175780	1706.55	292.04	0.6034
311.15	1.167260	1677.55	304.43	0.6084	298.90	1.175955	1707.16	291.78	0.6033
310.90	1.167437	1678.14	304.17	0.6083	298.65	1.176133	1707.77	291.53	0.6032
310.65	1.167617	1678.74	303.90	0.6082	298.40	1.176311	1708.39	291.28	0.6030

310.40	1.167794	1679.35	303.63	0.6081	298.15	1.176489	1709.00	291.02	0.6029
310.15	1.167970	1679.94	303.38	0.6080	297.90	1.176666	1709.61	290.77	0.6028
309.90	1.168149	1680.55	303.11	0.6079	297.65	1.176842	1710.22	290.52	0.6027
309.65	1.168325	1681.15	302.85	0.6078	297.40	1.177019	1710.84	290.27	0.6026
309.40	1.168502	1681.75	302.59	0.6077	297.15	1.177201	1711.45	290.02	0.6025
309.15	1.168680	1682.35	302.32	0.6076	296.90	1.177373	1712.06	289.77	0.6024
308.90	1.168859	1682.96	302.06	0.6075	296.65	1.177553	1712.67	289.52	0.6023
308.65	1.169036	1683.55	301.80	0.6074	296.40	1.177729	1713.28	289.27	0.6022
308.40	1.169213	1684.16	301.54	0.6073	296.15	1.177905	1713.90	289.01	0.6021
308.15	1.169391	1684.76	301.28	0.6072	295.90	1.178085	1714.52	288.76	0.6019
307.90	1.169567	1685.36	301.02	0.6071	295.65	1.178265	1715.13	288.51	0.6018
307.65	1.169742	1685.96	300.76	0.6070	295.40	1.178438	1715.75	288.26	0.6017
307.40	1.169922	1686.56	300.50	0.6069	295.15	1.178617	1716.36	288.01	0.6016
307.15	1.170102	1687.17	300.23	0.6068	294.90	1.178798	1716.97	287.76	0.6015
306.90	1.170276	1687.77	299.98	0.6067	294.65	1.178971	1717.58	287.52	0.6014
306.65	1.170454	1688.37	299.72	0.6066	294.40	1.179148	1718.20	287.27	0.6013
306.40	1.170632	1688.98	299.45	0.6065	294.15	1.179325	1718.81	287.02	0.6012
306.15	1.170810	1689.58	299.20	0.6064	293.90	1.179505	1719.42	286.77	0.6011
305.90	1.170986	1690.18	298.94	0.6063	293.65	1.179682	1720.04	286.52	0.6009
305.65	1.171165	1690.79	298.68	0.6062	293.40	1.179858	1720.66	286.27	0.6008
305.40	1.171343	1691.39	298.42	0.6060	293.15	1.180037	1721.27	286.03	0.6007
305.15	1.171518	1691.99	298.16	0.6059	292.90	1.180210	1721.88	285.78	0.6006
304.90	1.171699	1692.60	297.90	0.6058	292.65	1.180390	1722.50	285.53	0.6005
304.40	1.172053	1693.80	297.39	0.6056	292.15	1.180744	1723.72	285.04	0.6003
304.15	1.172230	1694.41	297.13	0.6055	291.90	1.180923	1724.34	284.80	0.6002
291.65	1.181104	1724.95	284.55	0.6000	279.40	1.189760	1755.38	272.77	0.5944
291.40	1.181278	1725.57	284.30	0.5999	279.15	1.189935	1756.03	272.53	0.5943
291.15	1.181453	1726.18	284.06	0.5998	278.90	1.190108	1756.62	272.31	0.5941
290.90	1.181631	1726.80	283.81	0.5997	278.65	1.190288	1757.23	272.08	0.5940
290.65	1.181809	1727.43	283.56	0.5996	278.40	1.190464	1757.88	271.83	0.5939
290.40	1.181990	1728.05	283.32	0.5995	278.15	1.190632	1758.50	271.60	0.5938
290.15	1.182162	1728.67	283.07	0.5994					
289.90	1.182339	1729.29	282.83	0.5993					
289.65	1.182515	1729.91	282.58	0.5991					

289.39	1.182700	1730.84	282.24	0.5990
289.15	1.182877	1731.59	281.95	0.5989
288.89	1.183052	1732.13	281.73	0.5988
288.64	1.183228	1732.78	281.48	0.5987
288.39	1.183407	1733.34	281.25	0.5986
288.15	1.183574	1733.91	281.03	0.5985
287.90	1.183753	1734.51	280.79	0.5983
287.64	1.183941	1735.04	280.58	0.5982
287.40	1.184107	1735.67	280.33	0.5981
287.15	1.184289	1736.27	280.10	0.5980
286.90	1.184462	1736.82	279.88	0.5979
286.65	1.184637	1737.45	279.63	0.5978
286.40	1.184815	1738.07	279.39	0.5977
286.15	1.184986	1738.68	279.16	0.5975
285.90	1.185168	1739.24	278.93	0.5974
285.65	1.185344	1739.86	278.69	0.5973
285.40	1.185519	1740.47	278.46	0.5972
285.15	1.185700	1741.08	278.22	0.5971
284.90	1.185886	1741.82	277.94	0.5970
284.64	1.186059	1742.42	277.71	0.5968
284.40	1.186228	1742.99	277.49	0.5967
284.15	1.186403	1743.61	277.25	0.5966
283.90	1.186582	1744.21	277.02	0.5965
283.65	1.186756	1744.84	276.78	0.5964
283.40	1.186933	1745.46	276.54	0.5963
283.15	1.187110	1746.08	276.30	0.5961
282.90	1.187288	1746.70	276.06	0.5960
282.65	1.187467	1747.32	275.82	0.5959
282.40	1.187641	1747.95	275.59	0.5958
282.15	1.187817	1748.57	275.35	0.5957
281.90	1.187991	1749.20	275.11	0.5956
281.65	1.188172	1749.83	274.87	0.5954
281.40	1.188344	1750.39	274.66	0.5953
281.15	1.188523	1751.00	274.42	0.5952

280.90	1.188699	1751.60	274.19	0.5951
280.40	1.189050	1752.86	273.72	0.5948
280.15	1.189231	1753.49	273.48	0.5947
279.90	1.189407	1754.12	273.24	0.5946
279.65	1.189580	1754.75	273.01	0.5945

2-hydroxy ethylammonium pentanoate (2-HEAPE)									
T	ρ	u	κ_s	$10^3 \cdot \alpha$	T	ρ	u	κ_s	$10^3 \cdot \alpha$
(K)	(gcm^{-3})	(ms^{-1})	(TPa^{-1})	(K^{-1})	(K)	(gcm^{-3})	(ms^{-1})	(TPa^{-1})	(K^{-1})
338.15	1.020672	1468.15	454.54	-3.6736	307.90	1.039467	1558.18	396.24	-3.8607
337.90	1.020820	1468.77	454.09	-3.6729	307.65	1.039618	1558.99	395.77	-3.8646
337.65	1.020969	1469.46	453.60	-3.6723	307.40	1.039772	1559.78	395.31	-3.8684
337.40	1.021126	1470.18	453.08	-3.6716	307.15	1.039925	1560.61	394.83	-3.8723
337.15	1.021280	1470.87	452.59	-3.6710	306.90	1.040077	1561.44	394.35	-3.8763
336.90	1.021436	1471.58	452.09	-3.6705	306.65	1.040230	1562.25	393.89	-3.8803
336.65	1.021593	1472.29	451.58	-3.6700	306.40	1.040384	1563.08	393.41	-3.8843
336.40	1.021745	1473.00	451.08	-3.6695	306.15	1.040533	1563.89	392.94	-3.8883
336.15	1.021898	1473.73	450.56	-3.6690	305.90	1.040687	1564.73	392.47	-3.8924
335.65	1.022205	1475.12	449.58	-3.6683	305.40	1.040991	1566.38	391.52	-3.9007
335.40	1.022364	1475.83	449.08	-3.6679	305.15	1.041143	1567.19	391.06	-3.9050
335.15	1.022520	1476.54	448.58	-3.6677	304.90	1.041297	1568.03	390.59	-3.9092
334.90	1.022671	1477.24	448.09	-3.6674	304.65	1.041450	1568.87	390.11	-3.9135
334.65	1.022828	1477.95	447.59	-3.6672	304.40	1.041602	1569.72	389.63	-3.9178
334.40	1.022986	1478.66	447.09	-3.6670	304.15	1.041753	1570.56	389.16	-3.9222
334.15	1.023146	1479.37	446.59	-3.6669	303.90	1.041907	1571.39	388.69	-3.9266
333.90	1.023305	1480.07	446.10	-3.6668	303.65	1.042059	1572.25	388.21	-3.9310
333.65	1.023463	1480.78	445.60	-3.6667	303.40	1.042209	1573.09	387.74	-3.9355
333.40	1.023622	1481.49	445.11	-3.6667	303.15	1.042363	1573.94	387.26	-3.9400
333.15	1.023780	1482.20	444.61	-3.6667	302.90	1.042516	1574.79	386.79	-3.9445
332.90	1.023940	1482.92	444.11	-3.6667	302.65	1.042668	1575.65	386.31	-3.9491
332.65	1.024100	1483.63	443.62	-3.6668	302.40	1.042820	1576.51	385.83	-3.9537
332.40	1.024257	1484.34	443.12	-3.6669	302.15	1.042972	1577.39	385.34	-3.9584
332.15	1.024414	1485.06	442.63	-3.6671	301.90	1.043124	1578.23	384.88	-3.9631
331.90	1.024574	1485.77	442.13	-3.6673	301.65	1.043277	1579.11	384.39	-3.9678
331.65	1.024732	1486.48	441.64	-3.6675	301.40	1.043429	1579.97	383.92	-3.9726
331.40	1.024890	1487.19	441.15	-3.6678	301.15	1.043579	1580.82	383.45	-3.9774

331.15	1.025050	1487.90	440.66	-3.6681	300.90	1.043732	1581.71	382.96	-3.9822
330.90	1.025207	1488.62	440.17	-3.6684	300.65	1.043883	1582.58	382.49	-3.9871
330.65	1.025363	1489.35	439.67	-3.6688	300.40	1.044037	1583.48	382.00	-3.9920
330.40	1.025523	1490.05	439.19	-3.6692	300.15	1.044188	1584.38	381.51	-3.9970
330.15	1.025679	1490.79	438.69	-3.6697	299.90	1.044340	1585.27	381.02	-4.0020
329.90	1.025838	1491.51	438.20	-3.6702	299.65	1.044492	1586.16	380.54	-4.0070
329.65	1.025997	1492.23	437.71	-3.6707	299.40	1.044644	1587.08	380.04	-4.0121
329.15	1.026310	1493.70	436.71	-3.6719	298.90	1.044973	1588.87	379.07	-4.0224
328.90	1.026467	1494.41	436.23	-3.6726	298.65	1.045148	1589.78	378.57	-4.0275
328.65	1.026627	1495.14	435.74	-3.6732	298.40	1.045311	1590.70	378.07	-4.0328
327.90	1.027097	1497.32	434.27	-3.6755	297.65	1.045807	1593.44	376.60	-4.0487
327.65	1.027255	1498.06	433.77	-3.6764	297.40	1.045975	1594.39	376.09	-4.0540
327.40	1.027411	1498.78	433.29	-3.6772	297.15	1.046142	1595.32	375.59	-4.0594
327.15	1.027568	1499.51	432.80	-3.6781	296.90	1.046304	1596.24	375.10	-4.0649
326.90	1.027725	1500.24	432.32	-3.6791	296.65	1.046470	1597.18	374.60	-4.0704
326.65	1.027883	1500.98	431.82	-3.6801	296.40	1.046642	1598.12	374.10	-4.0759
326.40	1.028039	1501.70	431.34	-3.6811	296.15	1.046804	1599.08	373.59	-4.0814
326.15	1.028194	1502.44	430.85	-3.6821	295.90	1.046975	1600.00	373.10	-4.0870
325.90	1.028352	1503.16	430.38	-3.6832	295.65	1.047135	1600.95	372.60	-4.0927
325.65	1.028508	1503.88	429.90	-3.6844	295.40	1.047303	1601.93	372.08	-4.0983
325.40	1.028665	1504.64	429.40	-3.6855	295.15	1.047465	1602.89	371.58	-4.1041
325.15	1.028822	1505.36	428.92	-3.6868	294.90	1.047628	1603.86	371.07	-4.1098
324.90	1.028976	1506.11	428.43	-3.6880	294.65	1.047795	1604.81	370.58	-4.1156
324.65	1.029135	1506.84	427.95	-3.6893	294.40	1.047960	1605.78	370.07	-4.1214
324.40	1.029289	1507.58	427.47	-3.6906	294.15	1.048125	1606.77	369.56	-4.1273
324.15	1.029445	1508.32	426.98	-3.6920	293.90	1.048288	1607.74	369.05	-4.1332
323.90	1.029602	1509.05	426.50	-3.6934	293.65	1.048451	1608.73	368.54	-4.1391
323.65	1.029757	1509.79	426.02	-3.6948	293.40	1.048614	1609.75	368.02	-4.1451
323.15	1.030071	1511.28	425.05	-3.6978	292.90	1.048944	1611.75	366.99	-4.1571
322.90	1.030226	1512.02	424.57	-3.6993	292.65	1.049105	1612.77	366.47	-4.1632
322.65	1.030381	1512.75	424.10	-3.7009	292.40	1.049271	1613.76	365.96	-4.1693
322.40	1.030537	1513.50	423.62	-3.7025	292.15	1.049433	1614.77	365.45	-4.1755
322.15	1.030693	1514.23	423.14	-3.7042	291.90	1.049593	1615.76	364.94	-4.1817
321.90	1.030846	1514.98	422.66	-3.7059	291.65	1.049759	1616.79	364.42	-4.1879

321.65	1.031002	1515.72	422.18	-3.7076	291.40	1.049921	1617.83	363.90	-4.1942
321.40	1.031159	1516.46	421.71	-3.7094	291.15	1.050082	1618.87	363.37	-4.2005
321.15	1.031314	1517.21	421.23	-3.7112	290.90	1.050244	1619.95	362.83	-4.2068
320.90	1.031468	1517.96	420.75	-3.7130	290.65	1.050407	1620.99	362.31	-4.2132
320.65	1.031625	1518.71	420.27	-3.7149	290.40	1.050566	1622.02	361.80	-4.2196
320.40	1.031780	1519.46	419.79	-3.7168	290.15	1.050730	1623.16	361.23	-4.2261
320.15	1.031934	1520.22	419.31	-3.7188	289.90	1.050889	1624.19	360.72	-4.2326
319.90	1.032088	1520.97	418.83	-3.7208	289.65	1.051050	1625.29	360.18	-4.2391
319.65	1.032243	1521.73	418.35	-3.7228	289.40	1.051211	1626.38	359.64	-4.2457
319.40	1.032399	1522.49	417.87	-3.7249	289.15	1.051372	1627.47	359.10	-4.2523
319.15	1.032553	1523.24	417.40	-3.7270	288.90	1.051531	1628.60	358.55	-4.2590
318.90	1.032709	1524.00	416.92	-3.7292	288.65	1.051691	1629.70	358.01	-4.2656
318.65	1.032862	1524.77	416.44	-3.7313	288.40	1.051853	1630.82	357.46	-4.2724
318.40	1.033016	1525.53	415.96	-3.7336	288.15	1.052010	1631.92	356.93	-4.2791
318.15	1.033171	1526.28	415.49	-3.7358	287.90	1.052170	1633.05	356.38	-4.2859
317.90	1.033327	1527.05	415.01	-3.7381	287.65	1.052330	1634.18	355.83	-4.2928
317.40	1.033635	1528.57	414.06	-3.7428	287.15	1.052647	1636.52	354.71	-4.3065
317.15	1.033790	1529.33	413.59	-3.7452	286.90	1.052803	1637.66	354.16	-4.3135
316.65	1.034098	1530.86	412.64	-3.7502	286.40	1.053121	1639.97	353.06	-4.3275
316.40	1.034253	1531.63	412.16	-3.7527	286.15	1.053282	1641.17	352.49	-4.3345
316.15	1.034406	1532.39	411.69	-3.7552	285.90	1.053440	1642.36	351.93	-4.3416
315.90	1.034559	1533.16	411.22	-3.7578	285.65	1.053595	1643.59	351.35	-4.3488
315.40	1.034867	1534.71	410.26	-3.7631	285.15	1.053914	1645.91	350.25	-4.3632
315.15	1.035022	1535.47	409.80	-3.7659	284.90	1.054069	1647.20	349.65	-4.3704
314.90	1.035175	1536.22	409.34	-3.7686	284.65	1.054227	1648.38	349.10	-4.3777
314.65	1.035330	1536.99	408.86	-3.7714	284.40	1.054384	1649.68	348.50	-4.3850
314.40	1.035483	1537.77	408.39	-3.7742	284.15	1.054542	1650.96	347.91	-4.3924
314.15	1.035638	1538.53	407.92	-3.7771	283.90	1.054697	1652.23	347.32	-4.3998
313.90	1.035792	1539.30	407.46	-3.7800	283.65	1.054853	1653.49	346.74	-4.4072
313.65	1.035945	1540.06	406.99	-3.7829	283.40	1.055012	1654.78	346.15	-4.4147
313.40	1.036100	1540.83	406.53	-3.7859	283.15	1.055166	1656.17	345.52	-4.4222
313.15	1.036252	1541.60	406.06	-3.7889	282.90	1.055325	1657.46	344.93	-4.4297
312.90	1.036406	1542.37	405.60	-3.7919	282.65	1.055479	1658.73	344.35	-4.4373
312.65	1.036558	1543.14	405.13	-3.7950	282.40	1.055637	1660.17	343.70	-4.4449

312.40	1.036711	1543.91	404.67	-3.7981	282.15	1.055795	1661.49	343.10	-4.4526
312.15	1.036865	1544.69	404.20	-3.8013	281.90	1.055948	1662.83	342.50	-4.4603
311.90	1.037019	1545.47	403.73	-3.8045	281.65	1.056104	1664.24	341.87	-4.4680
311.65	1.037171	1546.25	403.26	-3.8077	281.40	1.056260	1665.61	341.26	-4.4758
311.40	1.037325	1547.02	402.80	-3.8110	281.15	1.056416	1667.01	340.63	-4.4836
311.15	1.037479	1547.82	402.33	-3.8143	280.90	1.056572	1668.41	340.01	-4.4914
310.65	1.037785	1549.39	401.39	-3.8210	280.40	1.056883	1671.29	338.74	-4.5072
310.40	1.037938	1550.17	400.93	-3.8244	280.15	1.057038	1672.76	338.10	-4.5152
310.15	1.038089	1550.96	400.46	-3.8279	279.90	1.057192	1674.21	337.46	-4.5232
309.90	1.038244	1551.75	400.00	-3.8314	279.65	1.057349	1675.59	336.86	-4.5312
309.65	1.038396	1552.56	399.52	-3.8349	279.40	1.057504	1677.18	336.17	-4.5393
309.40	1.038550	1553.36	399.05	-3.8385	279.15	1.057659	1678.69	335.52	-4.5474
309.15	1.038704	1554.16	398.58	-3.8421	278.90	1.057816	1680.20	334.86	-4.5556
308.90	1.038856	1554.95	398.12	-3.8458	278.65	1.057971	1681.62	334.25	-4.5637
308.65	1.039008	1555.77	397.64	-3.8494	278.40	1.058124	1683.11	333.61	-4.5720
308.40	1.039161	1556.55	397.18	-3.8532	278.15	1.058279	1684.75	332.91	-4.5802
308.15	1.039313	1557.36	396.71	-3.8569					

Table 2. Densities (ρ), ultrasonic velocity (u), isentropic compressibilities (κ_s), isobaric expansibilities (α), 278.15-338.15K

The contrary effect is observed for conductivity. At the same temperature, higher viscosity was observed when the salt was of higher molecular weight. The effect of the temperature is similar for all salts.

A frequently applied derived property for industrial mixtures is the isobaric expansibility or thermal expansion coefficient (α), expressed as the temperature dependence of density. Thermal expansion coefficients are calculated by means of $(-\Delta\rho / \rho)$ as a function of temperature and assuming that α remains constant in any thermal range. As in the case of pure chemicals it can be computed by way of the expression:

$$\alpha = -\left(\frac{\partial \ln\rho}{\partial T}\right)_{P,x} \tag{2}$$

taking into account the temperature dependence of density. The results gathered in Table 2 showed that a minimum of isobaric expansibility is obtained (in terms of negative values) at approximately the same temperature for all ILs. The smaller the size of the cation (monoethylene cation), the lower the value of isobaric expansibility was obtained.

Temperature (K)	2-HEAF	2-HEAPE
278.15	2158.20	83.6
288.15	3069.00	143.3
298.15	4197.60	239.6
308.15	5623.20	453.4
318.15	6959.70	632.6
328.15	8563.50	910.8
338.15	10404.90	1202.9

Table 3. Values of ionic conductivity ($\mu S \bullet cm^{-1}$) of the 2-HEAF and 2-HEAPE in the range 278.15 – 338.15 K

The values of ionic conductivity are gathered in Table 3. These results show an increasing trend for higher temperatures in each case. This fact may be ascribed to the increasing mobility of the ions for increased temperatures. At the same time, the ionic conductivity values decrease when molecular weight increases, thus 2-HEAPE has a lower ionic conductivity than 2-HEAF, the shortest member of this IL family [23].

The factor studied in this work is the chain length of the anion. The influence of anion residue is higher in terms of steric hindrance, due to its longer structure [2, 23]. This factor produces a higher disturbation on ion package. This fact may be observed in terms of higher values of densities and ultrasonic velocities for those salts of the lighter anion [37].

The ILs studied in this work showed interesting properties for industrial use: low cost of preparation, simple synthesis and purification methods. Moreover, the very low toxicity and the degradability have been verified [38]. Thus, sustainable processes can be originated from their use.

With this in mind, we decided to test their catalytic potential for several aldol condensation reactions with interest for fine chemicals synthesis. At industrial level aldol condensations are catalyzed by homogeneous alkaline bases (KOH or NaOH) [39,40] but with this kind of catalysts numerous disadvantages arise such as loss of catalysts due to separation difficulties, corrosion problems in the equipment and generation of large amounts of residual effluents which must be subsequently treated to minimize their environmental impact. Consequently, new technological solutions have to be developed in order to generate new and more environmental friendly processes.

The condensation reaction between citral and acetone leads to the formation of pseudoionones which are precursors in the commercial production of vitamin A. In the last years, the aldol condensation between citral and acetone has been studied by several groups employing different types of catalysts: rehydrated hydrotalcites [41], mixed oxides derived from hydrotalcites [42, 43], organic molecules [44], ionic liquids [28] etc.

Using the mixed oxides derived from hydrotalcites Climent et al. [42, 43] obtained a conversion of 83% and selectivity to pseudoionones of 82% in 1 h. Abello et al. obtained a citral conversion of 81% in only 5 min employing rehydrated hydrotalcites as catalysts [41] highlighting that Brønsted basic sites are more active than Lewis sites for aldol condensation reactions. In the study of Cota et al. [44] it was shown that 1,8-diazabicyclo[5.4.0]undec-7-ene

(DBU) which has Lewis basic properties, is inactive for aldol condensation reactions; however when it reacts with equimolar amounts of water, this molecule transforms towards a complex that shows Brønsted basic properties and becomes active giving a conversion of 89.17% and a selectivity of 89.6% in 6 h. When choline hydroxide (ionic liquid) was used as catalyst a citral conversion of 93% and selectivity of 98.2% were obtained in 1 h [28].

Among the ILs studied in this work, for citral and acetone condensation (entry 1, Table 4) the most active IL is 2-HEAA, which gives a conversion of 52%, the less active is 2-HEAiB which gives a conversion of 10%. The selectivity obtained in this reaction ranges between 49-83%. No traces of diacetone alcohol derived from the self-condensation of acetone were found but other secondary products coming from the self-condensation of citral and oligomers derived from citral are detected in small quantities in the reaction mixture.

Entry	Substrate	Ketone	Product	Catalyst	Time	Conversion	Selectivity
					(h)	(%)	(%)
1				2-HEAF	7	35	83
				2-HEAP		40	63
				2-HEAA		52	74
				2-HEAB		33	60
				2-HEAiB		10	53
				2-HEAPE		38	49
2				2-HEAF	4	94	82
				2-HEAP	3	100	86
				2-HEAA	4	99	85
				2-HEAB	2	99	85
				2-HEAiB	2	93	85
				2-HEAPE	2	98	77

Table 4. Condensation reactions catalyzed by the studied ILs.

For the production of benzylideneacetone from the aldol condensation between acetone and benzaldehyde, Cota et al. [44] obtained a conversion of 99.9% and 93.97 selectivity in 2 h. When choline hydroxide was employed as catalyst [28] the total conversion was obtained in 0.1 hours but due to the production of dibenzylidenacetone the selectivity to benzylidenacetone decreased around 77%.

When ILs presented in this study were employed for this reaction (entry 2, Table 4), in 2 h of reaction, a conversion of 99% and a selectivity of 85% are obtained when using 2-HEAB as catalyst. Good conversion was also obtained with 2-HEAiB (93%) and 2-HEAPE (98%) with selectivity of 85% and 77% respectively. The decrease in the selectivity to benzylidenacetone is due to the formation of secondary products which include products of aldolisation of benzylidenacetone, like dibenzylidenacetone and other oligomers. The other studied ILs reached the maximum conversion in 3h (2-HEAP) and 4h (2-HEAF and 2-HEAA) and provided high selectivities between 82-86%.

For the repeated runs experiments, we used 2-HEAB in the condensation reaction between acetone and benzaldehyde. The catalyst was recycled 3 times, and in all runs a very good conversion was obtained. The results are presented in Figure 2.

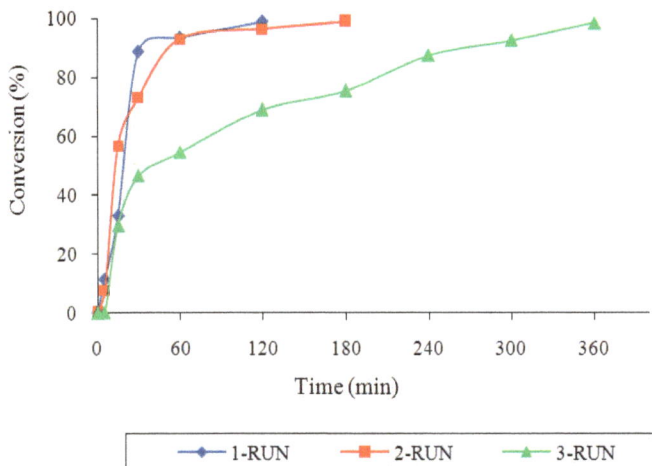

Figure 2. Repeated runs experiments using 2-HEAB in benzylideneacetone synthesis.

The loss of activity noticed in the second and third run can be attributed, on one hand to the loss of IL during the separation process and on the other hand due to the absorption of reaction products on the active sites of the catalyst. IL is partially soluble in the reaction product therefore during the separation procedure small quantities of IL can be dissolved in the organic phase and therefore lost during the separation process. This hypothesis is sustained by the evolution of the specific bands of the ILs which appear in the range 3500-2400 cm^{-1}, almost disappearing in the re-used sample as Figure 3 shows.

A weak band around 1591 cm^{-1} is present in the re-used sample accounting for the carbonyl stretching and N-H plane bending vibrations. On the other hand, deactivation of the catalyst, moreover exhibiting a dark yellow color, is probably due to the adsorption of oligomers and other secondary products on the surface of the catalyst during the reaction. This hypothesis is supported by the appearance of new bands in the re-used IL spectrum. The bands detected in the 1700-1200 cm^{-1} region corresponding to the symmetric and stretching vibrations of CH modes can be assigned to oligomeric species adsorbed on the surface. On the other hand in the 1260-700 cm^{-1} region bands which are normally weak appear and can be assigned to the C-C skeletal vibrations.

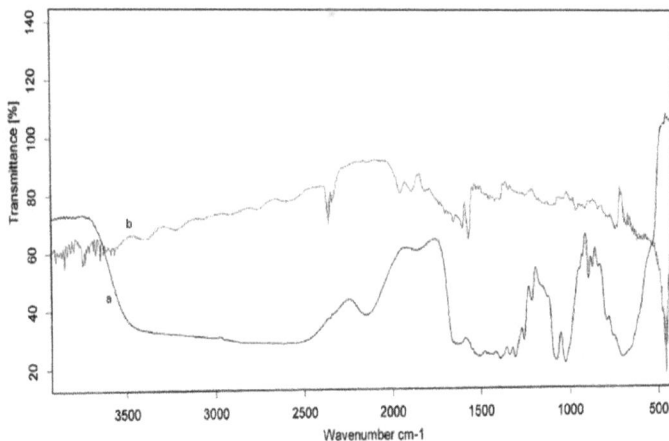

Figure 3. FT-IR spectra for (a) 2-HEAB before reaction, (b) 2-HEAB after reaction (3 consecutive runs).

In order to facilitate the recovery and re-use of the ILs we decided to immobilize them on a solid support. Immobilization and supporting of ILs can be achieved by simple impregnation, covalent linking of the cation or the anion, polymerization etc [45-47]. Compared to pure ILs, immobilized ILs facilitate the recovery and re-use of the catalyst. Previous reports describe the immobilization of ILs by adsorption or grafting onto silica surface and their use as catalysts for reactions like Friedel-Crafts acylation [45], hydrogenation [48] and hydroformilation [49]. Organic polymers [30], natural polymers [50] and zeolites [51] have been also used as supports for ILs.

For this purpose, the ILs were supported on alanine, a cheap readily available aminoacid. Their catalytic activity was tested in the same reactions as the pure ILs.

The catalytic activity results of the a-ILs for the citral-acetone condensation are presented in Table 5. After 6 h of reaction, the two isomers of citral can be converted into the corresponding pseudoionone with conversion between 30-56% except for a-HEAiB for which a conversion of 9% was obtained. The most active IL for this reaction is a-2-HEAA which provides a conversion of 56%. The selectivity obtained in this reaction ranges between 48-80%. No traces of diacetone alcohol derived from the self-condensation of acetone were found, but other secondary products coming from the self condensation of citral and oligomers derived from citral are detected in the reaction mixture. The support (entry 1) is not catalytically active.

In the condensation reaction of benzaldehyde and acetone the first step is the deprotonation of an acetone molecule to give the enolate anion whose nucleophilic attack on the C=O group of benzaldehyde leads to the β-aldol. This latter is easily dehydrated on weak acid sites and benzylidenacetone is obtained.

Entry	Catalyst	Conversion	Selectivity
		(%)	(%)
1	alanine	0	0
2	a-2-HEAF	30	61
3	a-2-HEAA	56	74
4	a-2-HEAP	49	80
5	a-2-HEAB	35	63
6	a-2-HEAiB	9	52
7	a-2-HEAPE	33	48

Table 5. Conversion at 6 h for citral-acetone condensation catalyzed by a-ILs

Entry	Catalyst	Conversion	Selectivity
		(%)	(%)
1	alanine	0	0
2	a-2-HEAF	99	83
3	a-2-HEAA	99	82
4	a-2-HEAP	99	85
5	a-2-HEAB	99	84
6	a-2-HEAiB	78	82
7	a-2-HEAPE	98	80

Table 6. Conversion at 2 h for benzaldehyde-acetone condensation catalyzed by a-ILs

In 2 hours of reaction a conversion of 98-99% is achieved for the majority of a-ILs, while a lower conversion (78%) is obtained for a-2-HEAiB (Table 6). The selectivity toward benzylidenacetone is around 80-86% due to the formation of dibenzylidenacetone as secondary product. The support, alanine (entry 1) is not active for citral acetone condensation.

It is noteworthy that, for both studied reactions, the conversions obtained with the a-ILs are in the same range as the ones obtained with free ILs (Figure 4 and 5).

The a-ILs are easily separated from the reaction mixture and reused. For the consecutive runs experiments we chose condensation between benzaldehyde and acetone as model reaction. The catalysts were recycled for 3 consecutive runs and in all runs a very good conversion was obtained. The results are presented in Figure 6.

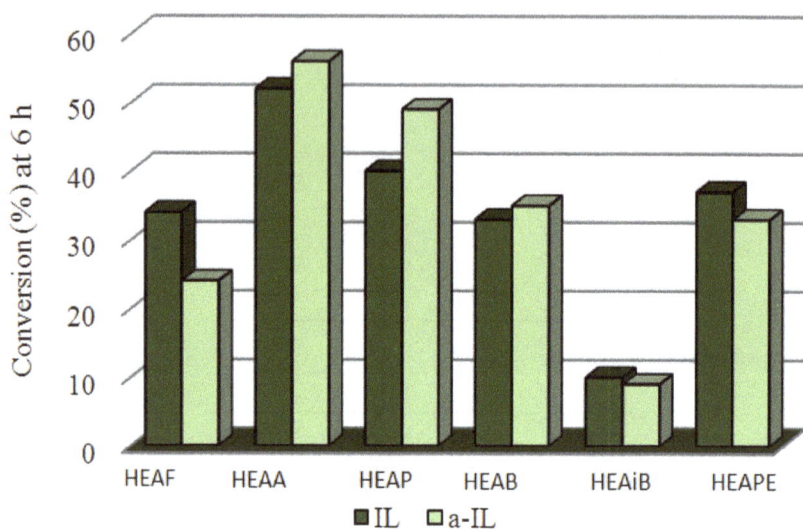

Figure 4. Conversion at 6 h for citral-acetone condensation for free ILs and a-ILs.

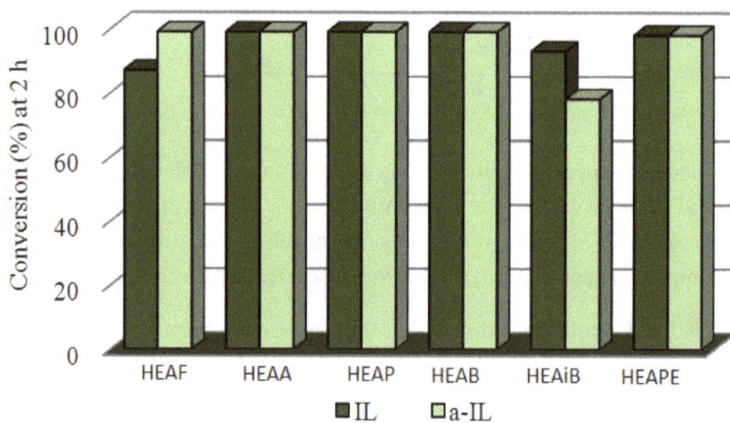

Figure 5. Conversion at 2 h for benzaldehyde-acetone condensation for free ILs and a-ILs.

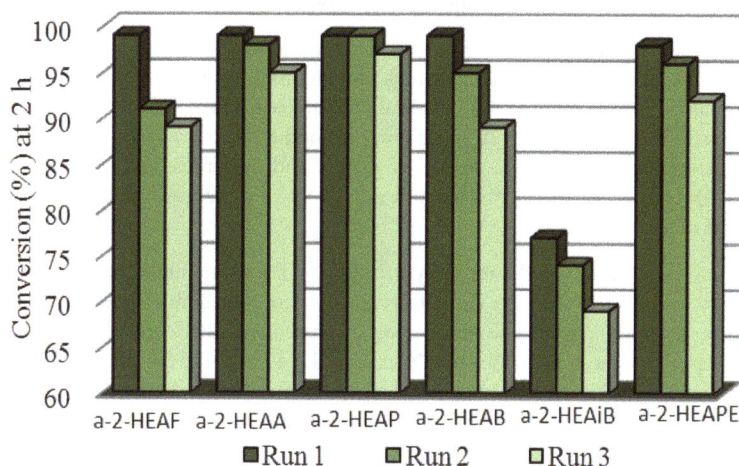

Figure 6. Consecutive runs experiments in benzaldehyde acetone condensation.

In the case of each IL, only a negligible loss of activity is detected in the second and third run which can be attributed to the possible adsorption of reactants or reaction products to the active sites of the catalyst.

From the comparison made with the aforementioned basic catalysts employed for these two aldol condensation reactions we can conclude that the ILs presented in this study are not the most active catalysts for these reactions but due to their green character and easy separation from the reaction media represent a convenient and environmental friendly alternative for the traditional homogeneous catalysts.

4. Conclusions

In this work, we present a simple and efficient synthesis protocol for protic ionic liquids and the experimental data for density, ultrasonic velocity and ionic conductivity of these liquid salts. It was found that increased temperature diminishes the interaction among ions and therefore lower values of density, ultrasonic velocity, viscosity, surface tension and refractive index are obtained for increased temperatures in each case. The contrary effect is observed for conductivity.

The influence of chain length of the anion on the physicochemical properties of the ILs has been also studied. The effect of the anion residue is higher in terms of steric hindrance, due to its longer structure. This factor produces a higher disturbance on ion package. The physicochemical data of ILs are important for both, designing cleaner technological processes and understanding the interactions in this kind of compounds

The catalytic potential of these new ILs was tested for two aldol condensation reactions with interest for fine chemistry industry. Conversions ranging from 35 to 52% and selectivities up to 83% are obtained for the condensation of citral with acetone. In the synthesis of benzilide-nacetone, conversions above 93% with selectivities around 85% are obtained. We also studied the optimization of the recovery process of the ILs and their reuse in repeated runs of experiments. The catalysts can be recycled and reused for three consecutive cycles without significant loss of activity.

In addition, in order to improve the recovery process, the ILs were immobilized on alanine, a cheap readily available aminoacid. The catalytic activity of the alanine supported ILs was tested for citral-acetone and benzaldyde-acetone condensations. It is noteworthy that, for both studied reactions, the conversions obtained with the a-ILs are in the same range as the ones obtained with free ILs; moreover the catalysts can be recycled and reused for three consecutive cycles without significant loss of activity.

The ILs studied in this work showed interesting properties for industrial use: low cost of preparation, simple synthesis and purification methods. Moreover, the very low toxicity and the degradability have been verified. Thus, sustainable processes can be originated from their use.

Acknowledgements

This work has been financed by the MEC of Spain and the Generalitat of Catalunya (ICREA ACADEMIA AWARD).

Author details

I. Cota[1], R. Gonzalez-Olmos[2], M. Iglesias[3] and F. Medina[1]

1 Departament d'Enginyeria Química, Escola Tècnica Superior d'Enginyeria Química, Universitat Rovira i Virgili, Avinguda Països Catalans 26, Campus Sescelades, 43007 Tarragona, Spain

2 Laboratory of Chemical and Environmental Engineering (LEQUiA), Institute of the Environment, University of Girona, Campus Montilivi s/n, Faculty of Sciences, E-17071 Girona, Spain

3 Departamento de Engenharia Química, Escola Politécnica, Universidade Federal da Bahia, 40210-630 Salvador-Bahia, Brazil

References

[1] Olivier-Bourbigou, H., Magna, L., & Morvan, D. (2010). Ionic liquids and catalysis: Recent progress from knowledge to applications. *Applied Catalysis A: General*, 1-56.

[2] Sheldon, R. (2001). Catalytic reactions in ionic liquids. *Chemical Communications*, 2399-2407.

[3] Bates, E., D., , Mayton, R. D., Ntai, I., & Davis, J. H. (2002). CO2 Capture by a Task-Specific Ionic Liquid. *Journal of the American Chemical Society*, 926-927.

[4] Huddleston, J. G., Willauer, H. D., Swatloski, R. P., Visser, A. E., & Rogers, R. D. (1998). Room temperature ionic liquids a novel media for'clean' liquid-liquid extraction. *Chemical Communications*, 1765-1766.

[5] Zhang, S., Zhang, Q., & Zhang, Z. C. (2004). Extractive Desulfurization and Denitrogenation of Fuels Using Ionic Liquids. *Industrial & Engineering Chemistry Research*, 614-622.

[6] Fuller, J., Carlin, R. T., & Osteryoung, R. A. (1997). The Room Temperature Ionic Liquid 1 Ethyl-3-methylimidazolium Tetrafluoroborate: Electrochemical Couples and Physical Properties. *Journal of the Electrochemical Society*, 3881-3886.

[7] Welton, T. (1999). Room-Temperature Ionic Liquids. Solvents for Synthesis and Catalysis. *Chemical Reviews*, 2071-2084.

[8] Dupont, J., de Souza, R. F., & Suarez, P. A. Z. (2002). Ionic Liquid (Molten Salt) Phase Organometallic Catalysis. *Chemical Reviews*, 3667-3692.

[9] Chauvin, Y. L., Mussmann, L., & Olivier, H. (1996). A Novel Class of Versatile Solvents for Two-Phase Catalysis: Hydrogenation, Isomerization, and Hydroformylation of Alkenes Catalyzed by Rhodium Complexes in Liquid 1, 3 Dialkylimidazolium Salts. *Angewandte Chemie. International Edition in English*, 34, 2698-2700.

[10] Brausch, N., Metlen, A., & Wasserscheid, P. (2004). New, highly acidic ionic liquid systems and their application in the carbonylation of toluene. *Chemical Communications*, 1552-1553.

[11] Jiang, T., Ma, X., Zhou, Y., Liang, S., Zhang, J., & Han, B. (2008). Solvent-free synthesis of substituted ureas from CO_2 and amines with a functional ionic liquid as the catalyst. *Green Chemistry*, 465-469.

[12] Earle, M. J., Mc Cormac, P. R., & Sheldon, K. R. (1999). Diels-Alder reactions in ionic liquids. A safe recyclable alternative to lithium perchlorate-diethyl ether mixtures. *Green Chemistry*, 1, 23-25.

[13] Doherty, S., Goodrich, P., Hardacre, C., Luo, H. K., Rooney, D. W., Seddon, K. R., & Styring, P. (2004). Marked enantioselectivity enhancements for Diels-Alder reactions in ionic liquids catalysed by platinum diphosphine complexes. *Green Chemistry*, 63-67.

[14] Wasserscheid, P., Sesing, M., & Korth, W. (2002). Hydrogensulfate and tetrakis(hydrogensulfato)borate ionic liquids: synthesis and catalytic application in highly Brønsted-acidic systems for Friedel-Crafts alkylation. *Green Chemistry*, 134-138.

[15] Adams, C. J., Earle, M. J., Roberts, G., & Seddon, K. R. (1998). Friedel-Crafts reactions in room temperature ionic liquids. *Chemical Communications*, 185-190.

[16] Song, C. E., Oh, C. R., Roh, E. J., & Choo, D. J. (2000). Cr(salen) catalysed asymmetric ring opening reactions of epoxides in room temperature ionic liquids. *Chemical Communications*, 1743-1744.

[17] Song, C. E., Shim, W. H., Roh, E. J., & Choo, J. H. (2000). Scandium (III) triflate immobilised in ionic liquids: a novel and recyclable catalytic system for Friedel-Crafts alkylation of aromatic compounds with alkenes. *Chemical Communications*, 1695-1696.

[18] Fraga-Dubreuil, J., Bourahla, K., Rahmouni, M., Bazureau, J. P., & Hamelin, J. (2002). Catalysed esterifications in room temperature ionic liquids with acidic counteranion as recyclable reaction media. *Catalysis Communications*, 185-190.

[19] Alleti, R., Oh, W. S., Perambuduru, M., Afrasiabi, Z., Simm, E., & Reddy, V. P. (2005). Gadolinium triflate immobilized in imidazolium based ionic liquids: a recyclable catalyst and green solvent for acetylation of alcohols and amines. *Green Chemistry*, 203-206.

[20] Bradaric, C. J., Downard, A., Kennedy, C., Rovertson, A. J., & Zhou, Y. H. (2003). Industrial preparation of phosphonium ionic liquids. *Green Chemistry*, 143-152.

[21] Wang, Y., Li, H., Wang, C., & Jiang, H. (2004). Ionic liquids as catalytic green solvents for cracking reactions. *Chemical Communications*, 1938-1939.

[22] Cota, I., Gonzalez-Olmos, R., Iglesias, M., & Medina, F. (2007). New Short Aliphatic Chain Ionic Liquids: Synthesis, Physical Properties, and Catalytic Activity in Aldol Condensations. *Journal of Physical Chemistry B*, 12468-21477.

[23] Iglesias, M., Torres, A., Gonzalez-Olmos, R., & Salvatierra, D. (2008). Effect of temperature on mixing thermodynamics of a new ionic liquid: {2Hydroxy ethylammonium formate (2-HEAF) + short hydroxylic solvents}. *Journal of Chemical Thermodynamics*, 119-133.

[24] Yuan, X. L., Zhang, S. J., & Lu, X. M. (2007). Hydroxyl Ammonium Ionic Liquids: Synthesis, Properties, and Solubility of SO_2. *Journal of Chemical & Engineering Data*, 596-599.

[25] Kurnia, K. A., Harris, F., Wilfred, C. D., Mutalib, M. I. A., & Murugesan, T. (2009). Thermodynamic properties of CO_2 absorption in hydroxyl ammonium ionic liquids at pressures of (100-1600) kPa. *Journal of Chemical Thermodynamics*, 1069-1073.

[26] Li, X. Y., Hou, M. Q., Zhang, Z. F., Han, B. X., Yang, G. Y., Wang, X. L., & Zou, L. Z. (2008). Absorption of CO2 by ionic liquid/polyethylene glycol mixture and the thermodynamic parameters. *Green Chemistry*, 879-884.

[27] Zhu, A., Jiang, T., Wang, D., Han, B., Liu, L., Huang, J., Zhang, J., & Sun, D. (2005). Direct aldol reactions catalyzed by 1,1,3,3-tetramethylguanidine lactate without solvent. *Green Chemistry*, 514-517.

[28] Abello, S., Medina, F., Rodriguez, X., Cesteros, Y., Salagre, P., Sueiras, J., Tichit, D., & Coq, B. (2004). Supported choline hydroxide (ionic liquid) as heterogeneous catalyst for aldol condensation reactions. *Chemical Communications*, 1096-1097.

[29] Kryshtal, G. V., Zhdankina, G. M., & Zlotin, S. G. (2005). Tetraalkylammonium and 1, 3Dialkylimidazolium Salts with Fluorinated Anions as Recoverable Phase-Transfer Catalysts in Solid Base-Promoted Cross-Aldol Condensations. *European Journal of Organic Chemistry*, 2822-2827.

[30] Lombardo, M., Pasi, F., Easwar, S., & Trombini, C. (2007). An Improved Protocol for the Direct Asymmetric Aldol Reaction in Ionic Liquids, Catalysed by Onium Ion-Tagged Prolines. *Advanced Synthis & Catalysis*, 2061-2065.

[31] Kotrusz, P., Kmentova, I., Gotov, B., Toma, S., & Solcaniova, E. (2002). Proline-catalysed asymmetric aldol reaction in the room temperature ionic liquid [bmim]PF$_6$. *Chemical Communications*, 2510-2511.

[32] Loh, T. P., Feng, L. C., Yang, H. Y., & Yiang, J. Y. l. (2002). l-Proline in an ionic liquid as an efficient and reusable catalyst for direct asymmetric aldol reactions. *Tetrahedron Letters*, 8741-8743.

[33] Wang, C., Zhao, W., Li, H., & Guo, L. (2009). Solvent-free synthesis of unsaturated ketones by the Saucy-Marbet reaction using simple ammonium ionic liquid as a catalyst. *Green Chemistry*, 843-847.

[34] Gu, Y., Zhang, J., Duan, Z., & Deng, Y. (2005). Pechmann Reaction in Non-Chloroaluminate Acidic Ionic Liquids under Solvent-Free Conditions. *Advanced Synthesis & Catalysis*, 512-516.

[35] Mallakpour, S., & Seyedjamali, H. (2009). Ionic liquid catalyzed synthesis of organosoluble wholly aromatic optically active polyamides. *Polymer Bulletin*, 605-614.

[36] Kim, D. W., & Chi, D. Y. (2004). Polymer-Supported Ionic Liquids: Imidazolium Salts as Catalysts for Nucleophilic Substitution Reactions Including Fluorinations. *Angewandte Chemie*, 483-485.

[37] Iglesias, M., Garcia-Muñoz, R., Gonzalez-Olmos, R., Salvatierra, D., & Mattedi, S. (2007). Analysis of methanol extraction from aqueous solution by n-hexane: Equilibrium diagrams as a function of temperatura. *Journal of Molecular Liquids*, 52-58.

[38] Peric, B., Marti, E., Sierra, J., Cruañas, R., Iglesias, M., & Garau, M. A. (2011). Terrestrial ecotoxicity of short aliphatic protic ionic liquids. *Environmental Toxicology and Chemistry*, 2802-2809.

[39] Gradeff, P. S. (1974). US Patent 3,840,601, to Rhodia Inc.

[40] Mitchell, P. W. D. (1989). US Patent 4,874,900, to Union Camp Corporation.

[41] Climent, M. J., Corma, A., Iborra, S., & Velty, A. (2002). Synthesis of pseudoionones by acid and base solid catalysts. *Catalysis Letters*, 157-163.

[42] Climent, M. J., Corma, A., Iborra, S., Epping, K., & Velty, A. (2004). Increasing the basicity and catalytic activity of hydrotalcites by different synthesis procedures. *Journal of Catalysis*, 316-326.

[43] Abello, S., Medina, F., Tichit, D., Perez-Ramirez, J., Groen, J. C., Sueiras, J., Salagre, P., & Cesteros, Y. (2005). Aldol Condensations Over Reconstructed Mg-Al Hydrotalcites: Structure-Activity Relationships Related to the Rehydration Method. *Chemistry a European Journal*, 728-739.

[44] Cota, I., Chimentao, R., Sueiras, J. E., & Medina, F. (2008). The DBU-H2O complex as a new catalyst for aldol condensation reactions. *Catalysis Communications*, 2090-2094.

[45] Valkenberg, M. H., de Castro, C. W., & Hölderich, F. (2002). Immobilisation of ionic liquids on solid supports. *Green Chemistry*, 88-93.

[46] Gadenne, B., Hesemann, P. J., & Moreau, J. E. (2004). Supported ionic liquids: ordered mesoporous silicas containing covalently linked ionic species. *Chemical Communications*, 1768-1769.

[47] Mehnert, C. P. (2005). Supported Ionic Liquid Catalysis. *Chemistry a European Journal*, 50-56.

[48] Mehnert, C. P., Mozeleski, E. J., & Cook, R. A. (2002). Supported ionic liquid catalysis investigated for hydrogenation reactions. *Chemical Communications*, 3010-3011.

[49] Mehnert, C. P., Cook, R. A., Dispenziere, N. C., & Afeworki, M. (2002). Supported Ionic Liquid Catalysis – A New Concept for Homogeneous Hydroformylation Catalysis. *Journal of the American Chemical Society*, 12932-12933.

[50] Baudoux, J., Perrigaud, K., Madec-J, P., Gaumont-C, A., & Dez, I. (2007). Development of new SILP catalysts using chitosan as support. *Green Chemistry*, 1346-1351.

[51] Hu, Y. Q., Wang, J. Y., Zhao, R. H., Liu, Y., Liu, R., & Li, Y. (2009). Catalytic Oxidation of Cyclohexane over ZSM-5 Catalyst in N-alkyl-N-methylimidazolium Ionic Liquids. *Chinese Journal of Chemical Engineering*, 407-411.

Chapter 2

Ionic Liquids: "Green" Solvent for Catalytic Oxidations with Hydrogen Peroxide

Liangfang Zhu and Changwei Hu

Additional information is available at the end of the chapter

1. Introduction

Over the past decade, ionic liquids (ILs) have received great deal of attention as possible "green" replacement for volatile organic solvent mainly due to their nonmeasurable vapor pressure and good dissolubility for other salts. [1-5] Reaction types successfully performed in ILs include Diels–Alder, [6] Friedel–Crafts, [7] olefin hydrogenation, [8] hydroformylation, [9] , [10] oligomerization, [11] and Heck and Suzuki coupling reactions. [12] , [13] In addition to solvent, ILs may have multiple functions in catalytic reactions. They may act as catalyst, co-catalyst, support, or ligands for the catalytic process. [14] In particular, some "unexpected" effects have been observed in affecting the catalytic reaction pathway. For example, the cations/anions in ILs may be involved in the formation of the active species changing the reaction mechanism. [15] - [20] Understanding the functions of ILs in the catalytic reaction is of critical importance for deliberately modifying existed reaction system and exploiting new types of synthetic route by using this "green" solvent.

Catalytic oxidation is a class of commercially important reaction. As an environmentally benign oxidant, hydrogen peroxide (H_2O_2) has been used for several catalytic oxidation. So far, significant improvements on the catalytic performance, in terms of yield and selectivity, have been observed using ILs as the solvent for the H_2O_2 oxidation reaction. [21] - [23] Actually, in many cases, ILs are active participant because the formation of radical species, stabilization of the charged reactive intermediate, and immobilization of the actual catalyst can be strongly affected by the presence of an ionic environment. In comparison with traditional organic solvent, the use of ILs in catalytic oxidation has been regarded as a new means for recycling the catalyst and enhancing the yield and selectivity of the product. Though a great number of catalytic oxidation have been performed in ILs, there are still rare examples which demonstrate how the ILs affect the reaction pathway and the reactivity.

This chapter aims at summarizing the examples that concern the H_2O_2 oxidation reactions in ILs, in particular the benzene hydroxylation, alcohol oxidation, and olefin oxidation. The effects of ILs on the reaction pathway and the selectivity are discussed, drawing to the conclusion that ILs are offering unique properties as solvent by recycling the catalyst and enhancing the yield and selectivity of the product. What should be pointed out is that the examples are limited as far as possible to those that inform the readers' understanding of the role of ILs in the H_2O_2 oxidation reaction. We apologize that some fine work is not covered, and we hope to stimulate more discussions in the future.

2. Benzene hydroxylation

Direct hydroxylation of benzene to phenol with H_2O_2 has been extensively investigated owing to the reduced reaction steps and environmentally benign byproduct of water when comparing to the commercial cumene process for phenol production. One of the fundamental targets in this intriguing investigation is to enhance the utilization efficiency of H_2O_2 and the selectivity of phenol. The low efficiency of H_2O_2 always derives from the fast decomposition of H_2O_2, and the low product selectivity is mainly originated from the over-oxidation of phenol. Studies have shown that solvents used in the hydroxylation play an important role on enhancing both the H_2O_2 efficiency and the product selectivity. For example, water was the solvent in the traditional Fenton's reagent (Fe^{II}-H_2O_2) catalyzed hydroxylation, [24] whereas, the decomposition of H_2O_2 was very fast. [25] The selectivity to phenol was rather poor in the aqueous solution since phenol is more reactive toward oxidation than benzene itself. Acetonitrile and acetic acid were then used as the solvents for most of the catalyzed hydroxylation of benzene, [26] - [28] and a biphasic water-acetonitrile (1:1) system was developed to decrease the over-oxidation of phenol. [29] In Bianchi et al.'s work, [30] sulfolane was believed to form complexes with phenolic compounds inducing increased selectivity to phenol.

In addition to organic solvents, Peng et al. [31] introduced a biphasic aqueous–imidazolium-based IL system for benzene hydroxylation in the presence of ferric tri(dodecanesulfonate) catalyst (Figure 1). In this aqueous- $[C_n mim]$ $[X]$ (n=4, 8, 10; X=PF_6, BF_4) IL biphasic system, both the catalyst and benzene were dissolved in the IL, whereas, H_2O_2 was mainly dissolved in aqueous phase. The produced phenol was extracted into water phase, minimizing the over-oxidation. A highest yield of 54% and selectivity of 100% to phenol was obtained in the aqueous-IL biphasic system. However, as H_2O_2 was existed in the IL-free aqueous phase, the active oxidizing agent were deemed to be radical species. Therefore, this biphasic system was not operative for the hydroxylation of toluene: only 1% of toluene was converted to benzaldehyde although its selectivity reached 100%.

In our work, [25] , [32] a benzene–triethylammonium acetate ([Et$_3$NH] [CH$_3$COO]) IL biphasic system was constructed for the benzene hydroxylation (Figure 2). The Fenton-like reagent (Fe^{III}-H_2O_2) existed in the IL phase and most of the phenol was extracted to the benzene layer. The [Et$_3$NH] [CH$_3$COO] IL was found to be stable in the water- and oxygen-rich environment. Benzene acted as both the substrate and the extractant in the hydroxylation reaction. In comparison with the aqueous-IL biphasic system, the continuous extraction of phenol by benzene from [Et$_3$NH] [CH$_3$COO] IL protected phenol from further oxidation by directly avoiding the contact of phenol with the catalyst and oxidant. As a result, moderate yield (20%, based on benzene converted, excluding evaporated) and high selectivity (> 99.5%) of phenol were obtained in the IL-benzene biphasic system.

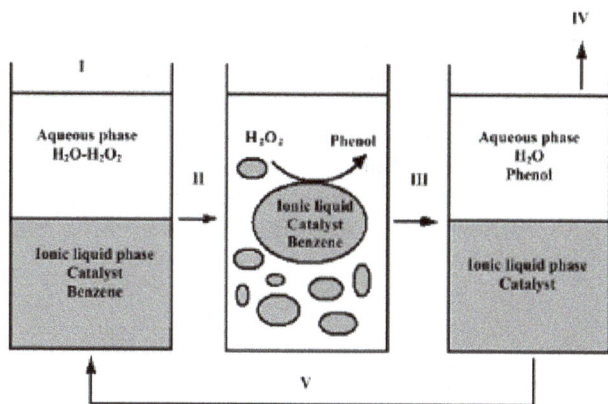

Figure 1. A schematic representation of the aqueous–IL biphasiccatalytic reaction system for benzene hydroxylation to phenol with H$_2$O$_2$.Step I charging; Step II reaction; Step III still; Step IV recovery of phenolvia extraction; Step V the IL and catalyst are reused for anotherreaction cycle. IL = [C$_n$mim] [X] (n = 4,8,10; X = PF$_6$, BF$_4$). [31] Reprinted with permission from ref. 31. Copyright ©2003, Royal Society of Chemistry.

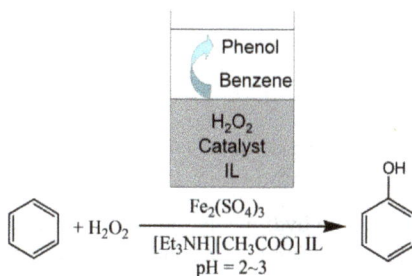

Figure 2. Schematic illustration of the benzene- [Et$_3$NH] [CH$_3$COO] IL biphasic system for benzene hydroxylation to phenol with H$_2$O$_2$. [25]

Moreover, the [Et$_3$NH] [CH$_3$COO] IL exhibited retardation performance for the decomposition of H$_2$O$_2$ and protection performance for the over-oxidation of phenol. From a molecular aspect, the CH$_3$COO- anions of [Et$_3$NH] [CH$_3$COO] IL were found to be coordinated with the Fe ions, forming Fe complexes, *by virtue of* the solution of Fenton-like reagent in the IL phase (Figure 2). [25] Such coordination "anchored" the catalyst in the IL, affecting the subsequent activation of benzene and H_2O_2. A higher electrophilicity of the Fe-complexes was favorable for the interaction of H_2O_2 with the Fe center, which may be the origin of the retardation role of IL on the decomposition of H_2O_2. High-valent Fe^{IV}-oxo species, formed from the O-O bond homolysis of a Fe^{III}-OOH intermediate, was found to be the main active oxidizing species in the ionic environment rather than the widely accepted oxidizing species of hydroxyl radical (OH•) in an aqueous Fenton system. The mechanism for hydroxylation of benzene in the [Et$_3$NH] [CH$_3$COO] IL was thus different from that occurred in aqueous solution (Figure 3) [32] : the activation of benzene was mainly achieved *via* the electrophilic attack by the Fe^{IV}-oxo species, rather than *via* hydrogen abstraction to form phenyl radical. Over-oxidation of phenol through H-abstraction from O-H of phenol by the Fe^{IV}-oxo species was partly prohibited by the hydrogen-bond interaction between phenol and the CH$_3$COO- anion. The electrophilic character of the Fe^{IV}-oxo species made the [Et$_3$NH] [CH$_3$COO] IL suitable for hydroxylation of other alkyl-benzenes. As an instance, the oxidation of toluene in [Et$_3$NH] [CH$_3$COO] IL resulted in the selective activation of benzene ring with the selectivity to methylphenols of about 62%. [33]

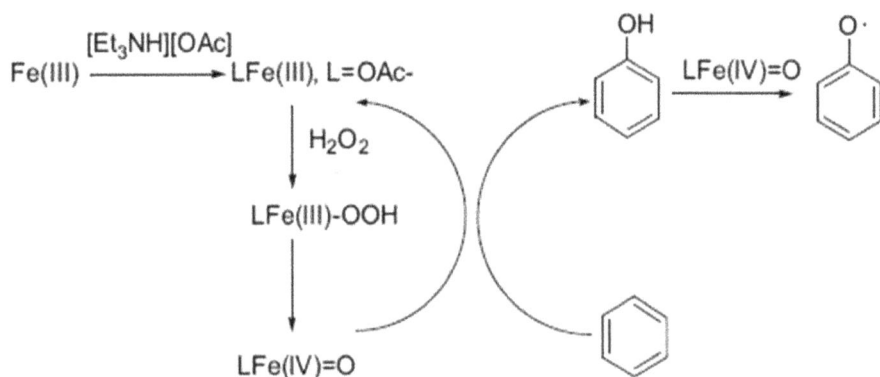

Figure 3. Mechanism for hydroxylation of benzene with H_2O_2 in [Et$_3$NH] [OAc] system. [32] Reprinted with permission from ref. 32. Copyright © 2011,Elsevier Science Ltd.

In the benzene hydroxylation, both hydrophobic and hydrophilic ILs have shown the feasibility of acting as solvent for enhancing both the yield and selectivity of phenol when compared with that in aqueous solution (*vide supra*). The nature of the active oxidizing species is mostly depended on the dispersion of both H_2O_2 and the catalyst in the ILs because the generation of active species is thus influenced by the ionic environment. Furthermore, the combination of ILs with a second solvent (co-solvent), either traditional organic solvent or water, may offer opportunity for breaking the thermodynamic equilibrium by mass-transfer of the product from phase to phase. Most importantly, the ILs-solvent biphasic system provides opportunity for the stabilization of charged reactive intermediate and the protection of unstable product from over-oxidation. Therefore, the ILs-co-solvent biphasic system can be tentatively developed as solvent for a wide range of H_2O_2 oxidation reaction to meet specific requirements, for example, yield, selectivity, or solubility, *etc.*

3. Alcohol oxidation

The partial oxidation of alcohols to aldehydes, or secondary alcohols to corresponding ke-tones is a fundamental synthetic transformation in organic chemistry and is industrially im-portant. [34] - [36] However, this transformation always suffers from drawbacks such as poor conversion and selectivity due to over-oxidation. Stable-free nitroxyl radicals such as TEMPO(2,2,6,6-tetramethylpiperidine-1-oxyl) has recently emerged as a catalyst or co-cata-lyst to promote the formation of the catalytically active species for selective oxidation of al-cohols to aldehydes or ketones where volatile organic solvents such as CH_2Cl_2 are frequently used. [37] - [44] However, the recycling of the quite expensive TEMPO is problematic due to the homogeneous character of the classic organic media. Replacement of organic solvents with ILs or immobilization of TEMPO on ILs provide alternative strategies for solving the above-mentioned problems. On the one hand, TEMPO can be anchored on ILs allowing the recycling of the catalyst; on the other hand, ILs provide advantages for increasing the selec-tivity of the product by promoting oxidation of alcohols to aldehydes but suppressing over-oxidation of these aldehydes to acids.

In Wang *et al.*'s work, [45] oxidation of alcohols with H_2O_2 was performed in pyridiniumtetra-fluoroborate([Bpy] BF$_4$) IL. The catalyst (vanadate) and co-catalyst (TEMPO and sulfonic acid) were both grafted on [Bpy] BF$_4$ IL. The functionalized IL showed good dissolubility in the [Bpy] BF$_4$ solvent. The as-formed homogeneous mixture of N-n-dodecyl pyridinium vana-date, N-(propyl-1-sulfonic acid) pyridiniumtetrafluoroborate, and 4-(propanoate-TEMPO) pyridiniumtetrafluoroborate exhibited good activity and recyclability for alcohols oxida-tion.Jiang *et al.* [46] used the acetamido-modified TEMPO as catalyst for the selective oxida-tion of benzylic alcohols to aldehydes in the 1-n-butyl-3-methylimidazolium hexafluorophosphate ([bmim] [PF$_6$]) IL (Figure 4). The [bmim] [PF$_6$] ILwas immiscible with H_2O_2, favoring the partial oxidation of alcohols to aldehydes but inhibiting the over-oxidation of aldehydes to acids by reducing the contact of the product with the oxidant. The miscibility of acetamido-TEMPO in [bmim] [PF$_6$] IL ensured good catalytic activity and efficient recycling of the catalyst. In comparison to the common organic solvents (ethyl acetate or chloroform), the yield of aldehydes was enhanced by about three times.

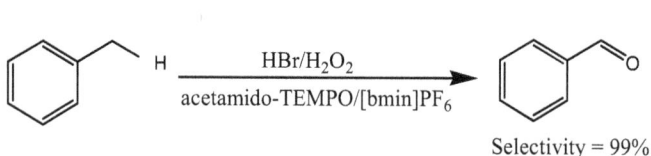

$$\text{Selectivity} = 99\%$$

Figure 4. Highly selective oxidation of benzyl alcohol to benzaldehyde with acetamido-TEMPO/HBr//H$_2$O$_2$ in [bmim] [PF$_6$] IL. [46]

The strategy of anchoring TEMPO on ILs has also been applied in other catalytic oxidation re-action. In Fall et al.'s work, [39] TEMPO was supported on ILs through click chemistry reaction (Figure 5a). The IL-supported catalyst exhibited good solubility in [HMIM] [BF$_4$] IL and high activity for alcohol oxidation using bis(acetoxy)iodobenzene (BAIB) as the terminal oxidant. The catalyst can be recycled together with the IL without loss of the efficiency for several cycles. Karimi et al. [47] grafted TEMPO on SBA-15 solid support and then synthesized IL@SBA-15-TEMPO catalyst by physically confining 1-methyl-3-butylimidazolium ([Bmim] Br) IL within the mesopores of the TEMPO-modified SBA-15 (Figure 5b). The catalyst showed high activity, improved selectivity, and good recyclability for the oxidation of alcohols to alde-hydesand ketones with t-butylnitrite (TBN) as oxidant in AcOH. Although the catalytic performance of the TEMPO-ILs and ILs@support-TEMPO catalysts in the H$_2$O$_2$ oxidation reaction is not investigated, we may expect the strategy of fixing TEMPO onto ILs or solid support applicable in recycling the expensive catalyst in a wide range of catalytic reactions.

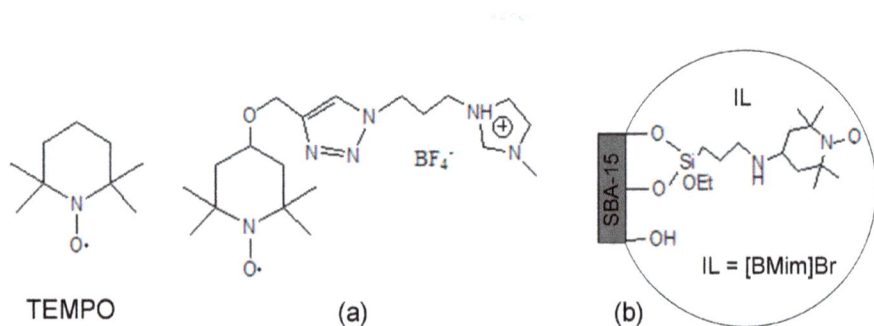

Figure 5. Strategies for immobilizing TEMPO on ILs. (a) IL supported TEMPO, [39] and (b) IL@SBA-15-TEMPO. [47]

In addition to TEMPO, the combination of various kinds of catalysts (i.e., inorganic salts, transition metal complexes, and oxides) with ILs has shown prospect as effective catalytic system for the H$_2$O$_2$ oxidation reaction. Chhikara et al. [48] synthesized imidazolium-based phosphotungstate catalyst (Figure 6, 1) by grafting phosphotungstate onto imidazolium-based IL, which showed good catalytic performance in the homogeneous oxidation of secondary alcohols in the 1-butyl-3-methylimidazolium tetrafluoroborate ([bmim] [BF$_4$]) IL (Figure 6a). All of the secondary alcohols were converted to corresponding ketones in good to excellent yields. Oxidation of primary alcohol, i.e. benzyl alcohol, produced benzaldehydein good yield (78%), or benzoic acid in high yield (96%) after increasing the H$_2$O$_2$ amount (Figure 6b). The IL and catalyst after the extraction of the products could be reused for further catalytic oxidation.

Bianchini et al. [49] described the oxidation of some secondary alcohols to their ketones with H$_2$O$_2$ catalyzed by methyltrioxorhenium(VII)(MTO) or polymer supported-MTO in [bmim] [PF$_6$] or 1-ethyl-3-methylimidazolium bis-triflic amide ([emim] [Tf$_2$N]) ILs. The supported catalyst was dispersed in the IL layer, which allowed the recycling of the catalyst and solvent system that could not be realized in a wide range of organic solvent. Moreover, in comparison with ethanol or acetic acid, the activity of the catalyst in the ILs was obviously improved.

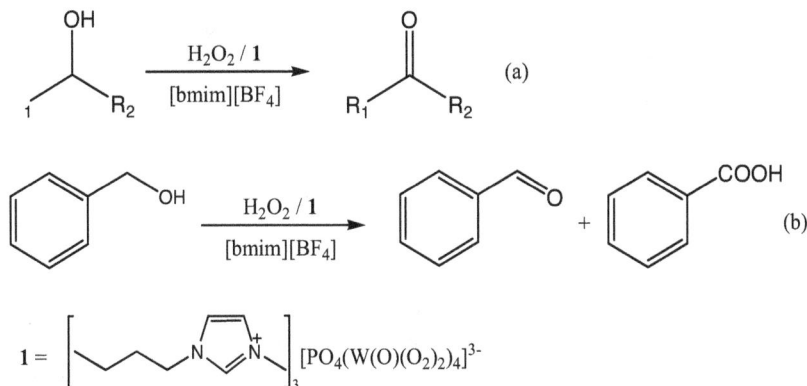

Figure 6. Oxidation of (a) secondary alcohols to ketones, and (b) benzyl alcohol to benzaldehyde and benzoic acid with imidazolium-based phosphotungstate (**1**) and H_2O_2 in [bmim] [BF$_4$] IL. [48]

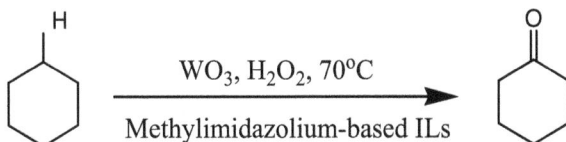

Figure 7. Selective oxidation of cyclohexanol to cyclohexanone with H_2O_2 in hydrophobic methylimidazolium-based ILs. [50]

Chen *et al.* [50] applied several hydrophobic methylimidazolium-based ILs in the oxidation of cyclohexanol to cyclohexanone with H_2O_2 using WO_3 as the catalyst (Figure 7). The oxidation of cyclohexanol in the absence of ILs produced cyclohexanone with a moderate yield (42%) or adipic acid at high cyclohexanol conversion. [51] In the biphasic cyclohexanol-ILs system, however, the ILs were found to effectively intensify cyclohexanol oxidation and resulted in 100% conversion of cyclohexanol with 100% selectivity to cyclohexanone. The high production of cyclohexanone can be explained by the fact that the oxidation of cyclohexanol occurred in aqueous phase contained H_2O_2 and WO_3, whereas, the produced cyclohexanone was abstracted into the organic phase, minimizing the further oxidation of the product. Among the three kinds of methylimidazolium-based ILs (1-hydroxyethyl-3-methylimidazoliumchloride ([HOemim] Cl); 1-hexyl-3-methylimidazoliumchloride ([Hmim] Cl); and 1-octyl-3-methylimidazolium chloride ([Omim] Cl,)) investigated, the [Omim] Cl IL exhibited the best solvent performance for enhancing the conversion of cyclohexanol when comparing with the traditional organic solvents (methanol, *n*-propanol, or acetone). Furthermore, a higher concentration of the [Omim] ClIL favored the transformation of cyclohexanol, giving evidence that the IL may be involved in the stabilization of the reaction intermediates in the catalytic process. Detailed investigation revealed that a longer alkyl chain of the IL increased the interaction between the solvent and the hydrophobic substrate, and a larger polarity of the IL increased the strength of coulombic forces arising in the solvation process. Both of these two factors improved the conversion of cyclohexanol considerably.

In this part, the anchoring of the actual catalyst on ILs offers opportunity for immobilizing catalyst with the solvent, allowing the recycling of the catalyst, especially some expensive reagent. This "anchoring" can be either chemical coupling or physical confinement. Chemical coupling requires special tailoring or functionalization of the ILs. The functionalized ILs show prospect as both catalyst and solvent for the H_2O_2 oxidation reaction. The physical confinement of ILs within some solid porous materials has dual effects on the catalytic oxidation: on the one hand, the ILs supply special microenvironment on affecting the reaction pathway; on the other hand, the micropores of the porous material allow the controlling of the selectivity of the product. In addition, we may expect structural modification of the functionalized ILs by deliberately varying the cations/anions to meet the distribution requirement of the substrate or product, which will be of great importance for increasing the yield and selectivity of the product. Beyond that, more synthetic method should be developed to support the actual catalyst in order to shed light on the effective utilization of the expensive reagent in future catalytic oxidation.

4. Olefin oxidation

Recently, significant improvements on the catalytic performance in some transition metal-catalyzed reactions have been observed using ILs as the solvent. [2] , [19] , [20] The room-temperature ILs have emerged as environmentally benign reaction media as well as new vehicles for the immobilization of transition metal-based catalysts. Singh et al. [52] reported the H_2O_2 epoxidation of substrates containing both electron rich and deficient olefins catalyzed with meso-tetraarylporphyrin iron(III) chlorides ([TAPFe(III)Cl]) in imidazolium ILs. The active oxidizing species depended on the substrate used (Figure 8): the ferric peroxy anions (TAP-Fe^{III}-OO$^-$) were effective intermediates in the epoxidation of electron deficient olefins, whereas the high valent oxoferrylporphyrin π-cation radicals (TAP-Fe^{IV}=O$^{\bullet+}$) were involved in the epoxidation of electron rich olefins. The ILs provide special microenvironment via the interactions between their cations and anions, where the active intermediates could be fast generated from TAPFe^{III}Cl/ [Bmim] [PF$_6$] and H_2O_2 and well stabilized in the [Bmim] [PF$_6$] IL. The nature of the anions in the ILs played an important role on the activity of the catalyst.

Han and coworkers [53] synthesized novel Ni^{2+}-containing 1-methyl-3- [(triethoxysilyl)propyl] imidazolium chloride (TMICl) IL immobilized on silica to catalytic oxidation of styrene to benzaldehyde with H_2O_2 under solvent-free condition (Figure 9). With the aid of the IL, both hydrophobic reactant and the hydrophilic reactant were accessible to the active sites of the catalyst: styrene and H_2O_2 are miscible with the IL, and the Ni^{2+} was coordinated by the immobilized IL that allowed both reactants to access to active sites of the catalyst effectively. Under solvent-free condition, the conversion of styrene reached 18.5%and the selectivity to benzaldehyde was as high as 95.9% on the IMM-TMICl-Ni^{2+} catalyst.

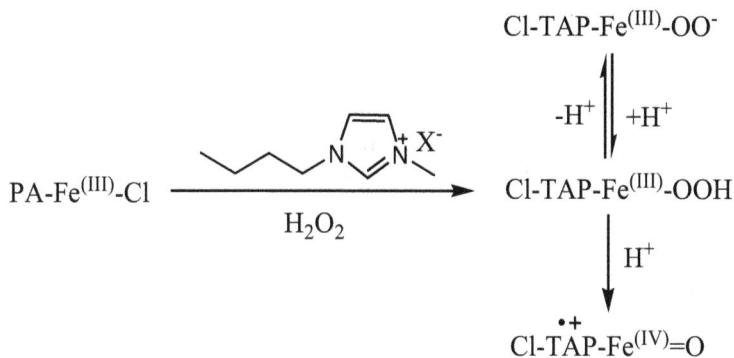

Figure 8. Mechanism for the generation of the reactive intermediate from TAPFe^{III}Cl/ [Bmim] [PF$_6$] and H_2O_2. [52]

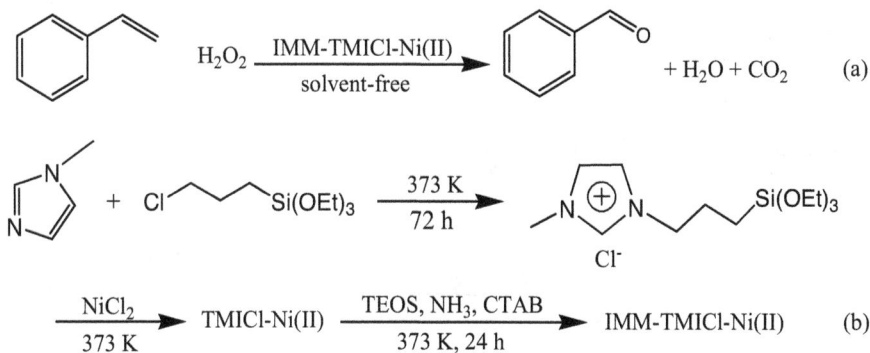

Figure 9. Schematic illustrations of (a) oxidation of styrene to benzaldehyde with H_2O_2 under solvent-free condition, (b) synthesis of IMM-TMICl-Ni(II) catalyst by grafting Ni^{2+} on IL-Silica. [53]

Some more examples are given by combining ILs with metal peroxides or polyoxometalates in the catalytic epoxidation of olefins with H_2O_2. In Yamaguchi et al.'s work, [54] peroxo-tungstate was immobilized on dihydroimidazolium-based IL-modified SiO_2 (Figure 10). The as-prepared catalyst showed high activity and selectivity for epoxidation of a wide range of olefins. Radical mechanism was excluded for the IL-involved epoxidation in CH_3CN solvent. Berardi et al. [55] embedded the catalytically active $[\gamma\text{-SiW}_{10}O_{36}(PhPO)_2]$ ^{4}polyanions in the hydrophobic IL ([bmim] [[PF$_6$] or hydrophilic IL [bmim] [(CF$_3$SO$_2$)$_2$N]). The catalyst gave out high yield and selectivity for epoxidation of olefins under microwave irradiation in the hydrophilic IL (Figure 11). The catalyst can be recycled with the catalytic IL phase. Liu *et al.* [56] demonstrated the role of [bmim] [PF$_6$] IL as an activator for efficient olefin epoxidation with H$_2$O$_2$catalyzed by Keggin polyoxometalate [bmim] $_3$PW$_{12}$O$_{40}$ (Figure 12). In the IL, the interaction between the anions and the cations supplied a special microenvironment, accelerating the generation of the active peroxotungstate [PO$_4${W(O)(O)$_2$}$_4$] $^{3-}$ species from [bmim] $_3$PW$_{12}$O$_{40}$ and H$_2$O$_2$. In some sense, the [bmim] [PF$_6$] IL could be considered as a co-catalyst to promote the formation of active species for epoxidation. Both of the TOF and selectivity for olefin epoxidation were significantly enhanced in the IL compared to that of traditional organic solvents, *e.g.*, 289 times TOF and 1.3 times selectivity as found in CH$_2$Cl$_2$ for the epoxidation of *cis*-cyclooctene. The utilization efficiency of H$_2$O$_2$ reached as high as 87%.

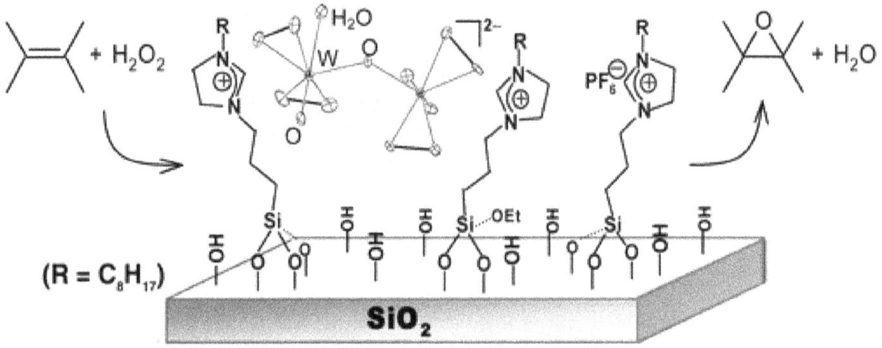

Figure 10. Epoxidation of Olefins with H$_2$O2 catalyzed by peroxotungstate immobilized on IL-modified SiO$_2$. [54] Reprinted with permission from ref.54. Copyright © 2005,AmericanChemical Society.

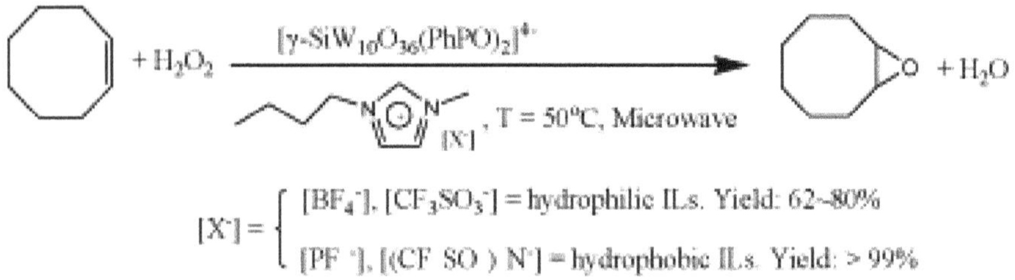

Figure 11. Epoxidaton of *cis*-Cyclooctene with H$_2$O$_2$ and polyoxometalates in both hydrophilic and hydrophobic ILs. [55]

Numerous examples have shown that ILs are offering unique properties in the transition metal-, metal peroxide- or polyoxometalates-catalyzed oxidation of olefins with H$_2$O$_2$. The

ILs supply special environment for the generation and stabilization of the active intermediate, or act as support for immobilizing and recycling the actual catalyst, both of which are necessary for performing effective catalytic oxidation. By delicately designing the combination of catalyst, support and ILs, the interactions between the hydrophobic substrate, hydrophilic oxidant, and the active site could be reinforced, intensifying the catalytic efficiency of the oxidation reaction.

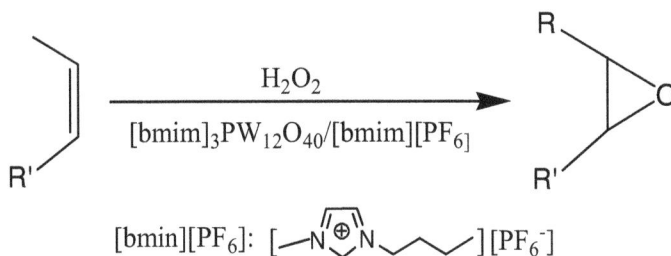

Figure 12. Epoxidation of olefins with H_2O_2 and polyoxometalates in [bmim] [PF$_6$] IL. [56]

5. Conclusion

Catalytic oxidations have been widely studied in ionic liquids, and much of this interest is centered on the possible use as "green" alternatives to traditionally used volatile organic solvents. This chapter summarizes limited examples that illustrate the applications of ILs in the catalytic oxidation using H_2O_2 as the oxidant, in particular benzene hydroxylation, alcohol oxidation, and olefin oxidation. We focus our discussion on understanding how the unusual solvent environment provide solute species that affect the reactions occurred in them.

As innocent and non-vaporized solvents, ILs provide good solubility to salts and most of the hydrophobic substrate, endowing them good solvent for the transition metal complexes-, peroxides-, oxides-, polyoxometalates-, or organic molecules-catalyzed oxidation. The miscibility of ILs with water and organic molecules can be elaborately tuned by varying the cations/anions (*i.e.*, length of alkyl chain, polarity, *etc.*). The interactions between ILs and the substrate, catalyst, oxidant, even reaction intermediate, make the ILs act as multi-functional solvent for the H_2O_2 involved catalytic oxidation. The cations/anions in ILs may influence the reaction pathway by stabilization of the charged transition state, active species or ligands which are necessary for many oxidation reaction. The structure of ILs can also be specially tailored to support actual catalyst and/or co-catalyst for effective recycling. The functionalized ILs have prospect as both catalyst and solvent for the H_2O_2 oxidation reaction. The grafting of ILs on some solid porous materials may improve the microenvironment of the reactive site, affecting the reaction pathway by increasing the selectivity of the product. Most importantly, the combination of ILs and a co-solvent (water or organic solvent) allows the mass transfer of unstable product from oxidative environment, protecting the product from over-oxidation by H_2O_2. As a conclusion, the integration of the multiple benefits from ILs will provide a greener scenario for recycling the catalyst and solvent, as well as improving the yield and selectivity of the product. We may expect novel synthetic strategy for functionalized ILs and their elaborate combination with prevailing catalytic materials for applications in a wide range of catalytic oxidation in future.

Acknowledgments

The financial support from the National Natural Science Foundationof China (No. 20901053 and 20872102) and PCSIRT (No. IRT0846) are greatly appreciated.

Author details

Liangfang Zhu and Changwei Hu*

*Address all correspondence to: changweihu@scu.edu.cn or chwehu@mail.sc.cninfo.net

Key Laboratory of Green Chemistry and Technology, Ministry of Education, College of Chemistry, Sichuan University, Chengdu, P.R. China

References

[1] Anastas, P. T.; Warner, J. C., *Green Chemistry: Theory and Practice*, Oxford University Press: New York, 1998.

[2] Welton, T., Room-Temperature Ionic Liquids. Solvents for Synthesis and Catalysis. *Chem. Rev.* 1999, 99, 2071-2084.

[3] Dupont, J.; De Souza, R. F.; Suarez, P. A. Z., Ionic Liquid (Molten Salt) Phase Organometallic Catalysis. *Chem. Rev.* 2002, 102, 3667-3692.

[4] Hernandez, O. R., ToTreat or Not to Treat? Applying Chemical Engineering Tools and a Life Cycle Approach to Assessing the Level of Sustainability of a Clean-up Technology. *Green Chem.* 2004, 6, 395-400.

[5] Clift, R., Sustainable Development and Its Implications for Chemical Engineering. *Chem. Eng. Sci.* 2006, 61, 4179-4187.

[6] Fischer, T.; Sethi, A.; Welton, T.; Woolf, J., Diels-Alder Reactions in Room-temperature Ionic Liquids. *Tetrahedron Lett.* 1999, 40, 793-796.

[7]   Boon, J. A.; Levisky, J. A.; L., P. J.; Wilkes, J. S., Friedel-Crafts Reactions in Ambient-Temperature Molten Salts. *J. Org. Chem.* 1986, 51, 480-483.

[8]   Brown, R. A.; Pollet, P.; McKoon, E.; Eckert, C. A.; Liotta, C. L.; Jessop, P. G., Asymmetric Hydrogenation and Catalyst Recycling Using Ionic Liquid and Supercritical Carbon Dioxide. *J. Am. Chem. Soc.* 2001, 123, 1254-1255.

[9]   Chauvin, Y.; Mussmann, L.; Olivier, H., A Novel Class of Versatile Solvents for Two-Phase Catalysis: Hydrogenation, Isomerization, and Hydroformylation of Alkenes Catalyzed by Rhodium Complexes in Liquid 1,3-Dialkylimidazolium Salts. *Angew. Chem., Int. Ed. Engl.* 1996, 34, 2698-2700.

[10] Favre, F.; Olivier-Bourbigou, H.; Commereuc, D.; Saussine, L., Hydroformylation of 1-Hexene with Rhodium in Non-aqueous Ionic Liquids : How to Design the Solvent and the Ligand to the Reaction. *Chem. Commun.* 2001, 1360-1361.

[11] Dullius, J. E. L.; Suarez, P. A. Z.; Einloft, S.; de Souza, R. F.; Dupont, J., Selective Catalytic Hydrodimerization of 1,3-Butadiene by Palladium Compounds Dissolved in Ionic Liquids. *Organometallics* 1998, 17 (5), 815-819.

[12] Kaufmann, D. E.; Nouroozian, M.; Henze, H., Molten Salts as an Efficient Medium for Palladium Catalyzed C-C Coupling Reactions. *Synlett.* 1996, 1091-1092.

[13] Matthews, C. J.; Smith, P. J.; Welton, T., Palladium Catalysed Suzuki Cross-coupling Reactions in Ambient Temperature Ionic Liquids. *Chem. Commun.* 2000, 1249-1250.

[14] Parvulescu, V. I.; Hardacre, C., Catalysis in Ionic Liquids. *Chem. Rev.* 2007, 107, 2615-2665.

[15] Hallett, J. P.; Welton, T., Room-Temperature Ionic Liquids: Solvents for Synthesis and Catalysis. 2. *Chem. Rev.* 2011, 111, 3508-3576.

[16] Olivier-Bourbigou, H.; Magna, L.; Morvan, D., Ionic Liquids and Catalysis: Recent Progress from Knowledge to Applications. *Appl. Catal. A* 2010, 373, 1-56.

[17] Lee, J. W.; Shin, J. Y.; Chun, Y. S.; Jang, H. B.; Song, C. E.; Lee, S., Toward Understanding the Origin of Positive Effects of Ionic Liquids on Catalysis: Formation of More Reactive Catalysts and Stabilization of Reactive Intermediates and Transition States in Ionic Liquids. *Acc. Chem. Res.* 2010, 43 (7), 985-994.

[18] Stark, A., Ionic Liquid Structure-Induced Effects on Organic Reactions. *Top Curr. Chem.* 2009, 290, 41-81.

[19] Wasserscheid, P.; Keim, W., Ionic Liquids-New "solution" for Transition Metal Catalysis. *Angew Chem, Int. Ed.* 2000, 39, 3772-3789.

[20] Sheldon, R., Catalytic Reactions in Ionic Liquids. *Chem. Commun.* 2001, 2399-2407.

[21] Gharnati, L.; Doering, M.; Arnold, U., Catalytic Oxidation with Hydrogen Peroxide in Ionic Liquids. *Curr. Org. Synth.* 2009, 6 (4), 342-361.

[22] Betz, D.; Altmann, P.; Cokoja, M.; Herrmann, W. A.; Kuehn, F. E., Recent Advances in Oxidation Catalysis Using Ionic Liquids as Solvents. *Coord. Chem. Rev.* 2011, 255 (13-14), 1518-1540.

[23] Muzart, J., Ionic Liquids as Solvents for Catalyzed Oxidations of Organic Compounds. *Adv. Synth. Catal.* 2006, 348, 275-295.

[24] Walling, C., Intermediates in the Reactions of Fenton Type Reagents. *Acc. Chem. Res.* 1998, 31, 155-157.

[25] Hu, X. K.; Zhu, L. F.; Guo, B.; Liu, Q. Y.; Li, G. Y.; Hu, C. W., Hydroxylation of Benzene to Phenol via Hydrogen Peroxide in Hydrophilic Triethylammonium Acetate Ionic Liquid. *Chem. Res. Chin. Univ.* 2011, 27 (3), 503-507.

[26] Zhang, J.; Tang, Y.; Li, G. Y.; Hu, C. W., Room Temperature Direct Oxidation of Benzene to Phenol Using Hydrogen Peroxide in the Presence of Vanadium-substituted Heteropolymolybdates. *Appl. Catal. A* 2005, 278, 251-261.

[27] Zhong, Y. K.; Li, G. Y.; Zhu, L. F.; Yan, Y.; Wu, G.; Hu, C. W., Low Temperature Hydroxylation of Benzene to Phenol by Hydrogen Peroxide over Fe/activated Carbon Catalyst. *J. Mol. Catal. A* 2007, 272, 169-173.

[28] Jian, M.; Zhu, L. F.; Wang, J. Y.; Zhang, J.; Li, G. Y.; Hu, C. W., Sodium Metavanadate Catalyzed Direct Hydroxylation of Benzene to Phenol with Hydrogen Peroxide in Acetonitrile Medium. *J. Mol. Catal. A* 2006, 253, 1-7.

[29] Bianchi, D.; Bertoli, M.; Tassinari, R.; Ricci, M.; Vignola, R., Ligand Effect on the Iron-catalysed Biphasic Oxidation of Aromatic Hydrocarbons by Hydrogen Peroxide *J. Mol. Catal. A* 2003, 204, 419-424.

[30] Bianchi, D.; Balducci, L.; Bortolo, R.; D' Aloisio, R.; Ricci, M.; Span, G.; Tassinari, R.; Tonini, C.; Ungarellia, R., Oxidation of Benzene to Phenol with Hydrogen Peroxide Catalyzed by a Modified Titanium Silicalite (TS-1B). *Adv. Synth. Catal.* 2007, 349, 979-986.

[31] Peng, J. J.; Shi, F.; Gu, Y. L.; Deng, Y. Q., Highly Selective and Green Aqueous-Ionic Liquid Biphasic Hydroxylation of Benzene to Phenol with Hydrogen Peroxide. *Green Chem.* 2003, 5 (2), 224-226.

[32] Hu, X. K.; Zhu, L. F.; Wang, X. Q.; Guo, B.; Xu, J. Q.; Li, G. Y.; Hu, C. W., Active Species Formed in a Fenton-Like System in the Medium of Triethylammonium Acetate Ionic Liquid for Hydroxylation of Benzene to Phenol. *J. Mol. Catal. A* 2011, 342-343, 41-49.

[33] Hu, X. K., Study on Hydroxylation of Benzene in Triethylammonium Acetate Ionic Liquid. *Chinese Doctoral Dissertation* 2011.

[34] Ley, S. V.; Madin, A., in: *Trost, B. M.; Flemming, I. (Eds.), Comprehensive Organic Synthesis*, Vol. 7, 305-327, Pergamon Press, Oxford, 1991.

[35] Hudlick, M., *Oxidations in Organic Chemistry*, American Chemical Society: Washington, DC, 1990.

[36] Sheldon, R. A.; Kochi, J. K., *Metal Catalyzed Oxidationd of Organic Compounds*, Academic Press, New York, 1984.

[37] Bobbitt, J. M.; Br °uckner, C., *Organic Reactions*, John-Wiley & Sons, New York, 2009.

[38] Anelli, P. L.; Biffi, C.; Montanari, F.; Quici, S., *J. Org. Chem.* 1987, 52, 2559-2562.

[39] Fall, A.; Sene, M.; Gaye, M.; Gomez, G.; Fall, Y., Ionic Liquid-Supported TEMPO as Catalyst in the Oxidation of Alcohols to Aldehydes and Ketones. *Tetrahedron Lett.* 2010, 51 (34), 4501-4504.

[40] Hoover, J. M.; Stahl, S. S., Highly Practical Copper(I)/TEMPO Catalyst System for Chemoselective Aerobic Oxidation of Primary Alcohols. *J. Am. Chem. Soc.* 2011, 133 (42), 16901-16910.

[41] Ma, S. M.; Liu, J. X.; Li, S. H.; Chen, B.; Cheng, J. J.; Kuang, J. Q.; Liu, Y.; Wan, B. Q.; Wang, Y. L.; Ye, J. T.; Yu, Q.; Yuan, W. M.; Yu, S. C., Development of a General and Practical Iron Nitrate/TEMPO-Catalyzed Aerobic Oxidation of Alcohols to Aldehydes/Ketones: Catalysis with Table Salt. *Adv. Synth. Catal.* 2011, 353 (6), 1005-1017.

[42] Hoover, J. M.; Steves, J. E.; Stahl, S. S., Copper(I)/TEMPO-catalyzed aerobic oxidation of primary alcohols to aldehydes with ambient air. *Nature Protoc.* 2012, 7 (6), 1161-1166.

[43] Gheorghe, A.; Chinnusamy, T.; Cuevas-Yanez, E.; Hilgers, P.; Reiser, O., Combination of Perfluoroalkyl and Triazole Moieties: A New Recovery Strategy for TEMPO. *Org. Lett.* 2008, 10 (19), 4171-4174.

[44] Liu, R.; Liang, X.; Dong, C.; Hu, X., Transition-Metal-Free: A Highly Efficient Catalytic Aerobic Alcohol Oxidation Process. *J. Am. Chem. Soc.* 2004, 126, 4112-4113.

[45] Wang, S. S.; Popovic, Z.; Wu, H. H.; Liu, Y., A Homogeneous Mixture Composed of Vanadate, Acid, and TEMPO Functionalized Ionic Liquids for Alcohol Oxidation by H2O2. *ChemCatChem* 2011, 3 (7), 1208-1213.

[46] Jiang, N.; Ragauskas, A. J., TEMPO-Catalyzed Oxidation of Benzylic Alcohols to Aldehydes with the H2O2/HBr/Ionic Liquid [bmim] PF6 System. *Tetrahedron Lett.* 2005, 46 (19), 3323-3326.

[47] Karimi, B.; Badreh, E., SBA-15-Functionalized TEMPO Confined Ionic Liquid: an Efficient Catalyst System for Transition-Metal-Free Aerobic Oxidation of Alcohols with Improved Selectivity. *Org. Biomol. Chem.* 2011, 9 (11), 4194-4198.

[48] Chhikara, B. S.; Chandra, R.; Tandon, V., Oxidation of Alcohols with Hydrogen Peroxide Catalyzed by a New Imidazolium Ion Based Phosphotungstate Complex in Ionic Liquid. *J. Catal.* 2005, 230 (2), 436-439.

[49] Bianchini, G.; Crucianelli, M.; De Angelis, F.; Neri, V.; Saladino, R., Highly Efficient C-H Insertion Reactions of Hydrogen Peroxide Catalyzed by Homogeneous and Het-

erogeneous Methyltrioxorhenium Systems in Ionic Liquids. *Tetrahedron Lett.* 2005, 46 (14), 2427-2432.

[50] Chen, L.; Zhou, T.; Chen, L.; Ye, Y.; Qi, Z.; Freund, H.; Sundmacher, K., Selective Oxidation of Cyclohexanol to Cyclohexanone in the Ionic Liquid 1-Octyl-3-Methylimidazolium Chloride. *Chem. Commun.* 2011, 47 (33), 9354-9356.

[51] Usui, Y.; Sato, K., A Green Method of Adipic Acid Synthesis: Organic Solvent- and Halide-Free Oxidation of Cycloalkanones with 30% Hydrogen Peroxide. *Green Chem.* 2003, 5, 373-375.

[52] Singh, P. P.; Ambika; Chauhan, S. M. S., Chemoselective Epoxidation of Electron Rich and Electron Deficient Olefins Catalyzed by Meso-Tetraarylporphyrin Iron(III) Chlorides in Imidazolium Ionic Liquids. *New J. Chem.* 2012, 36 (3), 650-655.

[53] Liu, G.; Hou, M. Q.; Song, J. Y.; Zhang, Z. F.; Wu, T. B.; Han, B. X., Ni2+-Containing Ionic Liquid Immobilized on Silica: Effective Catalyst for Styrene Oxidation with H2O2 at Solvent-Free Condition. *J. Mol. Catal. A* 2010, 316 (1-2), 90-94.

[54] Yamaguchi, K.; Yoshida, C.; Uchida, S.; Mizuno, N., Peroxotungstate immobilized on ionic liquid-modified silica as a heterogeneous epoxidation catalyst with hydrogen peroxide. *J. Am. Chem. Soc.* 2005, 127 (2), 530-531.

[55] Berardi, S.; Bonchio, M.; Carraro, M.; Conte, V.; Sartorel, A.; Scorrano, G., Fast Catalytic Epoxidation with H2O2 and [gamma-SiW10O36(PhPO)2]4 in Ionic Liquids under Microwave Irradiation. *J. Org. Chem.* 2007, 72 (23), 8954-8957.

[56] Liu, L. L.; Chen, C. C.; Hu, X. F.; Mohamood, T.; Ma, W. H.; Lin, J.; Zhao, J. C., A Role of Ionic Liquid as an Activator for Efficient Olefin Epoxidation Catalyzed by Polyoxometalate. *New J. Chem.* 2008, 32 (2), 283-289.

Chapter 3

Protic and Nonprotic Ionic Liquids in Polar Diels-Alder Reactions Using Properly Substituted Heterocycles and Carbocycles as Dienophiles. A DFT study

Pedro M. E. Mancini, Carla M. Ormachea,
Claudia D. Della Rosa, María N. Kneeteman and
Luis R. Domingo

Additional information is available at the end of the chapter

1. Introduction

The Diels-Alder (D-A) reaction is one of the most useful processes in preparative organic chemistry. Its potential in heterocyclic chemistry and natural products synthesis is very well known. It provides the chemist with one of his best tool for the preparation of cyclic compounds having a six-membered ring. The process is in one step inter or intramolecular from a diene and dienophile bearing an almost unlimited number of variants. It worth noting that these variants exist not only in the substitution of the reaction component but also in the electronic nature of these dienes and dienophiles. (Carruthers W, 1990; Fringelli F. et al 2002)

The D-A reaction has remained as one of the most powerful organic transformations in chemical synthesis, particularly in obtaining polycyclic rings. With the potential of forming carbon-carbon, carbon-heteroatom and heteroatom-heteroatom bonds, the reaction underlies the synthesis of diverse carbo- and heterocycle compounds. (Corey, 2002)

The design and discovery of ionic liquids (ILs) displaying a melting point lower than 100 ºC, mainly room temperature ionic liquids (RTILs), have been the subject of considerable research efforts over the past decade RTILs have attracted considerable attention because these are expected to be ideal solvents to provide novel reactions in green chemistry (Hitchcock, et al, 1986; Welton, 1999). The interest in this class of molecules arises from their use as liquid media for a variety of chemical transformations specially D-A reactions, as substitutes of organic molecular solvents. ILs has importance due to their unique properties. Thus, this class of molecules is increasingly employed in organic chemistry, material sciences and physical chemistry (Wasserscheid & Keim, 2000; Welton, 1999). An IL is a salt -substance composed exclusively of cations and anions-, and this fact differentiates them from simple ionic solutions, in which ions are dissolved in a molecular medium. They are also different from inorganic molten salts because their melting points are lower than 100 ºC (most of them exist in liquid form at or near room temperature).

RTILs exhibit a variety of desirable properties, such as negligible vapor pressure, which makes them interesting for various applications. In particular, the option of fine-tuning chemical and physical properties by an appropriate choice of cations and anions has stimu-lated much of the current excitement with respect to these compounds and has led to the term "design solvents". ILs have high solvation ability for a wide variety of polar, non polar, organic and inorganic molecules as well as organometallic compounds. Moreover, the possi-bility of changing their properties allowing the selective salvation of solvents and thus con-trol the mutual miscibility of particular organic compounds such as for instance alcohols and water. As a consequence, the characterization of the properties of different classes of ILs used as solvent for specific applications and for chemical reactions and catalysis, have been intensively investigated (Mancini, P.M.E., et al, 2012).

In base to their ionic nature, the structure of ILs incorporates different level of complexity. First and in order to maintain local electro neutrality, the high-charge density parts of cations and anions must create a three-dimensional network where the nearest neighbors of a given ion are of opposite sign. Second, the low-charge density residues that are often presents in the ions (generally as alkyl side chain) are segregated from the polar network, forming non polar do-mains. This nano-segregation/structuration between polar and non polar regions, first predict-ed by molecular dynamics simulation studies and later corroborated by diffraction techniques, implies the existence of differentiated and complex interactions not only in pure ionic liquid but also in their mixtures with molecular solutes or even other ionic liquids.

The imidazolium ILs were used to investigate the influence of the alkyl chain length and the presence of functional group on the cation on the polarity and the three Kamlet and Taft pa-rameters (π^*, α, and β) measured the solvent dipolarity/polarizability, hydrogen-bond do-nating (HBD) acidity and hydrogen-bond accepting (HBA) basicity. The results shown that the length of the alkyl chain on the cation has a significant influence on polarizability and on HBD, but only a small influence on HBA and π^*, indicating that in these type of IL's, HBD is a major contributor to polarity.

In the last years we reported the electrophilic behavior of different aromatic heterocyclopen-tadienes properly mono and disubstituted with an electron-withdrawing group such as ni-tro or carboxylate in their exposure to different dienes under thermal conditions, using molecular solvents and ionic liquids, respectively. Moreover, we use as dienophile in this type of polar D-A reactions (P-DA) with normal electron demand, nitrotosylindoles, nitro-benzofuranes and nitrobenzothiophenes, properly substituted.

In general, these reactions are domino processes which are initialized by a P-DA reaction to give the formal [4+2] cycloadduct, which undergo an irreversible elimination of nitrous acid; this elimination is the factor responsible for the feasibility of the overall process. An alternative way is the hetero D-A reaction which takes place when some thiophene derivatives act as electrophiles. This last behavior is probably due to the improved aromatic character of these heterocycles.

For P-DA reactions one of the most interesting aspects is its solvent dependence. Moreover, in recent years, these reactions have been subject of several studies in order to enhance the reactivity. For specific P-DA reactions was demonstrated that the aqueous solutions have remarkable increase in reactivity and selectivity, and these results were discussed in terms of hydrogen-bond (HB) formation. Protic ionic liquids (PILs) with similar properties to water, such as being highly ordered media and good hydrogen bonding donor, have also been shown to have potential influence the outcome of P-DA reactions. Also, in this direction it is interesting discuss which is the influence on these reactions of non protic ionic liquids.

Due to our interest in the cycloaddition chemistry of substituted aromatic heterocycles with electron-withdrawing groups, we have reported that 2- and 3-nitrotosylpyrroles, 2-nitrofurane, 2- and 3-nitrothiophenes, 1-tosyl-3-nitroindole, 2-nitrobenzofurane, 5- y 8-nitroquinolines and 1-nitronaphtalene, react as electrophiles in normal electron demand D-A reaction (Biolatto, B., et al, 2001) (Della Rosa, C., et al, 2004, 2005, 2007,2010, 2011; Brasca, R., et al, 2010, 2011; Cancian, S., et al, 2010; Paredes, E., et al, 2007). These dienophiles were exposed to different dienes strongly, moderately and poorly activated under thermal conditions using molecular solvents as reaction media and in the same cases using PILs. In these reactions, the best results were obtained with the PILs and with chloroform as molecular solvent, due to its potential character HBD which could be influence the reactivity of the reaction systems. The participation of N-tosyl-nitropyrroles in cycloaddition reactions made possible a one spot simple indole synthesis.

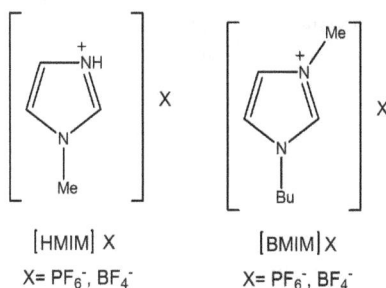

[HMIM] X
X= PF_6^-, BF_4^-

[BMIM] X
X= PF_6^-, BF_4^-

For to analyze the influence of room temperature ionic liquids (RTILs) in this type of polar cycloaddition reactions in which the dienophiles are relatively poor, ethylammoniun nitrate (EAN), 1-methylimidazolium tetrafluoroborate ([HMIM][BF_4]), and 1-methylimidazolium hexafluorophosphate ([HMIM][PF_6]), 1-n-butyl-3-methylimidazolium tetrafluoroborate

([BMIM][BF$_4$]) and 1-n-butyl-3-methylimidazolium hexafluorophosphate ([BMIM][PF$_6$]) were selected as reaction media. To explore the normal electron demand D-A dienophilicity of the proposed dienophiles we choose isoprene, 1-trimethylsilyloxy-1,3-butadiene, and 1-methoxy-3-trimethylsilyloxy-1,3-butadiene (Danishefsky diene) as dienes.

In general, in all cases studied the presences of ILs have two effects. On one hand improved the yields in reaction conditions softer than those when we used molecular solvents. Moreover manifest a clear tendency to the aromatization of adducts. In particular, when 2- and 3-nitrothiophene reacts in thermal conditions with isoprene in a molecular solvent we observed the corresponding hetero D-A adduct, however if an IL is the reaction media the reactions follow the normal D-A curse.

Considering that microwave irradiation has been used to enhance organic reactions in which an ionic liquid is used as the solvent, we have realized some experiences using a combination of microwave and PILs. In this case we noted that the microwave plus PILs constituted a synergetic mixture with strong effects on the reaction yields. ILs absorbed microwave irradiation extremely well and transfer energy rapidly by ionic conduction. At this point of the study we suppose that microwave irradiation has a major effect in a special range of energy activation barriers (ΔE). When the reaction ΔE is too low, the presence of microwave radiation is not especially important, and if the value is extremely high, this irradiation would not take effect.

Part of this work is specifically concerned with theoretically studies using DFT methods. We try to obtain information about the factors affecting reactivity and selectivity. Previous studies have been developed in this type of calculation including one molecule of the ILs corresponding cation coordinated with the dienofile. In this new generation of theoretical studies the aim is to get a better solvatation model including in the system some molecules (in this case, IL's cations and anions) obtaining a "solvent box" which involves the reactive molecule like a 3D electrostactic network.

2. Main Objetives

The aim of the present review is twofold. The first purpose is to analyze the influence of RTILs protic and nonprotic in polar cycloaddition reactions in which the dienophiles are relatively poor -aromatic carbocycles and heterocycles compounds substituted with electron withdrawing groups-. For this purpose alkylimidazolium and dialkylimmidazolium-based ILs have been selected because the differences in their HBD acidity. The second purpose concerned with theoretically studies using DFT methods. We try to obtain information about reaction mechanisms which would be affect reactivity and selectivity. In general, it would be possible demonstrated that the ILs solvent effect in these reactions is in general determined by the solvent hydrogen bond donation ability.

3. Results and Discussion

With the purpose of comparison and reference, we have taken in account the results obtained when the aromatic substrates heterocycles and carbocycles proposed, adequately substituted by electron with-drawing groups, take part in cycloaddition reactions with diverse dienes of nucleophilicity variable, development in molecular solvents or in PILs. These results will be compared with those reached when the cycloaddition reactions are developed in nonprotic ILs. If it is necessary the cited compilation will be included in the corresponding tables of results, looking for an appropriate comprehension of the influence of the reaction media in this type of Diels-Alder reactions.

3.1. Monocyclic five membered and benzofused five membered nitroheteroaromatic compounds, as dienophiles

The study of dienophilic character of substituted nitroaromatic heterocycles in the presence of ILs was carried out employing 2-nitrofuran (1a), 2,5-dimethyl-3-nitrofuran (1b), 1-tosyl-2-nitropyrrole (2a), 1-tosyl-3-nitropyrrole (2b), 2- and 3- nitrothiophene (3a, 3b), 1-tosyl-3-nitroindol (4) and 2-nitrobenzofurane (5). In addition, isoprene (6), 1- trimethylsiloxy-1,3-butadiene (7), 1-methoxy-3-trimethylsiloxy-1,3-butadiene (Danishefsky´s diene) (8) were selected as dienes, covering an attractive spectre of nucleophilic character (Figure 1).

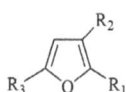 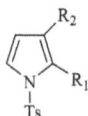

1a R 1= NO$_2$, R$_2$=R$_3$=H   2a R$_1$= NO$_2$, R$_2$ = H   3a R$_1$= NO$_2$, R$_2$=H
 b R$_2$= NO$_2$, R$_1$=R$_3$=Me   b R$_1$= H, R$_2$= NO$_2$ b R$_1$ = H, R$_2$= NO$_2$

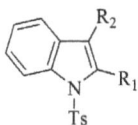 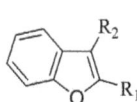

4 R$_1$= NO$_2$, R$_2$ = H   5 R$_1$= NO$_2$, R$_2$ = H

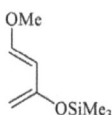

6   7   8

Figure 1. Dienophiles and dienes used in the different experiences.

When 1a was reacted with the less reactive isoprene 6 in a sealed ampoule at 60ºC for 24 h using [BMIM][BF$_4$] or [BMIM][PF$_6$] as solvent, respectively, the reactions proceeded to produce a mixture of isomeric benzofurans 10a and 10b (1:1) as the principal products with reasonable yield and dihydrobenzofurans 9a and 9b (1:1) (Figure 2). If the time of the reaction increased to 48 h we observe a 1:1 mixture of isomeric benzofurans 10a and 10b in reasonable yield and traces of the isomeric dihydrobenzofurans 9a and 9b. Similar results were observed when the reaction was development in EAN and [HMIM][BF$_4$] although with these solvents the yields were majors (Table 1).

In the same manner, in the case of 1-tosyl-2-nitropyrrole 2a, it reacted with isoprene in [BMIM][BF$_4$] or [BMIM][PF$_6$] (60ºC, 24 h) furnishing indole isomers 12a,b as the principal products in moderate yield, and dihydroindole isomers 11a,b. All addition products showed extrusion of the nitro group as nitrous acid. (Della Rosa, et al, 2007) (Table 1)

In contrast with the above mentioned behavior, when 2-nitrothiophene 3a was tested with 6, it gave traces of pyrrolyl-thiophene 13 formed by a heterocycloaddition followed by thermal rearrangement. (Della Rosa, et al, 2004) (Figure 2, Table 1). This unexpected behavior was also found with other compounds with stronger aromatic character. The observed low yield in this reaction would be attributed to the interaction between the nitro group and the ILs (Table 1).

Entry	Dienophile[a]	Conditions[b]	Products	Yield (%)[c]
1	1a	EAN	9a,b; 10a,b	40; 17
2		[HMIM][BF$_4$]	9a,b; 10a,b	30; 15
3		[HMIM][PF$_6$]	9a,b; 10a,b	28; 14
4		[BMIM][BF$_4$]	9a,b; 10a,b	25; 05
5		[BMIM][PF$_6$]	9a,b; 10a,b	26; 05
6	2a	EAN	11a,b;12a,b	40; 12
7		[HMIM][BF$_4$]	11a,b;12a,b	30; 10
8		[HMIM][PF$_6$]	11a,b;12a,b	28; 09
9		[BMIM][BF$_4$]	11a,b;12a,b	20; 04
10		[BMIM][PF$_6$]	11a,b;12a,b	21; 05
11	3a	EAN	13	18
12		[HMIM][BF$_4$]	13	12
13		[HMIM][PF$_6$]	13	12
14		[BMIM][BF$_4$]	13	Traces
15		[BMIM][PF$_6$]	13	Traces

[a] Diene/dienophile ratio 12:1
[b] Reaction´s time 24 h, reaction´s temperature 60 ºC.
[c] Based on consumed dienophile.

Table 1. Diels-Alder reactions of 2-nitroheterocycles with isoprene.

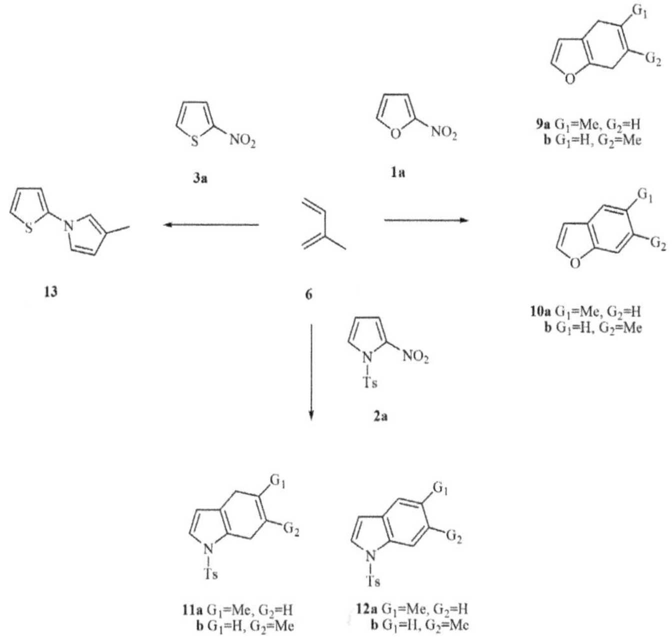

Figure 2. Diels-Alder reactions of aromatic 2- nitroheterocycles with isoprene.

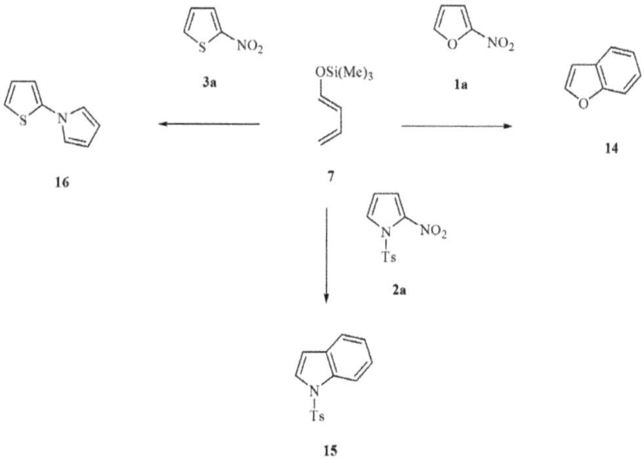

Figure 3. Diels-Alder reactions of aromatic 2- nitroheterocycles with diene 7.

The reactions of 1a with 1-trimethylsilyloxy-1,3-butadiene using [BMIM][BF$_4$] and [BMIM] [PF$_6$], respectively, as solvents, in seals ampoule (60ºC, 24 h), offered in all cases good yield in benzofuran 14. The yields were lower that those obtained with EAN, [HMIM][BF$_4$] and [HMIM][PF$_6$], as solvent. On the other hand, the reaction of 2a with this diene produced N-tosylindole 15, with reasonable yield little lower when PILs were used. In the reaction of 3a with 7 we observed again traces of the corresponding hetero Diels-Alder product 16 followed the same trend that the reaction with isoprene (Figure 3)(Table 2).

Entry	Dienophile[a]	Conditions[b]	Products	Yield (%)[c]
1	1a	EAN	14	62
2		[HMIM][BF$_4$]	14	58
3		[HMIM][PF$_6$]	14	57
4		[BMIM][BF$_4$]	14	42
5		[BMIM][PF$_6$]	14	43
6	2a	EAN	15	63
7		[HMIM][BF$_4$]	15	51
8		[HMIM][PF$_6$]	15	52
9		[BMIM][BF$_4$]	15	39
10		[BMIM][PF$_6$]	15	38
11	3a	EAN	16	15
12		[HMIM][BF$_4$]	16	12
13		[HMIM][PF$_6$]	16	12
14		[BMIM][BF$_4$]	16	Traces
15		[BMIM][PF$_6$]	16	Traces

[a] Diene/dienophile ratio 3:1
[b] Reaction´s time 24 h, reaction´s temperature 60 ºC.
[c] Based on consumed dienophile.

Table 2. Diels-Alder reactions of 2- nitroheterocycles with 1-trimethylsilyloxy-1,3-butadiene.

The reaction of Danishefsky´s diene with 1a using the nonprotic ILs cited in the before paragraphs yielded 5-hydroxyibenzofuran 17 in reasonable yield. Similarly to the reactions with isoprene and 1-trimethylsilyloxy-1,3-butadiene, the best yield was observed with EAN (Mancini, P.M.E., et al, 2011).In turn the reactions of 2a with diene 8 developed in these nonprotic ILs offered 1-tosyl-5-hydroxyindole in good yield. However, in the reactions of 3a with this diene and the neoteric solvents, we observed only traces of the aromatic product 19. The results derived from the aromatization of the nitro-adducts promoted by the loss of the nitro and methoxyl groups as nitrous acid and methanol, respectively. The intermediate that suffered nitrous acid extrusion and retained the methoxy group was not detected in any case. In all cases the presence of PIL´s as reaction media, improve the yields respect to the use of nonpolar ILs (Figure 4) (Table 3).

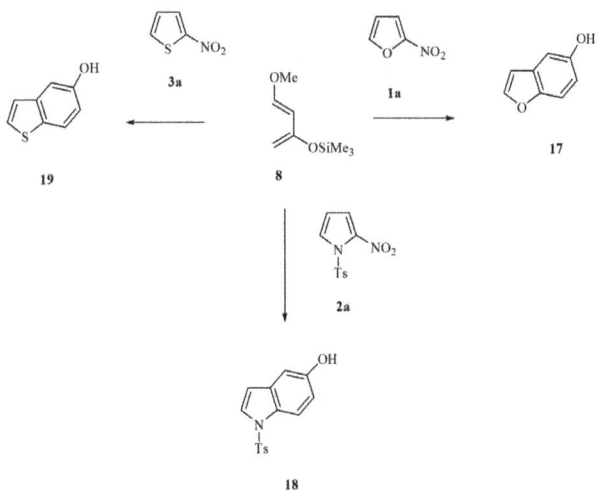

Figure 4. Diels-Alder reactions of aromatic 2- nitroheterocycles with Danishefsky´s diene.

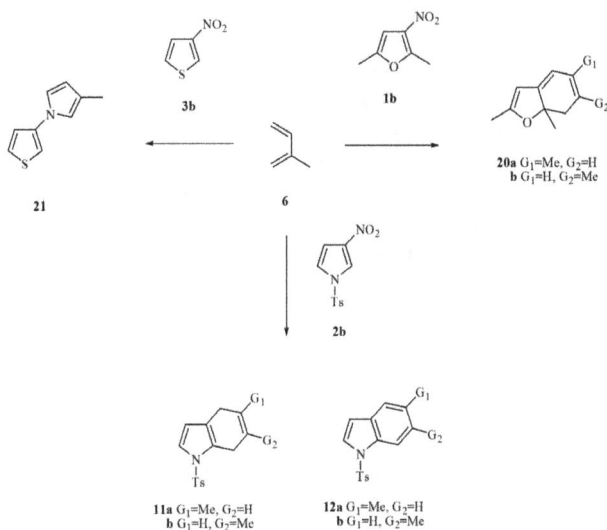

Figure 5. Diels-Alder reactions of aromatic 3- nitroheterocycles with isoprene.

Entry	Dienophile[a]	Conditions[b]	Products	Yield (%)[c]
1	1a	EAN	17	65
2		[HMIM][BF$_4$]	17	59
3		[HMIM][PF$_6$]	17	60
4		[BMIM][BF$_4$]	17	40
5		[BMIM][PF$_6$]	17	38
6	2a	EAN	18	57
7		[HMIM][BF$_4$]	18	55
8		[HMIM][PF$_6$]	18	54
9		[BMIM][BF$_4$]	18	42
10		[BMIM][PF$_6$]	18	41
11	3a	EAN	19	42
12		[HMIM][BF$_4$]	19	36
13		[HMIM][PF$_6$]	19	35
14		[BMIM][BF$_4$]	19	22
15		[BMIM][PF$_6$]	19	21

[a] Diene/dienophile ratio 2:1
[b] Reaction´s time 24 h, reaction´s temperature 60 ºC.
[c] Based on consumed dienophile.

Table 3. Diels-Alder reactions of 2-nitroheterocycles with Danishefsky´s diene.

When 1-tosyl-3-nitropyrrole 2b was tested with isoprene as diene and [BMIM][BF$_4$], [BMIM] [PF$_6$] as solvent (60ºC, 24 h), the reactions afforded a mixture of regioisomeric cycloadducts previously informed: 11 a,b; 12 a,b. In the same direction and due to the impossibility of obtaining 3-nitrofuran, when 2,5-dimethyl-3-nitrofuran 1b was exposed to 6, the D-A reaction (60ºC/24 h) in the nonprotic ILs proceeded to furnish the mixture of regioisomers 20a and 20b. Once again the reaction of 3b, and this diene development in ILs yielded only traces of the pyrrole derivative 21 in the same manner that 3a, formed by a heterocycloaddition followed by thermal rearrangement (Figure 5) (Table 4). In all cases the yields of these reactions in presence of the PILs are better

Entry	Dienophile[a]	Conditions[b]	Products	Yield (%)[c]
1	1b	EAN	20a,b	38; 15
2		[HMIM][BF$_4$]	20a,b	30: 15
3		[HMIM][PF$_6$]	20a,b	28; 14
4		[BMIM][BF$_4$]	20a,b	25; 05
5		[BMIM][PF$_6$]	20a,b	26; 05
6	2b	EAN	11a,b;12a,b	41; 12

Entry	Dienophile[a]	Conditions[b]	Products	Yield (%)[c]
7		[HMIM][BF$_4$]	**11a,b;12a,b**	30; 10
8		[HMIM][PF$_6$]	**11a,b;12a,b**	29; 09
9		[BMIM][BF$_4$]	**11a,b;12a,b**	20; 04
10		[BMIM][PF$_6$]	**11a,b;12a,b**	21; 05
11	**3b**	EAN	**21**	15
12		[HMIM][BF$_4$]	**21**	11
13		[HMIM][PF$_6$]	**21**	11
14		[BMIM][BF$_4$]	**21**	Traces
15		[BMIM][PF$_6$]	**21**	Traces

[a] Diene/dienophile ratio 3:1
[b] Reaction's time 24 h, reaction's temperature 60 °C.
[c] Based on consumed dienophile.

Table 4. Diels-Alder reactions of 3- nitroheterocycles with isoprene.

Exposure of 1-tosyl-3-nitropyrrole 2b to dienes 7 and 8 in the presence of [BMIM][BF$_4$], [BMIM][PF$_6$] yielded 1-tosyl-indole 15 and 1-tosyl-indole-6-ol 23, respectively with moderate yield. At the same time, mononitrated substrate 3b in its reaction with diene 7 afforded traces of the pyrrolyl derivative 22. However, 3b did not undergo cycloaddition with diene 8. Probably this behavior is a consequence of the special reactivity of this substrate connected with its aromaticity (Figures 6 and 7) (Tables 5 and 6).

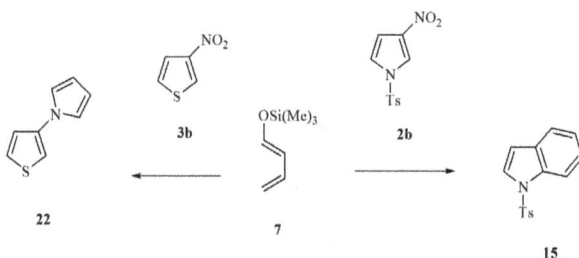

Figure 6. Diels-Alder reactions of aromatic 3- nitroheterocycles with diene 7.

Entry	Dienophile[a]	Conditions[b]	Products	Yield (%)[c]
1	**2b**	EAN	**15**	61
2		[HMIM][BF$_4$]	**15**	51
3		[HMIM][PF$_6$]	**15**	50
4		[BMIM][BF$_4$]	**15**	35

Entry	Dienophile[a]	Conditions[b]	Products	Yield (%)[c]
5		[BMIM][PF$_6$]	15	36
6	3b	EAN	22	12
7		[HMIM][BF$_4$]	22	11
8		[HMIM][PF$_6$]	22	11
9		[BMIM][BF$_4$]	22	Traces
10		[BMIM][PF$_6$]	22	Traces

[a] Diene/dienophile ratio 3:1
[b] Reaction´s time 24 h, reaction´s temperature 60 °C.
[c] Based on consumed dienophile.

Table 5. Diels-Alder reactions of 3- nitroheterocycles with 7.

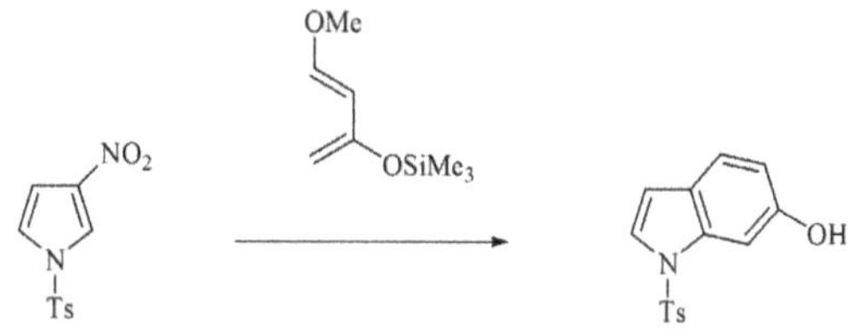

Figure 7. Diels-Alder reactions of aromatic 3- nitroheterocycles with Danishefsky´s diene.

Entry	Dienophile[a]	Conditions[b]	Products	Yield (%)[c]
1	2b	EAN	23	62
2		[HMIM][BF$_4$]	23	55
3		[HMIM][PF$_6$]	23	54
4		[BMIM][BF$_4$]	23	43
5		[BMIM][PF$_6$]	23	41

[a] Diene/dienophile ratio 3:1
[b] Reaction´s time 24 h, reaction´s temperature 60 °C.
[c] Based on consumed dienophile.

Table 6. Diels-Alder reactions of 3- nitroheterocycles with 9.

3.1.1. Nitroindole as dienophiles

When 1-tosyl-3-nitroindole 4 was tested with isoprene as diene and [BMIM][BF$_4$] or [BMIM][PF$_6$] as solvent (60°C, 24 h), the reactions afforded a mixture of regioisomeric cycloadducts 25a,b, as principal products in reasonable yield, and traces of the regioisomers 24a,b. On the

other hand the reactions of 4 with the dienes 7 and 8 in these ILs produced N-tosylcarbazole
26 and the hydroxycarbazole 27 in reasonable good yield. In all cases the yields were lower
than those obtained with PILs (EAN, [HMIM][BF$_4$] and [HMIM][PF$_6$]) (Figure 8) (Table 7).

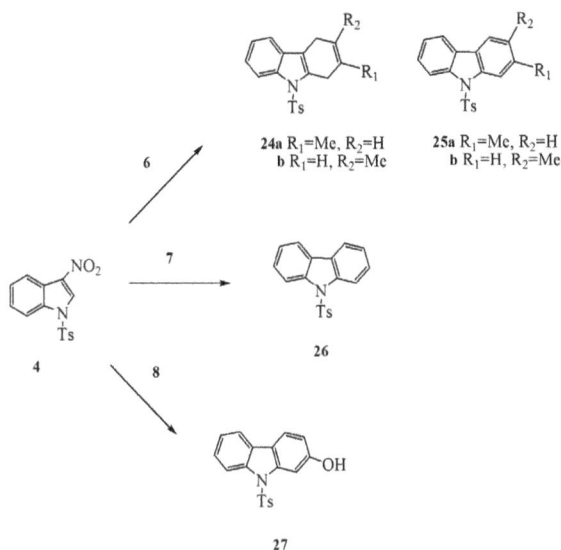

Figure 8. Diels-Alder reactions of 1-tosyl-3-nitroindol with dienes 6, 7 and 8.

Entry	Dienophile[a]	Conditions[b]	Products	Yield (%)[c]
1	6	EAN	24a,b; 25a,b	45; 05
2		[HMIM][BF$_4$]	24a,b; 25a,b	42: 03
3		[HMIM][PF$_6$]	24a,b; 25a,b	42; 04
4		[BMIM][BF$_4$]	24a,b, 25a,b	30; 02
5		[BMIM][PF$_6$]	24a,b; 25a,b	31; 02
6	7	EAN	26	62
7		[HMIM][BF$_4$]	26	55
8		[HMIM][PF$_6$]	26	56
9		[BMIM][BF$_4$]	26	40
10		[BMIM][PF$_6$]	26	41
11	8	EAN	27	72
12		[HMIM][BF$_4$]	27	65
13		[HMIM][PF$_6$]	27	63

Entry	Dienophile[a]	Conditions[b]	Products	Yield (%)[c]
14		[BMIM][BF$_4$]	27	51
15		[BMIM][PF$_6$]	27	48

[a] Diene/dienophile ratio 3:1
[b] Reaction´s time 24 h, reaction's temperature 60 °C.
[c] Based on consumed dienophile.

Table 7. Diels-Alder reactions of dienophile 4 with different dienes.

3.1.2. 2-NitroBenzofuran as dienophile

The reactions of 5 with isoprene employing [BMIM][BF$_4$] or [BMIM][PF$_6$] as solvent (60 °C, 24 h) afforded the mixture of aromatic regioisomeric cycloadducts 28a,b as principal products and traces of dihydrodibenzofurans 29a and 29b. On the other hand, reactions of 5 with 1-trimethylsilyloxy-1,3-butadiene (60[a]c, 24 h, [BMIM][BF$_4$] and [BMIM][PF$_6$]) yielded dibenzofuran 30 with loss of trimethylsilyloxy and nitro groups. The yield is good. In the same way, in the reaction of 2-nitrobenzofuran with the Danishefsky`s diene hidroxy aromatic cycloadduct 31 was obtained with very good yield and complete regioselectivity (Figure 9, Table 8). With these solvents the yields were lower than using PILs (EAN, [HMIM][BF$_4$] and [HMIM][PF$_6$])

Entry	Dieno[a]	Conditions[b]	Products	Yield (%)[c]
1	6	EAN	28a,b; 29a,b	42; 04
2		[HMIM][BF$_4$]	28a,b; 29a,b	38: 03
3		[HMIM][PF$_6$]	28a,b; 29a,b	38; 04
4		[BMIM][BF$_4$]	28a,b; 29a,b	27; 02
5		[BMIM][PF$_6$]	28a,b; 29a,b	27; 02
6	7	EAN	30	58
7		[HMIM][BF$_4$]	30	52
8		[HMIM][PF$_6$]	30	51
9		[BMIM][BF$_4$]	30	36
10		[BMIM][PF$_6$]	30	35
11	8	EAN	31	66
12		[HMIM][BF$_4$]	31	59
13		[HMIM][PF$_6$]	31	58
14		[BMIM][BF$_4$]	31	45
15		[BMIM][PF$_6$]	31	42

[a] Diene/dienophile ratio 3:1
[b] Reaction´s time 24 h, reaction´s temperature 60 °C.
[c] Based on consumed dienophile.

Table 8. Diels-Alder reactions of 2-nitrobenzofuran with different dienes.

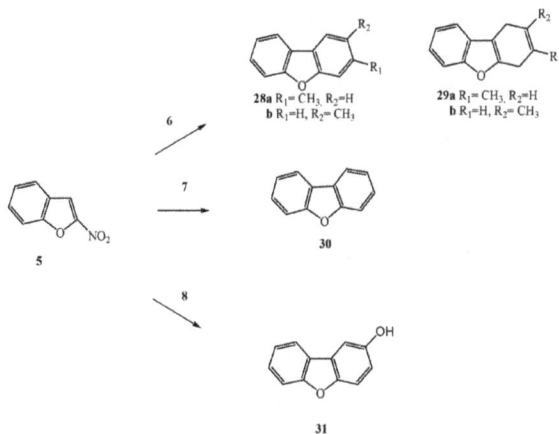

Figure 9. Diels-Alder reactions of 2-nitrobenzofuran with dienes 6, 7 and 8.

3.2. Azanitronaphathalenes as dienophiles

It was explored the cycloaddition reactions between 5-nitro and 8-nitroquinolines (32a and 32b) with the dienes 6, 7 y 8, respectively, in presence of [BMIM][BF$_4$] and [BMIM][PF$_6$] (60°C, 24h).

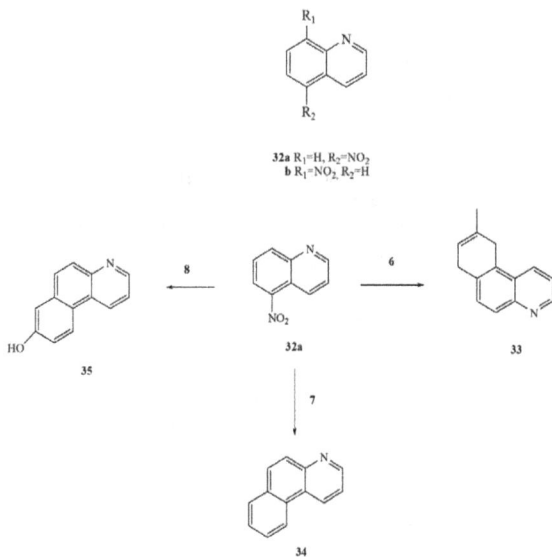

Figure 10. Diels-Alder reactions of 5-nitroquinoline with 6,7 and 8.

The reactions of 5-nitroquinoline with these dienes, yield the same products that using [HMIM][BF$_4$] as solvent but with lower yield. In the cases of dienes 7 and 8 the normal addition products 34 and 35, respectively, show complete aromatization due to the loss of the nitro and trimethylsyliloxy groups. The product 35 was obtained with complete regioselectivity. However, when the diene 6 was used the observed cycloaddition products was (9-Methyl-7,10-dihydro-benzo [f]quinoline) 33 with traces of its regioisomers (Cancian, et al, 2010) (Figures 10) (Tables 9).

Entry	Diene^a	Conditions^b	Products	Yield (%)^c
1	6	EAN	33	18
2		[HMIM][BF$_4$]	33	16
3		[HMIM][PF$_6$]	33	16
4		[BMIM][BF$_4$]	33	12
5		[BMIM][PF$_6$]	33	12
6	7	EAN	34	22
7		[HMIM][BF$_4$]	34	20
8		[HMIM][PF$_6$]	34	19
9		[BMIM][BF$_4$]	34	14
10		[BMIM][PF$_6$]	34	14
11	8	EAN	35	25
12		[HMIM][BF$_4$]	35	20
13		[HMIM][PF$_6$]	35	20
14		[BMIM][BF$_4$]	35	15
15		[BMIM][PF$_6$]	35	15

^{a} Diene/dienophile ratio 3:1
^{b} Reaction´s time 24 h, reaction´s temperature 60 ºC.
^{c} Based on consumed dienophile.

Table 9. Diels-Alder reactions of dienophile 32a with different dienes.

Entry	Diene^a	Conditions^b	Products	Yield (%)^c
1	6	EAN	36	17
2		[HMIM][BF$_4$]	36	15
3		[HMIM][PF$_6$]	36	15
4		[BMIM][BF$_4$]	36	12
5		[BMIM][PF$_6$]	36	12
6	7	EAN	37	20
7		[HMIM][BF$_4$]	37	18
8		[HMIM][PF$_6$]	37	19

Entry	Dieno[a]	Conditions[b]	Products	Yield (%)[c]
9		[BMIM][BF$_4$]	**37**	13
10		[BMIM][PF$_6$]	**37**	13
11	**8**	EAN	**38**	24
12		[HMIM][BF$_4$]	**38**	20
13		[HMIM][PF$_6$]	**38**	20
14		[BMIM][BF$_4$]	**38**	15
15		[BMIM][PF$_6$]	**38**	15

[a] Diene/dienophile ratio 3:1
[b] Reaction´s time 24 h, reaction´s temperature 60 °C.
[c] Based on consumed dienophile.

Table 10. Diels-Alder reactions of dienophile 32b with different dienes.

In the same way, the reactions of 8-nitroquinoline with these dienes, yield the same products that using [HMIM][BF$_4$] as solvent but with lower yield. When the were dienes 7 and 8
the normal addition products 37 and 38, respectively, show complete aromatization due to
the loss of the nitro and trimethylsyliloxy groups. The product 37 was obtained with complete regioselectivity. However, when the diene 6 was used the observed cycloaddition
products was (9-Methyl-7,10-dihydro-benzo[h]quinoline) 36 with traces of its regioisomers
(Cancian, et al, 2010) (Figures 11) (Tables 10).

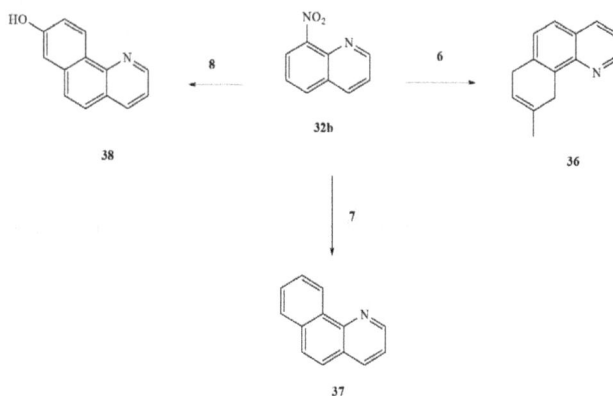

Figure 11. Diels-Alder reactions of 8-nitroquinoline with 6, 7 and 8.

3.3. Nitronaphthalenes as dienophiles

To explore the normal electron-demand D-A dienophilicity of nitronaphtahlenes in presence
of no-protic ILs ([BMIM][BF$_4$] and [BMIM][PF$_6$]) we selected 1-nitronaphthalene 39 as electrophile and 8 as diene.

When 39 and 8 were heated in a sealed ampoule (60°C, 24 h) using [BMIM][BF₄] and [BMIM] [PF₆], respectively, as solvents, in both cases ca de 50% of 2-hydroxy-phenanthrene 40, was regioselectively produced. The regioselectivity of these reactions was controlled by both the nitro group of the dienophile and the methoxyl group of Danishefsky´s diene. This product was obtained when a PILs was used (e.g. [HMIM][BF₄]), however in this case with major yield The preference for the normal D-A products in the presence of ILs respect the use of molecular solvent, probably is due to the increase of the electrophilicity of the dinenophile (Figure 12).

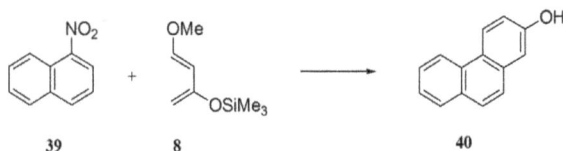

39   8   40

Figure 12. Diels- Alder reaction of 1-nitronaphthalene with Danishefsky's diene.

4. Diels-Alder reactions employing ionic liquids and microwave irradiation

Ionic liquids are becoming promising and useful substitutes for standard organic solvents. Not only they are environmentally benign, they also possess unique chemical and physical properties. Moreover, microwave irradiation has been used to enhance organic reactions in which an ionic liquid is used as the solvent. Ionic liquids absorb microwave irradiation extremely well and transfer energy rapidly by ionic conduction.

The D-A transformations usually require harsh conditions (high temperatures and pressures) and long reaction times. These cycloadditions were the first reaction type to be examined in conjunction with microwave irradiation. Microwave irradiation has also been used to enhance organic reaction in which an ionic liquid is used as the solvent.

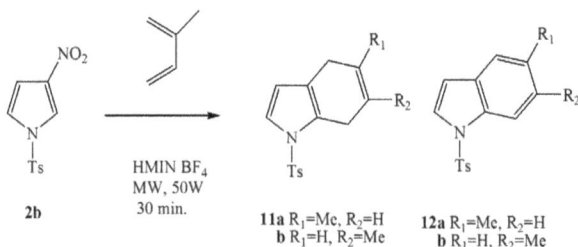

2b   HMIN BF₄
                MW, 50W
                30 min.

11a R₁=Me, R₂=H   12a R₁=Me, R₂=H
   b R₁=H, R₂=Me   b R₁=H, R₂=Me

Figure 13. Diels-Alder reaction of 1-tosyl-3-nitropyrrole with isoprene.

With microwave irradiation (50 W, 30 min.) and [HMIM] [BF$_4$] as reaction media, 1-tosyl-3-nitropyrrole reacts with isoprene yielding the mixture of isomeric dihydroindoles 11a and 11b and indoles 12a and 12b as the principal.products (global yield 95%). (Figure 13)

In turn, nitrobenzene react with isoprene using benzene as solvent to offer pyrroliybenzene as product in reasonable yield. If the diene is 8 we do not observe reaction probably due to the strong aromatic character of this substrate. It call our attention the absence of reactivity when nitrobenzene reacted with 6 in presence of [HMIM] [BF$_4$] and microwave irradiation. This result would be consequence of the strong interaction between the nitro group and the PIL.

5. Theoretical studies

5.1. General

The Density Functional Theory (DFT) is a model of quantum mechanics used to obtain electronic structure of different systems, in this case, molecules involved in the Diels-Alder reaction. Considering this theory, there are some parameters (or indexes) that can be used to explain the reactivity and regioselectivity of the cycloaddition reactions. The most significant ones are the chemical hardness (η),that describes the resistance of the chemical potential to a change in the number of electrons, and the electronic chemical potential (μ), which is usually associated with the charge-transfer (CT) ability of the system in its ground state geometry. Both quantities can be approximated in terms of the energies of the HOMO and LUMO frontier molecular orbitals (Eqs. 1 and 2) (Domingo, et al, 2002; Domingo & Aurell, 2002).

$$\eta = \left(\varepsilon_{LUMO} - \varepsilon_{HOMO} \right) \tag{1}$$

$$\mu = \frac{\left(\varepsilon_{LUMO} + \varepsilon_{HOMO} \right)}{2} \tag{2}$$

Based on these parameters Parr (Parr, et al, 1999) introduced the global electrophilicity index (ω), an useful descriptor of the reactivity that allows a quantitative classification of the electrophilicity character of the whole molecule in an unique scale. This index is defined as:

$$\omega = \frac{\mu^2}{2\eta} \tag{3}$$

Current studies based on the DFT have shown that this classification is a powerful tool to predict and justify the feasibility of the D-A process and the type of mechanism involved. The electrophilicity scale describes the effects of electron-donor and electron-withdrawing groups in the diene/dienofile pair. Reactants can be classified for the D-A cycloaddition as strong (>1.50 eV), moderate (1.49 - 0.90 eV) and poor (<0.90 eV) electrophiles.

The difference in electrophilicity between the diene/dienofile pair can be related to the electronic pattern expected in the transition state (TS) of a D-A process and, in consequence, its has been proposed as a measure of the polar character of the reaction.

Additionally, there are local reactivity indexes that are associated with site selectivity in a chemical reaction. They can be calculated from the Fukui function (Parr, R. G. *et al* 1984). Eq. (4) provides a simple and direct formalism to obtain it from an approach based on a relationship between the FMOs.

$$f_k^{\alpha} = \sum_{\mu \in k} |c_{\mu\alpha}|^2 + \sum_{v \neq \mu} c_{\mu\alpha} c_{v\alpha} S_{\mu v} \tag{4}$$

The condensed Fukui function for electrophilic attacks involves the HOMO FMO coefficients (c) and the atomic overlap matrix elements (S).

In this direction, to analyze at which atomic site of a molecule the maximum electrophilicity value is reached, Domingo (Domingo, *et al*, 2002) has introduced Eq. (5)

$$\omega_k = \omega f_k^+ \tag{5}$$

On the other hand, the first approach toward a quantitative description of nucleophilicity has also been reported by Domingo (Domingo, *et al*, 2008). The global nucleophilicity index, N, is defined in Eq. (6)

$$N = \left(\varepsilon_{HOMO,Nu} - \varepsilon_{HOMO,TCE}\right) \tag{6}$$

Where $\varepsilon_{HOMO,TCE}$ is the HOMO energy of tetracyanoethylene (TCE) (taken as a reference molecule due to the fact that it exhibits the lowest HOMO energy in a large series of molecules previously considered in D-A cycloadditions).

Its local counterpart, N_k -Eq. (7)- has been developed with the purpose of identifying the most nucleophilic site of a molecule. In this case Eq (4) is considered for a nucleophilic attack and involves LUMO FMO coefficients.

$$N_k = N f_k^- \tag{7}$$

This nucleophilicity index has been useful to explain the nucleophilic reactivity of some dienes with electrophiles in cycloaddition as well as substitution reactions (Domingo, *et al*, 2008).

In general, the polarity of the normal electron demand D-A process is studied trough global electrophilicity indexes difference between reactants. And the regioselectivity of the normal electron demand D-A reaction, using the local electrophilicity index for dienophiles (electrophiles in the reaction) and the local nucleophilicity index for dienes (nucleophiles in the reaction).

In this stage, we show different theoretical studies related to the polar D-A reactions experimentally described, where the dienophiles are aromatic heterocycles or carbocycles. In the
same way, the mechanism of these reactions, specially respect to regio-, site- and stereochemistry have been analyzed in detail.

5.2. Dienes

Global electronic properties of the dienes experimentally used in the cycloadditions previously described are exposed in Table 11.

Molecule	Global properties			
	μ (a.u.)	η (a.u.)	ω (eV)	N (eV)
Isoprene	-0.1209	0.1962	1.01	2.93
1-trimethylsilyloxy-1,3-butadiene	-0.0911	0.1977	0.75	3.59
Danishefsky's diene	-0.0945	0.1851	0.66	3.55

Table 11. Dienes.

As the substitution of the dienes with electron-donor groups increases, its global electrophilicity decrease. This indicates that the diene became a better nucleophile for D-A reaction.

The effect of this kind of substitutive groups is also reflected in the difference between local
nucleophilicity indexes of the extreme carbon atoms. Considering these values, regioselectivity of D-A reactions is evaluated.

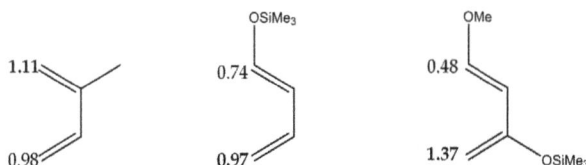

Figure 14. local nucleophilicity index (N_k) in eV.

In Figure 14, local nucleophilicity indexes for carbon atoms that would react in this type of
cycloadditions (C1 and C4), are shown. A higher value of nucleophilicity is observed in C4
for 1-trimethylsilyloxy-1,3-butadiene and 1-methoxy-3-trimethylsilyloxy-1,3-butadiene
(Danishefsky diene). For isoprene, C1 is the most nucleophilic atom.

The electrophilicity of isoprene falls in the range of moderate electrophiles within the electrophilicity scale proposed by Domingo *et al*. When electron-donating substituents, -OCH_3
and –$OSi(CH_3)_3$, are incorporated into the structure of butadiene, a decrease in the electrophilicity power is observed. Therefore, the electrophilicity of Danishefsky's diene falls in the
range of marginal electrophiles, good nucleophiles, within the electrophilicity scale. This be-

havior indicates that the nucleophilic activation in Danishefsky´s diene is better than in iso-prene, in clear agreement with the high nucleophilicity index of the diene. 1-trimethylsilyloxy-1,3-butadiene have a intermediate behavior.

5.3. Dienophiles

The electrophilicity power (ω) of the different dienophiles is shown in Tables 12, 13, 14, 15, 16 and 17. In the tables we also included some global properties such as the chemical poten-tial and the chemical hardness. A good electrophile is characterized by a high absolute value of μ and a low value of η.

We also analyze its local counterpart (ω_k), the values are exposed in the Figures 15, 16, 17, 18, 19 and 20 below each table.

5.3.1. Five-membered heterocycles

• Furan and derivatives.

Molecule	Global properties			
	μ (a.u.)	η (a.u.)	ω (eV)	N (eV)
Furan	-0.1024	0.2441	0.99	3.23
2-Nitrofuran	-0.1810	0.1775	2.51	1.99
3-Nitrofuran	-0.1767	0.1808	2.35	2.07

Table 12. global electronic properties.

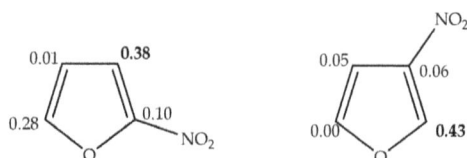

Figure 15. local electrophilicity indexes (ω_k) in eV.

• Tiophene and derivatives.

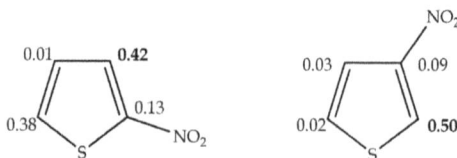

Figure 16. local electrophilicity indexes (ω_k) in eV.

Molecule	Global properties			
	μ (a.u.)	η (a.u.)	ω (eV)	N (eV)
Thiophene	-0.1545	0.1566	0.87	3.01
2-Nitrothiophene	-0.1845	0.1738	2.66	1.95
3-nitrothiophene	-0.1794	0.1821	2.40	1.98

Table 13. *global electronic properties*

• Selenophene and derivatives.

Molecule	Global properties			
	μ (a.u.)	η (a.u.)	ω (eV)	N (eV)
Selenophene	-0.1220	0.2195	0.49	3.16
2-Nitroselenophene	-0.1829	0.1695	2.68	2.05
3-Nitroselenophene	-0.1776	0.1803	2.38	2.05

Table 14. *global electronic properties*

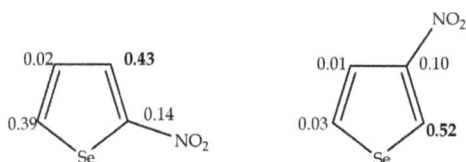

Figure 17. local electrophilicity indexes (ω_k) in eV.

• 1-Tosyl-pyrrole and derivatives.

Molecule	Global properties			
	μ (a.u.)	η (a.u.)	ω (eV)	N (eV)
1-Tosylpyrrole	-0.1348	0.1752	1.41	3.28
1-Tosyl-2-nitropyrrole	-0.1655	0.1739	2.31	2.43
1-Tosyl-3-nitropyrrole	-0.1668	0.1765	2.14	2.39

Table 15. global electronic properties.

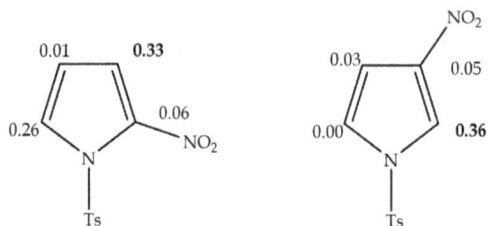

Figure 18. local electrophilicity indexes (ω_k) in eV.

5.3.2. Benzofused heterocycles

• Benzofuran and derivatives.

Molecule	Global properties			
	μ (a.u.)	η (a.u.)	ω (eV)	N (eV)
Benzofuran	-0.1180	0.2032	0.93	3.36
2- nitrobenzofuran	-0.1757	0.1536	2.73	2.46
3-nitrobenzofuran	-0.1694	0.1613	2.42	2.53

Table 16. global electronic properties.

Figure 19. local electrophilicity indexes (ω_k) in eV.

• N-Tosyl-indole and derivatives.

Molecule	Global properties			
	μ (a.u.)	η (a.u.)	ω (eV)	N (eV)
1-Tosyl-indole	-0.1316	0.1683	1.40	3.47
1-Tosyl-2-nitroindole	-0.1657	0.1455	2.57	2.84
1-Tosyl-3-nitroindole	-0.1606	0.1562	2.25	2.84

Table 17. global electronic properties.

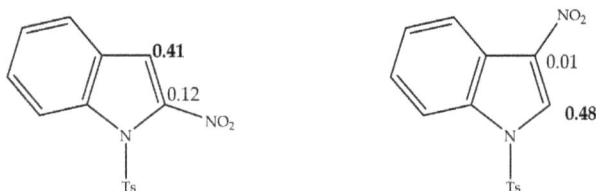

Figure 20. local electrophilicity indexes (ω_k) in eV.

We can assume that high nucleophilicity and high electrophilicity corresponds to opposite extremes of this scale (Della Rosa, et al, 2011; Brasca, et al, 2009).

The substitution of one hydrogen atom in all the dienophiles by the nitro group, one of the most powerful electron-withdrawing groups, produces an increment in the electrophilicity character and therefore an increase in the reaction rate is expected. The 2-nitro-substituted heterocycles show high electrophilicity power respect to the 3-nitro-substituted ones. Experimentally we obtained higher yields when the nitro group is place in the position 2 of the thiophene's ring than when it is in the 3-position. So these last results support the tendency observed in the tables.

The differences in the global electrophilicity power between the dienophile/diene pair $(\Delta\omega)$ are higher for the Danishefsky's diene than for isoprene. Therefore, we can expect a high reactivity for the pair nitrosubstituted-dienophile/Danishefsky's diene.

As a consequence of the high electrophilic character of these substituted dienophiles and the high nucleophilic character of the dienes, it is expected that these D-A reactions proceed with polar character. The polarity of the process is assessed comparing the electrophilicity index of the interacting pairs. Evidently, the differences in the global electrophilicity power $(\Delta\omega)$ are higher for the Danishefsky's diene than 1-trimethylsilyloxy-1,3-butadiene and isoprene.

On the other hand, a high regioselectivity is expected for Danishefsky's diene, due to the fact that the difference between local nucleophilicity indexes of C1 and C4 (ΔN_k) presents the highest value in this diene $(\Delta N_k=0.89)$. For isoprene and 1-trimethylsilyloxy-1,3-butadiene, as this difference is low $(\Delta N_k=0.13$ and $\Delta N_k=0.23$ respectively), both isomers are expected as D-A products. For dienophiles, the carbon atom adjacent to the nitro-substituted one is the most electrophile site in all the cases $(\Delta\omega_k=0.49\text{-}0.27)$.

When asymmetric reactants participate in this polar D-A cycloaddition, the most favorable interaction will take place between the most nucleophile site of the diene and the most electrophile site of the dienophile. This fact is also consistent with the experimental researches in ionic liquids as solvents.

When 2-nitrobenzofuran and 3-nitrobenzofuran reacted with isoprene, 1-trimethylsilyloxy-1,3-butadiene and the Danishesfky's diene, under different reaction conditions they showed their dienophilic character taking part in a normal demand polar D-A cycloaddition reactions.

Finally, the flux of the electron-density in these polar cycloaddition reactions is also support-ed by means of a DFT analysis besed on the electronic chemical potentials of the reagents. The electronic chemical potentials of the substituted heterocyclic dienophiles, nearly -5 eV, are higher than those of the dienes, nearly -3 eV, thereby suggesting that the net charge transfer will take place from these electron-rich dienes towards the aromatic dienophiles.

According to the global electrophilicity index ω showed that the dienes will act as nucleo-philes and the dienophiles as electrophiles. To study the regioselectivity we used the local electrophilicity and nucleophilicity indexes for dienophiles and dienes respectively. The more favored adducts are the ones where the most electrophilic and nucleophilic sites inter-act first. In the reactions in which it is possible discussed the regioselectivity of the experi-mental data agree with the computational results. (Della Rosa, el al, 2011)

The 2-nitrosubstituted benzofuran show higher electrophilicity power than the 3-nitrosub-stituted benzofuran probably due to the proximity of the nitro group with the heteroatom.

5.4. Ionic Liquids effect

The influence of ILs has also been considered in this theoretical study. For the calculation of electronic properties of the "complexes" dienophile-IL we use a model that includes the di-enophile molecule interacting with the anion-cation IL's system. In Table 18 we show an ex-ample of the different global electronic properties (ω and N) for these five-membered heterocycle-IL systems in comparison with the same study in gas phase.

	ω(eV)	N(eV)			ω(eV)	N(eV)	
	2.51	1.78	Gas Phase		2.35	1.85	Gas Phase
	4.31	1.83	[HMIM][BF4]		4.02	2.00	[HMIM][BF4]
2-nitrofuran	4,18	1.26	[HMIM][PF6]	3-nitrofuran	3.87	1.34	[HMIM][PF6]
	3.02	1.52	[BMIM][BF4]		2.78	1.64	[BMIM][BF4]
	2.95	1.60	[BMIM][PF6]		2.63	1.49	[BMIM][PF6]

Table 18. Global electronic properties for 2- and 3- nitrofuranes.

It can be observed that for nitrofuran dienophiles the electrophilicity index reach higher val-ues in presence of IL ([HMIM][BF$_4$], [HMIM][PF$_6$], [BMIM][BF$_4$], [BMIM][PF$_6$]). The stron-gest hydrogen-bond donation ability of [HMIM] cation is reflected in the significant increment of ω in about 2 eV for all the heterocycles in study. The [PF$_6$]⁻ anion present little lower values than [BF$_4$]⁻.

Taking into account that the reactivity of a D-A reaction depends on the HOMO-LUMO en-ergy separation of the reactants, and that in a normal electron demand D-A reaction the strongest interaction takes place between the HOMO of the diene and the LUMO of the di-enophile, we compared the corresponding energies of the reacting partners in order to ex-plain the experimental tendency observed.

	HOMO (a.u.)
Isoprene	-0.2272
1-trimethylsilyloxy-1,3-butadiene	-0.2046
Danishefsky's diene	-0.2045

Table 19. HOMO energy values.

	LUMO (a.u.)
2-nitrofuran	-0.0922
2-nitrofuran+[HMIM][BF$_4$]	-0.1379
2-nitrofuran+[HMIM][PF$_6$]	-0.1358
2-nitrofuran+[BMIM][BF$_4$]	-0.1016
2-nitrofuran+[BMIM][PF$_6$]	-0.1004

Table 20. LUMO energy values.

When the FMO of the reacting pairs are closer in energies, the interaction is higher. Thus, the FMO energies of the reactants were evaluated.

Therefore, in Tables 8 and 9 it can be observed that the expected higher reactivities for the dienophile-IL complexes is due to the fact that LUMO's energy of the dienophile gets closer to the HOMO's energy of the diene, which is consistent with the experimental results. This effect is also revealed by an increase in the yield of the D-A reaction. The energy difference between the FMOs is lower for [HMIM] based IL, what we attribute to the formation of the hydrogen-bond between the nitro group and the IL cation. The tendency is the same in all the cases of nitroheterocycle compounds.

5.5. Theoretical Mechanistic Approach

These D-A reactions could be considered domino processes that are initialized by a polar cycloadittion, and the latter concerted elimination of nitrous acid from the [4+2] cycloadduct yields the corresponding products (Della Rosa, et al, 2011).

5.5.1. Monocyclic five membered nitroheteroaromatic compounds as dienophiles.

Specifically the reactions of nitro-substituted five-membered heterocycles with Danishefsky´s diene were studied using the hardness, the polarizability and the electrophilicity of the corresponding D-A primary adducts as global reactivity indexes. The experimentally observed products for these D-A reactions using different conditions were indicated in the Figure 21 and related experiments. It has been demonstrated that both the hardness as well as the electrophilicity power of the adducts are appropriate descriptors to predict the major product of the reactions at least in the cases in study. (Brasca, et al, 2011)

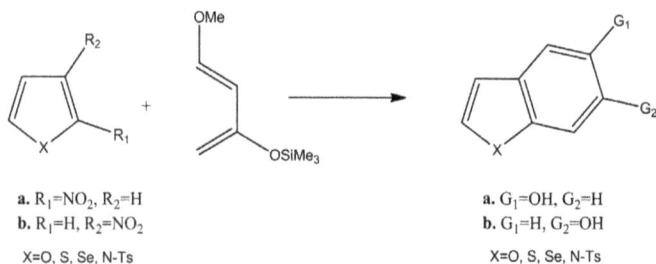

a. $R_1=NO_2$, $R_2=H$
b. $R_1=H$, $R_2=NO_2$

X=O, S, Se, N-Ts

a. $G_1=OH$, $G_2=H$
b. $G_1=H$, $G_2=OH$

X=O, S, Se, N-Ts

Figure 21. Reactions of nitrofuran whit Danishefsky diene.

For each reaction four channels, which lead to the regioisomers I, I', II and II' are feasible (Figure 9). As we can observe, depending on the orientation of the nitro group, two stereo-isomers can be obtained in each channel (i.e. *endo* and *exo* adducts).

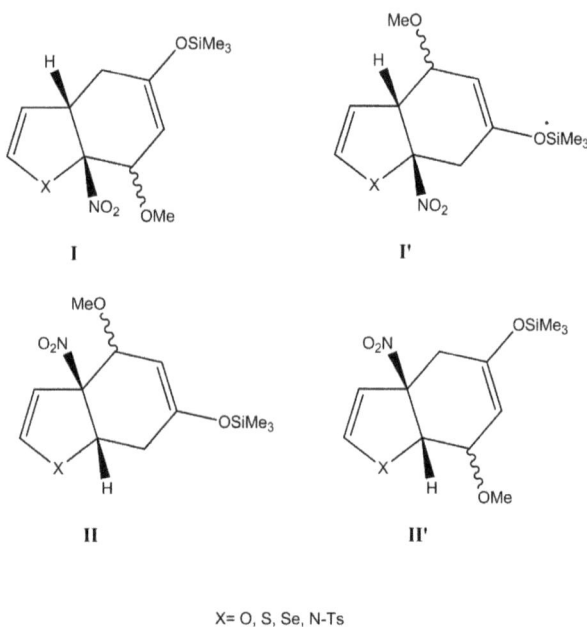

X= O, S, Se, N-Ts

Figure 22. Possible regioisomeric D-A adducts.

The regioisomer that have the higher value of η and the lower values of α and ω, should correspond to the major product. The calculated hardness and electrophilicity power correctly predict the regioisomers I.a-I.c as the main adducts of the D-A reactions.

The results obtained in gas phase revealed the same tendency as in ionic liquids as solvents.

We can conclude that the predominant regioisomeric adduct of the reactions between five-membered heterocycles derivatives and Danishefsky's diene have always the less electro-philicity and high hardness values. Moreover, the regioselectivity experimentally observed can be confirmed by this approach.

The obtained energies show that the I.*endo* isomer is more stable than the I.*exo* one. Moreover, the more stable I.*endo* isomer has lower electrophilicity value than the I.*exo* isomer in all cases.

The investigation by DFT theory, in which we include solvent effects (considering cation and anion of the ionic liquids), show that these cycloadditions proceeded by a concerted but asynchronus reaction mechanism. The lowest activation energies for concerted reactions are obtained. However, the stepwise additions have significantly lower activation energy lead to substantially less stable products. Moreover, the primary cycloadducts could never be isolated but were converted into 5-hydroxybenzofused heterocycles by subsequent extru-sion of nitrous acid, hydrolysis of the silyl enol ether, and elimination of methanol. Elimina-tion of nitrous acid is calculated to have lower overall barriers than cycloaddition reactions and is strongly exothermic, thus explaining the preferred reaction channel.

5.5.2. Nitronaphthalenes as dienophiles

The reactions of 1-nitronaphthalene with a serie of dienes were evaluated with the the fron-tier molecular orbitals (FMO) theory which provide qualitative information about the feasi-bility of this D-A reaction. Besides, the global electrophilicity index (ω) is employed to estimate the electrophilic character of the dienophiles used in the cycloaddition reactions.

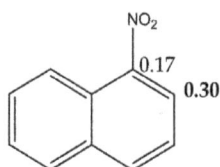

1-nitronaphtalene	
μ (a.u.)	-0.1650
η (a.u)	0.1485
ω (eV)	2.49
N (eV)	2.83

Table 21. 1-nitronaphthalene with the Danishefsky diene.

The reaction of 1-nitronaphthalene with the Danishefsky diene to obtain 3-hydroxyfenan-trene has been theoretically studied using DFT methods. This reaction is a domino proc-ess that is initialized by a polar Diels–Alder reaction between the par dienophile/diene to give the formally [2 + 4] cycloadduct. The subsequent concerted elimination of nitrous acid from the primary adduct yields the precursor of the fenatrene derivative. An analy-sis of the global reactivity indices as well as the thermodynamic data for this domino process indicate that while the large electrophilic carácter of 1-nitronaphthalene together with the large nucleophilic character of Danishefsky diene are responsible for the partici-pation of these reagents in a polar D-A reaction. The D-A reaction has a two-step non-in-termediate mechanism characterized by the nucleophilic attack of the non-substituted methylene of the diene to the electrophilically activated C2 position of 1-nitronaphtalene. The subsequentring-closure affords the primary cycloadduct. The latter concerted elimina-tion of nitrous acid yielded the precursor of the tricyclic aromatic final product. Spite of the large activation free energy associated with the D-A reaction and the endergonic char-acter in the primary adduct, the irreversible extrusion of nitrous acid make feasible ther-modynamically the domino reaction. (Domingo, et al, 2008)

5.5.3. Nitroquinolines

Although the global electrophilicity for the 5-nitroquinoline indicated a lightly major reac-tivity than the 8-nitro isomer (Figure 23). This result does not agree with the experimental data. In this respect it is possible think that the attack of the dienophile to the *para* posi-tion would be a reversible process, meanwhile the attack to the *orto* position to the nitro group evolve in form irreversible to the cycloaddition product. The major reactivity of the 8-nitro derivative could be occur due to the presence of electroelectronic factors more fa-vourables which are produced during the nucleophilic attack of the diene, for instance a better stabilization of the negative charge in the nitro group. These effects are not consid-ered in the reagents. (Cancian, *et al*, 2010)

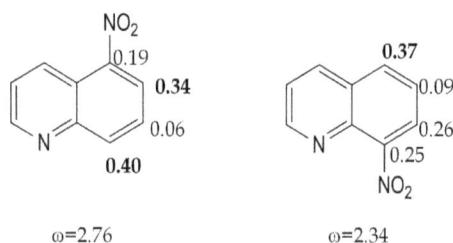

Figure 23. Global and local electrophilic indexes for 5- and 8-nitroquinolines.

Analysis of the local electrophilicities ω_k at 5-nitroquinoleine indicates that the C8 carbon, $\omega_{C8} = 0.40$ eV, and the C6 carbon, $\omega_{C6} = 0.34$ eV, are the electrophilic centres of the quino-leine moiety, while at 8-nitroquinoleine these centers are the C5 carbon, $\omega_{C5} = 0.37$ eV,

and the C7 carbon. They correspond with the carbon that contains the nitro group and
that located at the *ortho* position.

Formation of the HB between the acidic N10 hydrogen of HMIM$^+$ and the O9 oxygen of the nitro group does not only increase the electrophilicity index of the nitroquinoleine-PIL complexes, but also polarizes nitroquinoleine system. At the nitroquinoleine-PIL complexes, the
most electrophilic centres are the C8 carbon, $\omega_{C5} = 0.55$ eV, and the C6 carbon, $\omega_{C7} = 0.54$ eV, at 5-
nitroquinoleine-PIL, and at the C5 carbon, $\omega_{C5} = 0.45$ eV, and the C7 carbon, $\omega_{C7} = 0.40$ eV, at 8-
nitroquinoleina-PIL. These results are similar to those found in 5- and 8-nitroquinoleines.

Using the 5 and 8-nitroquinoline we showed specifically the interactions models with the
lowest energy between the dienophile and [HMIM][BF4] acting as reaction media.

Isoprene vs. 5-nitroquinoleine and 8-nitroquinoleine. A theoretical mechanism study.

In order to understand the catalytic role of polar ionic liquids (PILs) in P-DA reactions between isoprene and nitroquinoleines, in the absence and in the presence of [HMIM][BF4] as
a model of PILs, were theoretically studied using DFT methods. These reactions are domino
processes that comprises two consecutive reactions:

1. a P-DA reaction between isoprene, acting as diene, and nitroquinoleines acting as dien-
    ophiles, to yield the formal [4+2] cycloadducts (CA) and;

2. a concomitant nitroso acid extrusion at these intermediates to yield the final products
    (Figure 24).

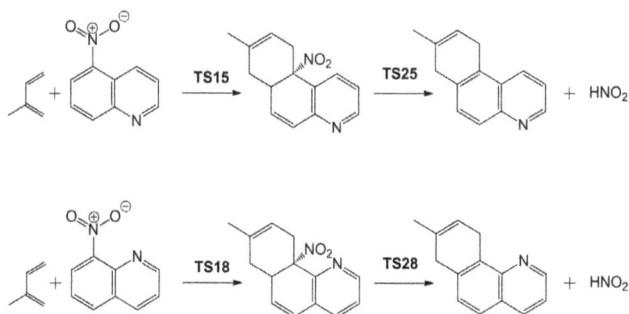

Figure 24. Reaction of isoprene with 5- and 8-nitroisoquinoline

Four reactive channels for the initial attack of isoprene on these nitroquinoleines are feasible:
two pairs of stereoisomeric channels, the *endo* and the *exo* ones, and two pairs of regioisomeric
channels, the *meta* and the *para* ones. Since both *endo* and *exo* channels yield the same final products after extrusion of nitroso acid, and as P-DA reactions involving isoprene present low regioselectivity, only the channels associated with the *endo/para* approach mode of isoprene,
respect to the electron-withdrawing nitro group of nitroquinoleines were considered.

An analysis of the gas-phase potential energy surfaces (PES) associated with these P-DA reactions indicates that the cycloadditions take place through a one-step mechanism via high asynchronous transition states (TS). Therefore, in both cases one TS, TS15 and TS18, and the formal [4+2] CAs, were located and characterized. The second reaction of these domino processes also takes place via a one-step mechanism via a high asynchronous TSs. Thus, one TS, TS25 and TS28, and the corresponding final products were located and characterized. Total and relative gas-phase energies are given in Table 11.

	E	ΔE
Isoprene	-195.3055	
5-nitroquinoleine	-606.4260	
TS15	-801. 6914	25.2
Cycloadduct	-801.7496	-11.4
TS25	-801.7129	11.6
Product	-596.0775	
HNO_2	-205.6952	
Product + HNO_2	-801.7727	-25.9
8-nitroquinoleine	-606.4195	
TS18	-801.6819	27.0
Cycloadduct	-801.7499	-15.6
TS28	-801.7136	7.2
Product	-596.0798	
Product + HNO_2	-801.7750	-31.4

Table 22. Gas phase total energies (E, in au) and relative energies ($ΔE$, in kcal/mol) of the stationary points involved in the domino reaction between isoprene and nitroquinoleines (5- and 8- nitrosubstituted).

In gas phase, the activation energies associated with the nucleophilic attack of the C1 carbon of isoprene on the C6 carbon of nitroquinoleines via TS15 or TS18 present high values, 25.2 and 27.0 kcal/mol; formation of the formal [4+2] CAs are exothermic by -11.4 and -15.6 kcal/mol, respectively. Although these P-DA reactions are thermodynamically favorable, these high activation energies associated with these processes prevent the cycloaddition reactions.

The [4+2] CAs suffer a nitroso acid extrusion regenerating the aromatic system present in quinoleine. The activation energies associated with the nitroso acid extrusion via TS25 and TS28 are 23.0 and 28.8 kcal/mol; formation of the tricyclic compounds plus nitroso acid is exothermic by -14.5 and -15.8 kcal/mol, respectively. Taking into account the favorable reaction entropies associated with the extrusion processes, we can consider these reactions thermodynamically irreversible. Since TS25 and TS28 are located below TS15 and TS18, the P-DA reactions between isoprene 1 and nitroquinoleines 2 or 3 via TS15 and

TS18 become the rate-determining steps of these domino processes. The high activation energy associated with TS15 and TS18 are in agreement with the drastic reaction conditions demanded for the reactions to take place.

These activation energies are higher than those associated with the P-DA reaction and nitroso acid extrusion associated with the domino reaction between nitronaphtalene and Danishefsky's diene, 16.5 and 23.9 kcal/mol, respectively (Figure 25).

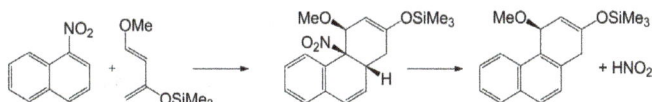

Figure 25. 1-nitronaphtalene and Danishefsky's diene

However, the lower activation energy associated with the P-DA reaction between nitronaphtalene and Danishefsky's diene, and the endothermic character of the formation of the formal [4+2] CA turn the nitroso acid extrusion into the rate-determining step in this domino reaction.

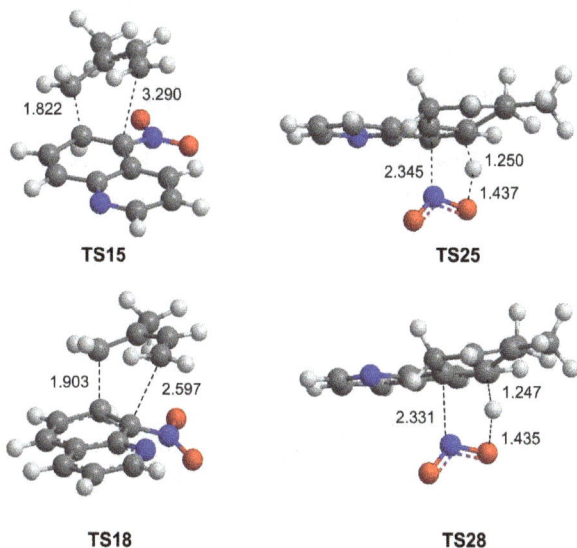

Figure 26. TS of the reaction between 5- and 8-nitroquinoline and isoprene

The gas-phase geometries of TS15, TS18, TS25 and TS28 are given in Figure 13. The lengths of the C1-C5 and C2-C8 forming bond at TSs associated with the P-DA reactions between isoprene and nitroquinoleines are 1.822 and 3.290 Å at TS15 and 1.903 and 2.597 Å at TS18, respectively. These values suggest asynchronous bond-formation processes in which the C-C bond formation between the most nucleophilic center of isoprene, the C5 carbon, and one of the most electrophilic centers of nitroquinoleines, the C1 carbon, is more advanced than the C-C bond formation between the C2 and C8 carbons. The short C1-C5 distance indicates that the cycloaddition processes are very advanced, in clear agreement with the high activation energies associated with TS15 and TS18.

At TSs associated with the nitroso acid extrusion the length of the C2-N3 breaking bond is 2.345 Å at TS25 and 2.331 Å at TS28, while the lengths of the C1-H1 breaking- and O4-H1' forming-bonds are 1.250 and 1.437 Å at TS25, and 1.247 and 1.435 Å at TS28, respectively. These values suggest asynchronous processes in which the C2-N3 breaking bond is more advanced than the H1' proton transfer process to the O4 oxygen.

The polar nature of these D-A reactions was evaluated analyzing the charge transfer (CT) at TS15 and TS18. The natural charges at these TSs were shared between the isoprene and the nitroquinoleine frameworks. At TS15 and TS18, the CT that flows from isoprene to nitroquinoleines is 0.29 eV and 0.26 eV, respectively. These values point at the zwitterionic character of these TSs. They are lower than that obtained at the TS associated with the nucleophilic attack of Danishefsky's diene on nitronaphtalene, 0.39 eV, as a consequence of the stronger nucleophilic character of Danishefsky's diene than that of isoprene.

Reaction mechanism pattern in presecence of the [HMIM][BF4] PIL.

The effects of the [HMIM][BF$_4$] PIL on the domino reactions between isoprene and nitroquinoleines were evaluated considering two computational models. In *Model I*, the implicit effects of the PIL were considered by forming a hydrogen bond (BH) between the acidic H10 hydrogen of HMIM and the O9 oxygen of the nitro group of nitroquinoleines (Figure 27).

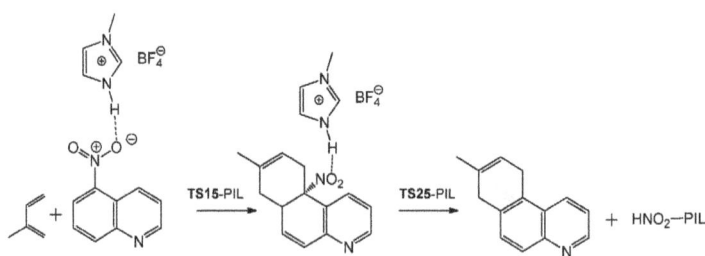

Figure 27. Reaction between 5-nitroquinoline and isoprene in presence of PIL's

While in *Model II*, the solvent effect of the PIL is completed including electrostatic interactions modeled by the polarizable continuum model (PCM) of Tomasi's group. For the PCM calculations, 1-heptanol was considered as solvent since it has a dielectric constant closer to [HMIM][BF4]; ε= 11.3. The energy results are given in Table 12.

	gas phase		1-heptanol	
	E	ΔE	E	ΔE
5-nitroquinoleine-PIL	-1297.0092		-1297.0411	
TS15-PIL	-1492.2827	20.1	-1492.3173	19.4
cycloadduct-PIL	-1492.3293	-9.2	-1492.3636	-9.7
TS25-PIL	-1492.3020	8.0	-1492.3395	5.5
HNO$_2$ -PIL	-896.2688		-896.30298	
Product + HNO$_2$ -PIL	-1492.3463	-19.8	-1492.3857	-23.5
8-nitroquinoleine-PIL	-1297.0040		-1297.0391	
TS18-PIL	-1492.2736	22.5	-1492.3117	21.6
cycloadduct-PIL	-1492.3341	-15.4	-1492.3673	-13.3
TS2-PIL	-1492.3021	4.6	-1492.3397	4.1
Product + HNO$_2$ -PIL	-1492.3486	-24.5	1492.3873	-25.8

Table 23. Total energies (E, in au) and relative energies (ΔE, in kcal/mol), in gas phase and in 1-heptanol, of the stationary points involved in the domino reaction between isoprene and nitroquinoleines in [HMIM][BF4].

In Figure 27, the stationary points involved in the domino reactions between isoprene and [HMIM][BF$_4$]:nitroquinoleines complexes are given. Formation of the HB between HMIM and an oxygen atom of the nitro group decreases the activation energies associated with these P-DA reactions significantly. Now, TS1-PIL and TS2-PIL are located 20.1 and 22.5 kcal/mol above the separated reagents. In spite of this behaviour, the exothermic character of the cycloadditions, -9.2 and -15.4 kcal/mol, remains unmodified. The large acceleration found in the presence of the PIL ionic pair can be understand as an increase of the polar character of the reactions as a consequence of the increased electrophilic character of the dienophile-PIL complexes, which favors the CT process.

The second reactions of these domino processes are also slightly catalyzed by the presence of [HMIM][BF$_4$] PIL, since it remains hydrogen-bonded at the intermediate cycloadducts-PIL. Now, the activation energies associated with the extrusion of HNO$_2$-PIL are 17.2 and 20.0 kcal/mol. These reactions are exothermic by -10.6 and -9.1 kcal/mol.

Inclusion of the solvent effects by means of the PCM calculations in *Model II* stabilizes all species between 21.0 and 25.0 kcal/mol, as a consequence of the charged [HMIM][BF$_4$] PIL. TS15 and TS18 are slightly more stabilized than the reagents due to their zwitterionic character. As a consequence, the activation energies associated with the P-DA reactions decrease by 1.7 and 0.9 kcal/mol relative to the gas-phase calculations. Consequently, a comparison between the gas-phase relative energies associated with the domino reactions between isoprene and nitroquinoleines with those obtained in *Model I* and *II* for the domino reactions in PILs indicate that the HB formation at nitroquinoleines 2 and 3 is the main factor responsible for the acceleration of these domino reactions in PILs.

The gas-phase geometries of TS15-PIL, TS18-PIL, TS25-PIL and TS28-PIL are given in Figure 15. The lengths of the C1-C5 and C2-C8 forming bonds at the TSs associated with the P-DA reactions between isoprene and the dienophile-PIL complexes are 1.879 and 3.338 Å, and 1.835 and 3.141 Å, respectively. These values suggest a two-stage one-step mechanism channel an asynchronous TS. At TS15-PIL and TS18-PIL, the distances between the nitro O9 oxygen and the acidic H10 hydrogen of HMIM, 1.769 and 1.776 Å, suggest a strong HB interaction.

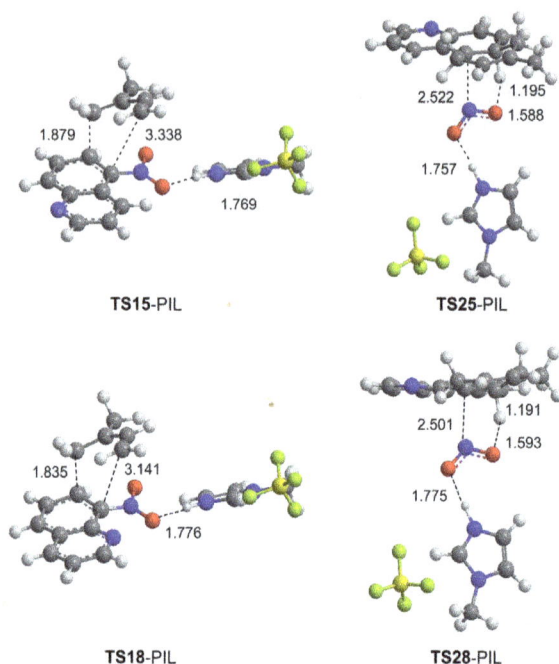

Figure 28. TS of the reaction between 5-nitroquinoline and isoprene in presence of PIL`s

At TSs associated with the extrusion of the 6-PIL complex, the length of the C2-N3 breaking bond is 2.522 Å at TS25-PIL and 2.501 Å at TS28-PIL, while the lengths of the C1-H1' breaking- and O4-H1' forming-bonds are 1.195 and 1.588 Å at TS25-PIL, and 1.191 and 1.593 Å at TS28-PIL, respectively. These values also suggest a highly asynchronous process in which the C2-N3 breaking bond is very advanced with respect to the H1' proton transfer process.

6. Conclusions

It was possible to demonstrate again the influence of the solvent in these particular type of D-A reactions. A series of aromatic carbocyclic and hetrocyclic substituted by electron withdrawing groups can act as dienophiles in polar cycloaddition reactions besides different dienes in the presence of PIL's. However, D-A reactions proceed at an appreciable rate only when either the diene or the dienophile are activated by an electron donating or electron withdrawing group, normally characterized by the presence of a heteroatom that can therefore efficiently interact with the solvent. IL's, with their peculiar properties such as high polarizability/dipolarity, good hydrogen bond donor ability, were straight away considered to have the potential to influence the outcome of these D-A reactions, accelerating them.

In general, the products of the reactions development in PIL's are similar to those in molecular solvents. However, the presence of PIL's improved the reaction rate probably due to the hydrogen bonding interactions between the neoteric solvent and the dienophile. Only in a few cases we can note differences in the product distribution. For the reactions in which are possible to observe a competition between normal and hetero D-A process, the PIL's favor the normal pathway because they improve the electrophilicity of the dienophiles.

The DFT analysis of the global properties of the interacting pair diene/dienophile illustrates the normal electron demand character of these D-A reactions. It is possible to show that the local indexes provide useful clues about the regiodirector effects, particularly of the nitro group. The presence of a solvent (molecular or neoteric) as the reaction media does not impart a prominent influence on the relative reactive sites. In few cases among those studied we can note that the relative reactive sites are affected by the solvent and the basis set (e.g. methyl 5-nitrofuran-2-carboxylate).

The site-, regio-, and stereochemistry of some of these D-A reactions has been investigated by the density functional theory, including solvent effects. Generally, these cycloadditions proceed by a concerted but asynchronous reaction mechanism. The *endo* stereochemistry is in most case preferred.

In general, the normal D-A reaction mechanism is a domino process that is initialized by the polar reaction between the diene and the dienophile to give the primary cicloadduct. These D-A reactions have a two-step non-intermediate mechanism characterized by the nucleophilic attack on the non-substituted methylene of the diene to the electrophilically activated position of the dienophile. The subsequent ring-closure affords the primary cicloadduct. This behavior makes the reaction to be regioselective. The latter concerted elimination of the nitrous acid from the primary cicloadduct yields the precursor of the final aromatic product. Spite of the large activation free energy associated with the D-A reaction and the endergonic character of formation of the primary cicloadduct, the irreversible extrusion of the nitrous acid make feasible thermodynamically the domino reaction.

DFT calculations of the electrophilicity and nucleophilicity indexes in general agree with the experimental results and they are a good reactivity and regioselectivity predictors in this type of polar cycloaddition reactions

The presence of a PIL's in the reaction media improves significantly the electrophilic character of the dienophile. However, the differences between the experimental results using PIL's or molecular solvents are not so bigger how the calculated electrophilic values indicated.

Author details

Pedro M. E. Mancini[1*], Carla M. Ormachea[1], Claudia D. Della Rosa[1],
María N. Kneeteman[1,2] and Luis R. Domingo[3]

*Address all correspondence to: pmancini@fiq.unl.edu.ar

1 Área de Química Orgánica-Departamento de Química-Facultad de Ingeniería Química-Universidad Nacional del Litoral (UNL), Santa Fe, Argentina

2 Consejo Nacional de Investigaciones Científicas y Técnicas (CONICET), de la República Argentina

3 Departamento de Química Orgánica, Facultad de Química, Universidad de Valencia, España

References

[1] Arnó, M., & Domingo, L. R. (2002). Density functional theory study of the mechanism of the proline-catalyzed intermolecular aldol reaction. *Theoretical Chemistry Accounts*, 108(4), 232-239.

[2] Biolatto, B., Kneeteman, M., Paredes, E., & Mancini, P. M. E. (2001). Reactions of 1-tosyl-3-substitutedindoles with conjugated dienes under thermal and/or hyperbaric conditions. *Journal of Organic Chemistry*, 66, 3906-3912, 0022-3263.

[3] Brasca, R., Kneeteman, M. N., Mancini, P. M. E., & Fabian, W. M. F. (2011). Comprehensive DFT Study on Site-, Regio-, and Stereoselectivity of Diels-Alder Reactions Leading to 5 -Hydroxybenzofurans. *Eur. J. Org. Chem*, 721-729.

[4] Brasca, R., Kneeteman, M. N., Mancini, P. M. E., & Fabian, W. M. F. Diels-Alder reactions for the rational design of benzo[b]thiophenes. DFT-based guidelines for synthetic chemists. *Journal of Molecular Structure*, THEOCHEM, 1010, 158.

[5] Brasca, R., Della Rosa, C., Kneeteman, M., & Mancini, P. (2011). Five-membered aromatic heterocycles in polar cycloaddition reactions: theoretical studies as a complement of the experimental researches. *Letters in Organic Chemistry*, 8(2), 82-87.

[6] Brasca, R., Della Rosa, C., Kneeteman, M., & Mancini, P. M. E. (2011). Five-Membered aromatic heterocycles in diels-alder cycloaddition reactions: Theoretical Studies as a Complement of the Experimental Researches. *Letters in Organic Chemistry*, 8(2), 82-87.

[7] Brasca, R., Kneeteman, M. N., Mancini, P. M. E., & Fabian, W. M. F. (2009). Theoretical explanation of the regioselectivity of polar cycloaddition reactions between furan derivatives and Danishefsky's diene. *Journal of Molecular Structure: THEOCHEM*, 911(1-3), 124-131.

[8]   Brasca, R., Kneeteman, M. N., Mancini, P. M. E., & Fabian, W. M. F. (2011). Compre-
      hensive DFT study on site-, regio-, and stereoselectivity of diels-alder reactions lead-
      ing to 5-hydroxybenzofurans. *European Journal of Organic Chemistry*, 4, 721-729.

[9]   Cancian, S., Kneeteman, M., & Mancini, P.M.E. (2010). Nitroquinolines as dienophiles
      in Polar Diels-Alder reactions. Influence of molecular solvent and ionic liquids. *14th
      International Electronic Conference on Synthetic Organic Chemistry (ECSOC-14)*,
      0095-2338.

[10] Carruthers, W. (1990). *Cycloaddition Reactions in Organic Synthesis*, Pergamon Press,
      Oxford, UK.

[11] Corey, E.J. (2002). Catalytic enantioselective Diels-Alder reactions: methods, mecha-
      nistic fundamentals, pathways, and applications. *Angew.Chem. Int. Ed.*, 41(10),
      1650-67.

[12] Della Rosa, C., Ormachea, C., Kneeteman, M. N., Adam, C., & Mancini, P. M. E.
      (2011). Diels-Alder reactions of N-tosylpirroles developed in protic ionic liquids.
      Theoretical studies using DFT methods. *Tetrahedron Lett.*, 52, 6754-6757.

[13] Della Rosa, C., Sanchez, J.P., Kneeteman, M.N., & Mancini, P.M.E. (2011). 2-Nitroben-
      zofuran and 3-Nitrobenzofurans as dienophiles in Polar Diels-Alder Reactions. A
      Simple Dibenzofurans Synthesis. Theoretical studies using DFT Methods. *Tetrahedron
      Lett.*, 52, 2316-2319.

[14] Della Rosa, C., Kneeteman, M., & Mancini, P. M. E. (2007). Behavior of selenophenes
      substituted with the electron withdrawing groups in polar Diels-Alder reactions. *Tet-
      rahedron Letters*, 48, 7075-7078, 0040-4039.

[15] Della Rosa, C., Kneeteman, M., & Mancini, P. M. E. (2007). Comparison of the reactiv-
      ity between 2- and 3-nitropyrroles in cycloaddition reactions. A simple indole syn-
      thesis. *Tetrahedron Letters*, 48, 1435-1438, 0040-4039.

[16] Della Rosa, C. D., Sanchez, J. P., Kneeteman, M. N., & Mancini, P. M. E. (2011). Diels-
      Alder reactions of nitrobenzofurans: A simple dibenzofuran Synthesis. Theoretical
      studies using DFT methods. *Tetrahedron Letters*, 52(18), 2316-2319.

[17] Della Rosa, C., Kneeteman, M., & Mancini, P. M. E. (2005). Nitrofurans as dienophiles
      in Diels-Alder reactions. *Tetrahedron Letters*, 46, 8711-8714, 0040-4039.

[18] Della Rosa, C., Paredes, E., Kneeteman, M., & Mancini, P. M. E. (2004). Behavior of
      thiophenes substituted with electron-withdrawing groups in cycloaddition reactions.
      Letters in Organic Chemistry, 1, 148-150, 1687-6865.

[19] Della Rosa, C., Sanchez, J. P., Kneeteman, M., & Mancini, P. M. E. (2010). Nitrobenzo-
      furan as dienophile in Polar Diels-Alder reactions. A simple Dibenzofuran Synthesis.
      Journal of Chemistry and Chemical Engineering, 4(11), Serial 36, 54-59.

[20] Fringelli, F., & Tatichi, A. (2002). *The Diels-Alder Reaction*, J. Wiley & Sons, Chichester,
      UK.

[21] Domingo, L. R. (2002). A density functional theory study for the Diels-Alder reaction between N-acyl-1-aza-1,3-butadienes and vinylamines. Lewis acid catalyst and solvent effects. *Tetrahedron*, 58(19), 3765-3774.

[22] Domingo, L. R., Arnó, M., Contreras, R., & Pérez, P. (2002). Density functional theory study for the cycloaddition of 1,3-butadienes with dimethyl acetylenedicarboxylate. Polar stepwise vs concerted mechanisms. *Journal of Physical Chemistry A*, 106(6), 952-961.

[23] Domingo, L. R., & José, Aurell M. (2002). Density functional theory study of the cycloaddition reaction of furan derivatives with masked o-benzoquinones. Does the furan act as a dienophile in the cycloaddition reaction. *Journal of Organic Chemistry*, 67(3), 959-965.

[24] Domingo, L. R., Chamorro, E., & Pérez, P. (2008). An understanding of the electrophilic/nucleophilic behavior of electro-deficient 2,3-disubstituted 1,3-butadienes in polar Diels-Alder reactions. A density functional theory study. *Journal of Physical Chemistry A*, 112(17), 4046-4053.

[25] Hitchcock, P. B., Mohammed, T. J., Seddon, K. R., Zora, J. A., Hussey, C. L., & Ward, E. H. (1986). 1-methyl-3-ethylimidazolium hexachlorouranate(IV) and 1-methyl-3-ethylimidazolium tetrachlorodioxo-uranate(VI): Synthesis, structure, and electrochemistry in a room temperature ionic liquid. *Inorganic Chimica Acta*, 113, L25-L26, 19-191, 0020-1693.

[26] Jaramillo, P., Domingo, L. R., Chamorro, E., & Pérez, P. (2008). A further exploration of a nucleophilicity index based on the gas-phase ionization potentials. *Journal of Molecular Structure: THEOCHEM*, 865(1-3), 68-72.

[27] Mancini, P. M. E., Fortunato, G., Bravo, M. V., & Adam, A. (2012). Ionic Liquids:Binary Mixtures with Selected Molecular Solvents. Characterization of its Molecular-Microscopic Properties. Reactivity. Chapter 13, 335-362, *in Green Solvents Book 2: Properties and Applications in Chemistry, Eds. Ali Mohammad and M.P. Inamuddin*, 978-94-007-2890-5, 978-94-007-2891-2-eBook-, Springer, United Kingdom.

[28] Mancini, P. M. E., Kneeteman, M., Della Rosa, C., Bravo, V., & Adam, C. (2011). Ionic Liquids in Polar Diels-Alder Reactions using Carbocycles and Heterocycles as Dienophiles. *Ionic Liquid/Book 1. Ed. Scott Hardy*, INTECH Open Access, cap. 13, 311-344, 978-953-308-66-6.

[29] Paredes, E., Brasca, R., Kneeteman, M., & Mancini, P. M. E. (2007). A novel application of the Diels-Alder reaction: nitronaphthalene as normal electron demand dienophiles. *Tetrahedron*, 63, 3790-3799, 0040-4020.

[30] Parr, R. G., Szentpály, L. V., & Liu, S. (1999). Electrophilicity index. *Journal of the American Chemical Society*, 121(9), 1922-1924.

[31] Parr, R. G., & Yang, W. (1984). Density functional approach to the frontier-electron theory of chemical reactivity. *Journal of the American Chemical Society*, 106(14), 4049-4050.

[32] Welton, T. (1999). Room-Temperature Ionic Liquids. Solvents for Synthesis and Catalysis. *Chemical Review*, 99(2071), 0010-8545.

Chapter 4

Multicomponent Reactions in Ionic Liquids

Ahmed Al Otaibi and Adam McCluskey

Additional information is available at the end of the chapter

1. Introduction

In our group, we place a premium on the rapid access to a wide rage of diverse small molecules. Our current focus spans the inhibition of dynamin GTPase, protein phosphatases 1 and 2A and the development of anti-cancer lead compounds.[1]-[10] While rapid access is paramount, we also strive to develop high levels of diversity in an environmentally friendly manner. That is, we are keen to apply the tenants of green chemistry at all stages of our drug development programs. To satisfy this need we have developed a particular interest in multicomponent reactions in benign solvents.

A multicomponent reaction (MCR) can be simply classified as a reaction in which three or more components are combined together in a single reaction vessel to produce a final product or products displaying features of all inputs and thus offers greater possibilities for molecular diversity per step with a minimum of synthetic time and effort. Products from such MCRs result in high atom and step economy.

The first reported MCR was Strecker's synthesis of racemic amino acids in the 1850's.[11] Strecker's amino acid synthesis combined an aldehyde, hydrogen cyanide and ammonia in a one pot procedure leading to a range of amino acids. With over 150 years history and development, MCRs have recently seen a resurgence, in part due to the ease of access to a wide range of diverse, highly functionalized molecules, in particular the synthesis of small heterocyclic rings of medicinal chemistry importance. A discourse on the physical properties of ionic liquids is out with the scope of this work. The chemistry, reactions and properties of ionic liquids has been addressed in a number of excellent review articles in this area.[12]-[20] In this chapter we describe the current state of play associated with MCRs in ionic liquids, with a focus on 3- and 4- component MCRs (3CRs and 4CRs respectively).[21]

For ease of discussion the application of MCRs in ionic liquids is broken down into the type of product generated: heterocyclic rings containing various numbers of heteroatoms and a classification of the reaction as either a 3CR or a 4CR.

2. Three component MCRs (3CRs)

2.1. Synthesis of acyclic products

MCRs are not only applicable to the synthesis of heterocyclic systems, but represent a very facile entry point to a range of acyclic compounds such as the amido substituted naphthols shown in Scheme 1. The treatment of β-naphthol with a wide range of aldehydes (aliphatic and aromatic), substituted amides in the presence of conventional ionic liquids (Ils), such as those based on the N-methyl, N-sulfonic acid imidazolonium [MSIM] cation, afforded rapid access to 1-amidoalkyl-2-naphthols and 1-amidoaryl-2-naphthols in good to excellent yields. In this area Zolfigol et al. have had particular success with the functionalised ILs such as [MSIM][Cl], [DSIM][Cl] and [MSIM][AlCl₄], under solvent-free conditions (Scheme 1).[22]

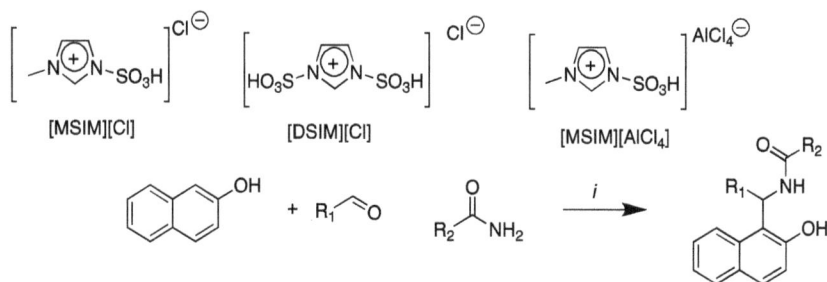

Scheme 1. Synthesis of 1-amidoalkyl-2-naphthols: (i) [MSIM][Cl] or [DSIM][Cl] or [MSIM][AlCl₄], 120°C, ~40 min.

Hajipour et al. and Hervai et al. effected the same transformations, and extended the methodology to allow the use of urea as the amide source using a range of Brønstead acid based Ils (BAILs).[23],[24] Hajipor et al.'s approach used N-(4-sulfonic acid)butyltriethyl ammonium hydrogen sulfate [TEBSA][HSO₄], while Hervai et al. applied two BAILs: 3-methyl-1-(4-sulfonic acid)-butylimidazolium hydrogen sulfate [MIM-(CH₂)₄SO₃H][HSO₄]) and N-(4-sulfonic acid)butylpyridinium hydrogen sulfate [Py-(CH₂)₄SO₃H][HSO₄] to effect the same transformations. [TEBSA][HSO₄] has been used previously as an efficient and reusable catalyst for nitration of aromatic compounds and esterification of various alcohols by different acids.[25]-[27] The acidic nature of BAILs has been exploited as catalysts for many other significant organic reactions, which proceed with excellent yields and selectivities and demonstrate the great potential of these ILs in catalytic technologies for chemical production.[23]

Kotadia et al. and Zhang independently reported the synthesis of similar 1-amidoalkylnaphthols using solid supported ionic liquids (SSILs).[28] Kotadia used a benzimidazolium based ionic liquid immobilized on silica based solid support, while Zhang conducted the reaction in the presence of polyethylene glycol (PEG)-based dicationic acidic ionic liquid as a catalyst under solvent-free conditions.[29],[30] Supported reagents offer the advantages of simple and safe catalyst recycling.[28] All MCR-IL based approaches to 1-amidoalkylnaphthols where highly substituent tolerant with excellent yields observed ith both electron donating and electron withdrawing aldehydes.

Yadav and Rai reported a three component MCR approach to β-nitrocarbonitriles and β-nitrothiocyanates in [BMIM][OH] or [BMIM][BF$_4$]. The reaction proceeds via a Henry reaction to yield the β-nitrostyrenes follwed by Michael addition of trimethylsilyl cyanide (TMSCN) or ammonium thiocyanate to yield the β-nitrocarbonitriles and β-nitrothiocyanates in modest overall yields of 53-58% (Scheme 2).[31],[32]

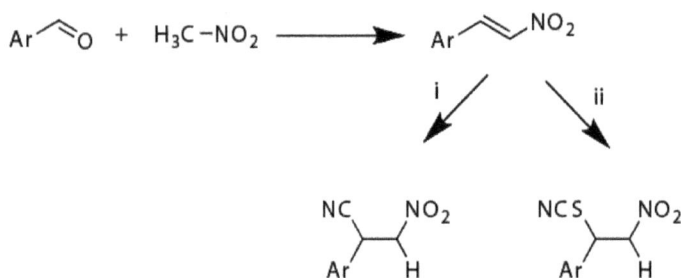

Scheme 2. Synthesis of β-nitrocarbonitriles and β–nitrothiocyanates: (i) [BMIM][OH] or [BMIM][BF$_4$], TMSCN, CH$_3$CN, 85-90°C, 6-9h; (ii) [BMIM][OH] or [BMIM][BF$_4$], NH$_4$SCN, CH$_3$CN, 85-90°C, 6-9h.

Zaho et al. has reported the combined Barbier / Friedel-Crafts alkylation of unsusbtituted benzaldehydes with allylbromide and phenols to yield 4-(2-hydroxyphenyl)-4-phenylbut-1-enes promoted by BuPyCl/SnCl•2H$_2$O and their subsequent application to the synthesis of 4-(substituted phenyl)chromans (Scheme 3).[33]

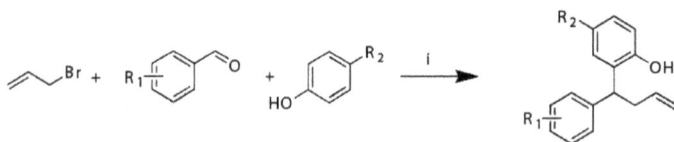

Scheme 3. Synthesis of 4-(2-hydroxyphenyl)-4-substituted phenylbut-1-enes: (i) BuPyCl/SnCl•2H$_2$O.

The use of an enolisable ketone facilitated the synthesis of a family of β-amido ketones (Scheme 4). The reaction of an enolizable ketone, aryl aldehyde and acetonitrile or benzonitrile in the presence of TMSCl using a Brønsted-acidic ionic liquid 3-methyl-1-(4-sulfonic acid) butylimidazolium hydrogen sulfate [MIM-(CH$_2$)$_4$SO$_3$H][HSO$_4$] as catalyst gave a family of β-amido ketones in good yield.[34]

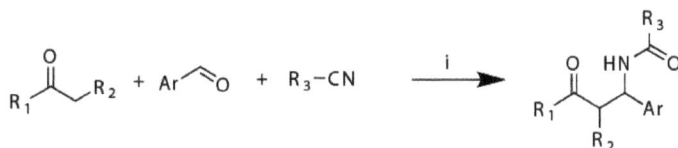

Scheme 4. Synthesis of β-amidoketones: (i) [MIM-(CH$_2$)$_4$SO$_3$H][HSO$_4$], TMSCl, 80°C.

This enolisable ketone approach also allowed Fang *et al.* to conduct a three-component Mannich-type reaction (Scheme 5) with aromatic aldehydes, aromatic amines, and ketones catalyzed by a novel functionalized ionic liquid, 3-(N,N-dimethyldodecylammonium)propanesulfonic acid hydrogen sulfate ([DDPA][HSO$_4$]) at room temperature to give various β-aminocarbonyl compounds in good yields.[35] [DDPA] [HSO$_4$] was recycled and after six cycles, no loss in catalytic activity was reported. Gong *et al.* conducted the same reaction using cyclohexanone and [BMIM][OH] as the IL catalyst to afford the β-aminocarbonyl compounds in excellent yields.[36]

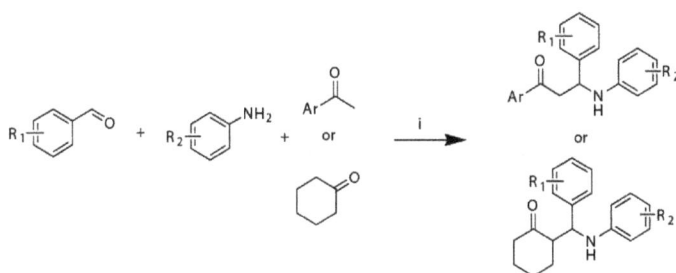

Scheme 5. Mannich-type approach to β-amidoketones: (i) [DDPA][HSO$_4$] or [BMIM][OH].

Liu *et al.* also explored the IL mediated Mannich reaction, but utilised a series task-specific ionic liquids in developing an asymmetric of β-aminoketones from isovaleraldehyde, methyl ketones, and aromatic amines in excellent yields (ca. 90%) and %ee's (~95%).[37] L-proline was used as the chiral catalyst (Scheme 6). The reaction with [DEMIm][BF$_4$], [DEEIm][BF$_4$], [BEIm][BF$_4$], [MEIm][BF$_4$] and [PEIm][BF$_4$] typically gave the desired product with an ee >95%. While the chemical yield dropped marginally on re-use from 96 to 85% over four cycles of IL use, the %ee remained constant.

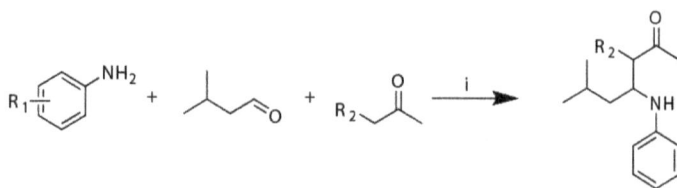

Scheme 6. Catalysed asymmetric Mannich reaction: (i) task specific ionic liquid chosen from [DEMIm][BF₄], [DEEIm] [BF₄], [BEIm][BF₄], [MEIm][BF₄] and [PEIm][BF₄] / L-proline.

In a similar reaction sequence, Akbari and Heydari, replaced the activated ketone with tri-methyl phosphite, in the presence of [MIM-(CH₂)₄SO₃H][CF₃SO₃], to affect rapid access to α-aminophosphonates (Scheme 7).[38] The reaction proceded via protonation of the carbonyl moiety, imine formation and attack at the protonated imine by trimethylphosphite. The re-action was highly tolerant of substituents on the carbonyl containing compound with pyrid-yl, cinnamyl, etc., affording excellent yields of the corresponding α-aminophosphonates. The ionic liquid could be recycled with no observable loss of efficacy after six cycles.

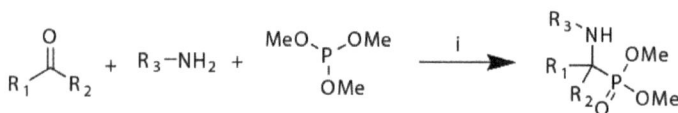

Scheme 7. Synthesis of α–aminophosphonates: (i) [MIM-(CH₂)₄SO₃H][CF₃SO₃] (10 mol%), H₂O.

Reddy *et al.* also reported the synthesis of α-aminophosphonates via a three-component re-action of 5-amino-2,2-difluoro-1,3-benzodioxole, aromatic aldehydes, and diethylphosphite catalysed by silica-supported boron trifluoride (BF₃.SiO₂) in [BMIM][HCl] at room tempera-ture (Scheme 8).[39] Yields were good to excellent and reaction times were typically 5 min versus 3h using conventional solvents.

Scheme 8. Synthesis of α-aminophosphonate catalysed: (i) BF₃.SiO₂ / [BMIM][HCl].

O-Protected cyanohydrins are versatile synthetic intermediates in organic synthesis for the preparation of a wide variety of organic compounds such as α-hydroxyacids, α-hydroxy ke-tones, α-amino acids, and β-amino alcohols.[40]-[43] Shen and Ji developed a mild synthesis of these key intermediates via the condensation of an aldehyde, trimethylsilyl cyanide (TMSCN), and Ac₂O in [BMIM][BF₄] (Scheme 9).[44] In addition, the recovered ionic liquid could be reused for subsequent runs without the loss of activity.

Scheme 9. One-pot synthesis of O-acetyl cyanohydrin: (i) [BMIM][BF$_4$].

2.1.1. 3CRs yielding heterocycles with one ring nitrogen

Arguably the major utility of MCRs is in the synthesis of highly decorated heterocyclic compounds. In our group we are interested in the synthesis of heterocyclic scaffolds that can be used in medicinal chemistry programs to instill the correct level of biological activity. Davoodnia et al. reported an efficient procedure for preparation of 2,4,6-triarylpyridines by treatment of acetophenones, aryl aldehydes, and NH$_4$OAc in the presence of [MIM-(CH$_2$)$_4$SO$_3$H][HSO$_4$] (Scheme 10).[45] Aromatic aldehydes with both electron donating and electron withdrawing substituents were well tolerated.

Scheme 10. Preparation of 2,4,6-triarylpyridines: (i) [MIM-(CH$_2$)$_4$SO$_3$H][HSO$_4$], 120°C.

In a related study, Heravi and Fakhr, reported a high yielding ultrasonic promoted synthesis of 2-amino-6-(arylthio)-4-arylpyridine-3,5-dicarbonitrile derivatives (Scheme 11), by the reaction of aryl aldehydes, thiols and malononitrile catalyzed by ZrOCl$_2$.8H$_2$O/NaNH$_2$ in [BMIM][BF$_4$] at room temperature.[46] Access to the same type of pentasubstituted pyridines was also possible using [BMIM][OH] as described by Ranu (Scheme 11).[47]

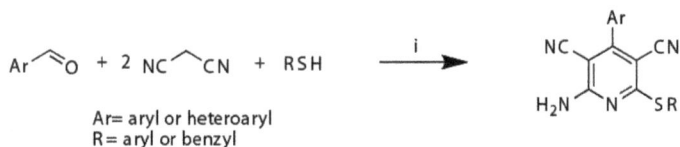

Ar= aryl or heteroaryl
R = aryl or benzyl

Scheme 11. Synthesis of penta substituted pyridines: (i) [BMIM][OH] / EtOH, rt; or ZrOCl$_2$.8H$_2$O/NaNH$_2$, [BMIM][BF$_4$], ultrasound.

The related 2,4,6-triaryl-1,4-dihydropyridines were generated in a Aldol-Michael-addition reaction cascade involving an aromatic aldehyde, acetophenone and NH$_4$OAc in [BMIM] [BF$_4$] (Scheme 12).[48] The resulting 2,4,6-triaryl-1,4-dihydropyridines were then examined as potential catalysts for the the Diels-Alder reaction of p-quinone and cyclopentadine, and maleic anhydride and cyclopentadiene.

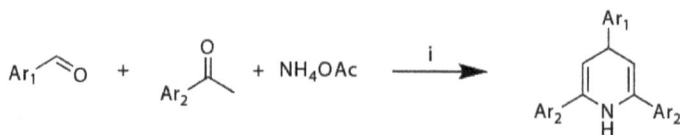

Scheme 12. Synthesis of 2,4,6,-triaryl-1,4-dihydropyridines: (i) [BMIM][BF$_4$], 80°C.

Wu used essentially the same reaction cascade described above, replacing acetophenone with acetoacetates which yielded, from [Bpy][BF$_4$], a series of 2,6-dimethyl-4-aryl-1,4-dihy-dropyridine-3,5-dicarboxylate esters (Scheme 13).[49] Compared with classical Hantzsch re-action conditions towards this type of product, this IL mediated reaction had the advantage of excellent yields, short reaction time, and easy workup.

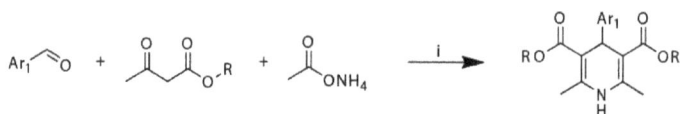

Scheme 13. Synthesis of 1,4-dihydropyridines: (i) [BPy][BF$_4$], 100-110°C.

Quinolin-4(1H)ones constitute an important class of heterocyclic compounds because of their important pharmaceutical properties, such as anti-viral,[50],[51] anti-platelet,[52] and anti-tumor effects.[53] These compounds have been exploited as precursors for anti-cancer and anti-malarial agents.[54],[55] Yadav et al. decribed an efficient two step synthesis of qui-nolin-4(1H)ones, 5H-thiazolo[3,2-a]pyrimidine, and 4H-pyrimido[2,1-b]benzothiazoles at room temperature.[56] The initial reaction in reaction was conducted arylamine with Mel-drum's acid) and trimethylorthoformate in [BMIM][Br] at 40°C giving the corresponding ar-ylaminomethylene-1,3-dioxane-4,6-diones. Cyclization occured in [BMIM][BF$_4$] / OTf at 80°C under nitrogen to the quiniolin-4(1H)ones in excellent yields (Scheme 14).[56]

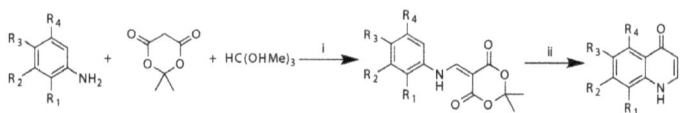

Scheme 14. Synthesis of 4(1H)-quinolones: (i) [BMIM][Br], 40°C, N$_2$; (ii) [BMIM][BF$_4$]/OTf, 80°C.

Wang et al. reported a novel reaction of 2-(1-substitutedpiperidin-4-ylidene)malononitrile, benzaldehyde, and malononitrile or cyanoacetate in the synthesis of highly substituted iso-quinoline derivatives (Scheme 15).[57] The three-component reaction of benzaldehyde, ma-lononitrile, and ethyl 4-(dicyanomethylene)piperidine-1-carboxylate was reacted in [BMIM] [BF$_4$] at 50°C, delivering ethyl 6-amino-5,7,7-tricyano-3,4,7,8-tetrahydro-8-arylisoquinoline-2(1*H*)-carboxylate derivatives being obtained in excellent yields. The highest yields were obtained with [BMIM][BF₄].

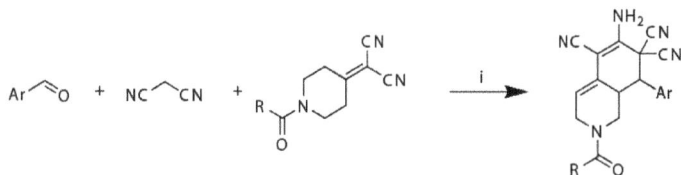

Scheme 15. Synthesis of ethyl 6-amino-5,7,7-tricyano-3,4,7,8-tetrahydro-8-arylisoquinoline-2(1*H*)-carboxylate derivatives: (i) [BMIM][BF₄], 50°C.

There are many methods for the synthesis of acridine compounds containing 1,4-dihydropyridine moieties from aldehydes, dimedone, and anilines or ammonium acetates via heating in organic solvents, or catalysis by triethyl(benzyl)ammonium chloride (TEBAC) in water, or under microwave irradiation.[58] In a much more efficient approach, Li *et al.* utilised the three component MCR in [BMIM][BF₄] at room temperature of an aromatic aldehyde, 2-hydroxy-1,4-naphthoquinone and naphthalen-2-amine giving rise to a series of 14-aryl-1,6,7,14-tetrahydrodibenzo[*a,i*]-acridine-1,6-dione derivatives (Scheme 16).[59]

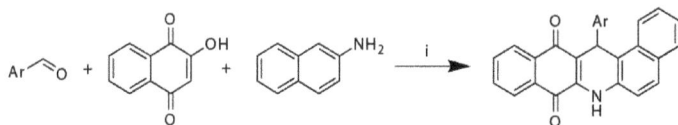

Scheme 16. Synthesis of 14-aryl-1,6,7,14-tetrahydrodibenzo[*a,i*]acridine-1,6-diones: (i) [BMIM][BF₄], rt.

Shi reported an efficient and green synthetic route to 3,3'-benzylidenebis(4-hydroxy-6-methylpyridin-2(1*H*)-ones) via condensation, addition and ammonolysis of an aldehyde, aniline and 6-methyl-4-hydroxypyran-2-one (Scheme 17).[60] Different solvents including [BMIM][Br], [BMIM][BF₄] and [BMIM][PF₆] were examined, with [BMIM][Br] giving the most favourable outcome (high yield and ease of product isolation).

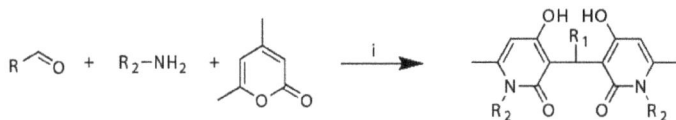

Scheme 17. Synthesis of 3,3'-benzylidenebis(4-hydroxy-6-methylpyridin-2(1*H*)-ones): [BMIM][Br], 95°C.

2.1.2. 3CRs yielding heterocycles with two ring nitrogens

Pyrimidine derivatives are important biologically active heterocyclic compounds which possess antimalarial.[61] Gholap's synthesis of 3,4-dihydropyrimidin-2-(1*H*)-ones from aromatic
or aliphatic aldehydes with ethyl acetoacetate and urea (or thiourea), was promoted by ultrasound in [BMIM][BF$_4$] at room temperature affording the target compounds in excellent
yields and short reaction times (Scheme 18).[62]

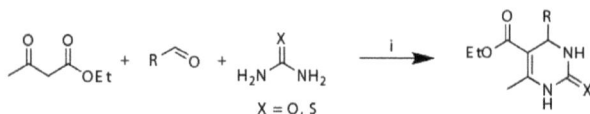

Scheme 18. Synthesis 3,4-dihydropyrimidin-2-(1*H*)-ones: (i) [BMIM][BF$_4$], 30°C, ultrasound.

Similar pyrimidine analogues were accessed by Gui *et al.* through the use of acidic ionic liquids such as [MIM-CH$_2$COOH][HSO$_4$], [MIM-CH$_2$COOH][H$_2$PO$_4$], [MIM-(CH$_2$)$_2$COOH]
[HSO$_4$] and [MIM-(CH$_2$)$_2$COOH][H$_2$PO$_4$]) which successfully promoted the Biginelli coupling of an aldehyde, 1,3-dicarbonyl compound, and urea giving easy access to 3,4-dihydro-
pyrimidin-2(1*H*)-ones.[63] Peng and Deng, used [BMIM][BF$_4$] and [BMIM][PF$_6$] as catalysts
for the same Biginelli condensation reaction at 100ºC (Scheme 19).[64]

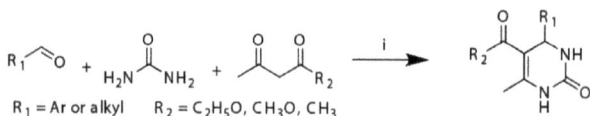

Scheme 19. The synthesis of 3,4-dihydropyrimidin-2(1*H*)-ones: (i) IL chosen from [MIMCH$_2$COOH][HSO$_4$] or
[MIMCH$_2$COOH][H$_2$PO$_4$] or [MIM(CH$_2$)$_2$COOH][HSO$_4$] or [MIM(CH$_2$)$_2$COOH][H$_2$PO$_4$], 75°C.

Brønsted acidic ionic liquids have designed to replace solid acids and traditional mineral liquid acids like sulfuric acid and hydrochloric acid in chemical procedures.[65],[66] Using 3-
carboxypyridinium hydrogensulfate [HCPy][HSO$_4$], 1,3-dicarbonyl compounds, aromatic
aldehydes and urea or thiourea, Hajipour and Seddighi, successfully removed the traditional acid requirement in the synthesis of 3,4-dihydropyrimidin-2(1*H*)-ones (Scheme 20).[67]

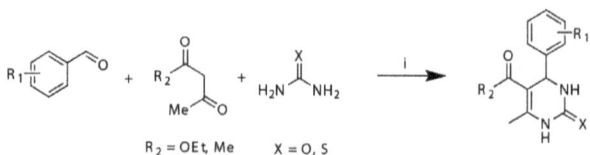

Scheme 20. Synthesis of 3,4-dihydropyrimidin-2-(1*H*)-ones: (i) [HCPy][HSO$_4$], 120°C.

Fang *et al.*'s dicationic acidic IL catalytic approach, in what amounted to a modified Biginelli approach, resulted in the synthesis of 3,4-dihydropyrimidin-2(1H)-one and 3,4-dihydropyrimidin-2(1H)-thione derivatives (Scheme 21), in good yields.[68] The products could be separated simply from the catalyst–water system, and the catalysts could be reused at least six times without noticeably reducing catalytic activity.

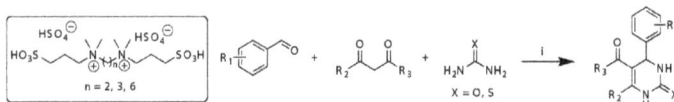

Scheme 21. Synthesis of 3,4-dihydropyrimidin-2(1H)-(thi)ones: (i) Dicationic acidic IL (shown in box), 90°C.

Rather than use an acidic IL approach to dihydropyrimidinones (above) Mirzai and Valizadeh developed a microwave assisted Biginelli route using the weakly Lewis basic nitrite based ionic liquid, IL-ONO (Scheme 22).[69] These nitrite based Ils have also been used to carry out nitrosations of aromatic compounds in aqueous media.[70] Valizadeh, have reported the nitrozation of aromatic compounds using the same nitrite ionic liquid in aqueous media.[70]

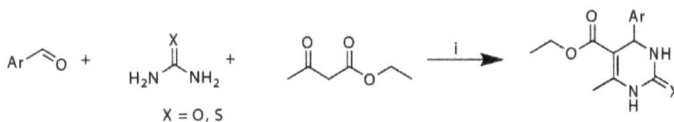

Scheme 22. Synthesis of dihydropyrimidinones: (i) IL-ONO, MW, 80°C.

Trisubstituted imidazoles can be rapidly accessed from a one-pot condensation of 1,2-diketone or α-hydroxyketone, aldehyde, and NH_4OAc in 1,1,3,3-N,N,N,N-tetramethylguanidinium trifluoroacetate (TMGT) at 100°C (Scheme 23).[71] The synthesis of trisubstituted imidazoles in TMGT as promoter and solvent for the synthesis of trisubstituted imidazoles not only represented a dramatic improvement (15-40 min, 81-94%) over conventional thermal heating but the reaction times were comparable to the recently reported microwave irradiation (20 min in HOAc, 180–200°C). There are related routes in a four-component approach (see section 2.2.3).

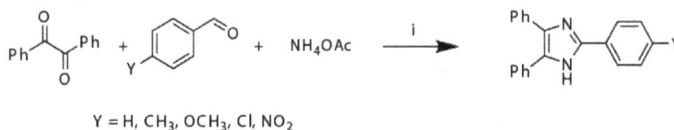

Y = H, CH_3, OCH_3, Cl, NO_2

Scheme 23. Synthesis of trisubstituted imidazoles: (i) TMGT, 100°C.

Khurana and Kumar have reported a simple and convenient synthesis of octahydroquinazo-linone and biscoumarin derivatives (Scheme 24).[72] Despite prolonged heating (10 h) at 100ºC, the reaction of benzaldehyde, dimedone and urea in [BMIM][Br] gave only 30% of the expected product, 4,6,7,8-tetrahydro-7,7-dimethyl-4-phenyl-1H,3H-quinazoline-2,5-di-one. Addition of TMSCl saw a reduction in reaction duration to 2.5 h wand a 92% isolated yield. Of the Ils examined [BMIM][Br] and [BMIM][BF$_4$] gave higher yields than [BMIM][Cl] and [BMIM][PF$_6$].

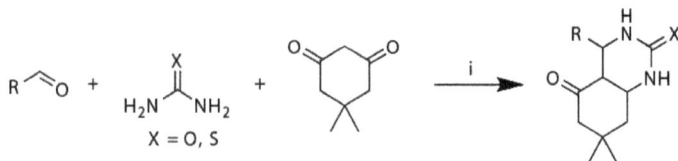

Scheme 24. Synthesis of octahydroquinazolinones: (i) [BMIM][Br], TMSCl, 100ºC.

Omprakash et al. used catalytic [BMIM][BF$_4$] and ultrasonics to obtain excellent yields of quinazolin-4(3H)ones from anthranilic acid, primary aromatic amine and carboxylic acids (Scheme 25). Of the anilines examined, only 4-nitroaniline required elevated temperature (50°C), but this reaction was complete after 20 minutes.[73]

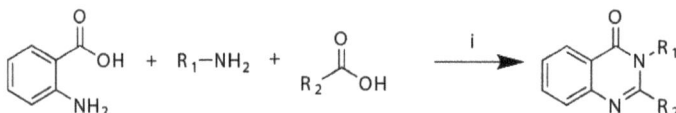

Scheme 25. Synthesis of 4(3H)-quinazolinones: (i) [BMIM][BF$_4$], ultrasound.

The related quinazoline nucleus has been prepared by Dabiri et al. from 2-aminobenzophe-none derivatives, aldehydes and ammonium acetate in the presence of an protic ionic liquid, 1-methylimidazolium triflouroacetate, [HMIM][TFA] (Scheme 26).[74]

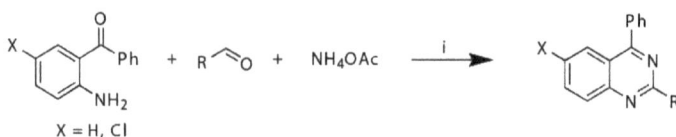

Scheme 26. Synthesis of quinazoline derivatives: (i) [HMIM][TFA], 80ºC.

An analogous 6, 5-ring system, the 3-aminoimidazo[1,2-a]pyridine, was accessed via con-densation of an aldehyde, 2-aminoazine and trimethylsilylcyanide, as an isocyanide equiv-alent under by simple heating in [BMIM][Br] in high yields with rather short reaction times (1-2 h) (Scheme 27).[75] Shaabani took a slighly different route to 3-arylsubstituted 3-aminoimidazo[1,2-*a*]pyridines using an isocyanide rather than an isocyanide equivalent. The use of substituted aminopyridines also allowed for the introduction of an additional C6 substituent.[76]

Scheme 27. Synthesis of 3-aminoimidazol[1,2-a]pyridines: (i) [BMIM][Br], 80°C

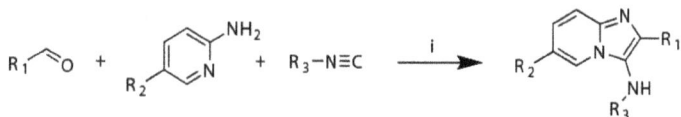

Scheme 28. Synthesis of 3-aminoimidazo[1,2-a]pyridines: (i) [BMIM][Br], 80°C.

Using the Brønsted acidic ionic liquid triethylammonium hydrogen sulfate [TEBSA][HSO$_4$] Hajipour *et al.* synthesied pyrimidinone derivatives from a range of aromatic aldehydes, cyclopentanone, and urea or thiourea (Scheme 28).[77]

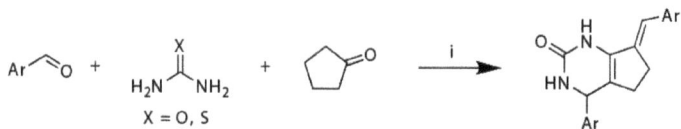

Scheme 29. Synthesis of pyrimidinone derivatives: (i) [TEBSA][HSO$_4$], 100°C.

The reaction of phthalhydrazide, aromatic aldehydes, and malononitrile using controlled under microwave irradiation in the presence of [BMIM][OH] at an ambient temperature of 45°C allowed facile access to 1*H*-pyrazolo[1,2-*b*]phthalazine-5,10-dione (Scheme 29).[78]

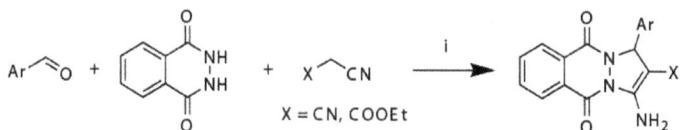

Scheme 30. Synthesis 1H-pyrazolo[1,2-b]phthalazine-5, 10-dione derivatives: (i) [BMIM][OH], microwaves, 45°C.

Mixed solvent systems comprising [BMIM][BF$_4$], water and ethanol were used in the synthesis of 2H-indazolo[2,1-b]phthalazine-triones by condensation of phthalhydrazide, aromatic aldehydes, and cyclic 1,3-dicarbonyl compounds. Interestingly, the reaction required the addition of a catalytic quantity of sulfuric acid to effect the desired transformation (Scheme 30).[79]

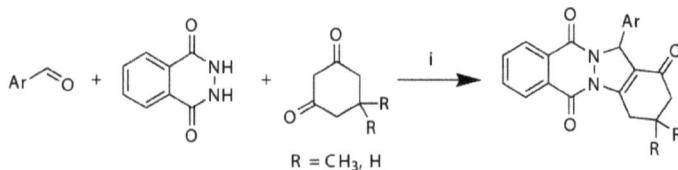

Scheme 31. Synthesis of a series of 2H-indazolo[2,1-b]phthalazinetriones: (i) [BMIM][BF$_4$] / H$_2$O-EtOH / H$_2$SO$_4$.

2.1.3. 3CRs yielding heterocycles with three ring nitrogen atoms

While not strictly speaking the synthesis of a new ring system by MCR in ILs, Wang has exploited the MCR approach in an elegant synthesis of N-(α-alkoxyalkyl)benzotriazoles (Scheme 31) via the condensation of benzotriazole with various aldehydes and alcohols catalysed by acidic ionic liquid [HMIM][HSO$_4$] at room temperature.[80] The yield was up to 99%. Wang's approach was effective when triethoxymethane was utilized instead of alcohols. Moreover, the [HMIM][HSO$_4$] was recyclable with no loss in catalytic activity.

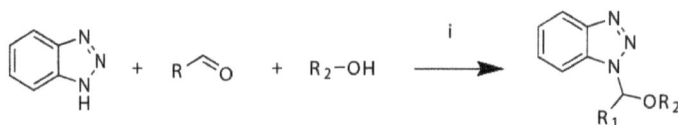

Scheme 32. N-(α-alkoxyalkyl)benzotriazoles: (i) [HMIM][HSO$_4$], rt.

Pyrazolo[3,4-b]pyridines possess a wide range of biological activities such as psychotropic and cytotoxic, and are thus a very important scaffold in medicinal chemistry.[81],[82]

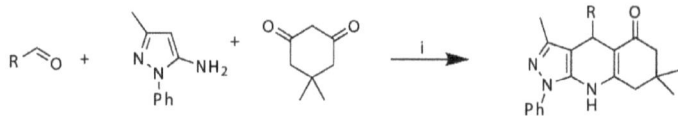

Scheme 33. Preparation of pyrazolo[3,4-b]pyridinone and pyrazolo[3,4-b]quinolinone derivatives: (i) [BMIM][BF$_4$], 80°C.

Judicious choice of the 1,3-dicarbonyl source allows the synthesis of either pyrazolo[3,4-b]pyridinone derivatives or pyrazolo[3,4-b]quinolinone (Scheme 32). The combination of an aldehyde, 5-amino-3-methyl-1-phenylpyrazole, and Meldrum's acid in [BMIM][BF$_4$] affords the pyrazolo[3,4-b]pyridinone, while the use of dimedone affords the pyrazolo[3,4-b]quinolinones.[83]

Zhang showed that the 1,3-dicarbonyl source was not limited to Meldrum's acid or dimedone derivatives with the introduction of 1H-indene-1,3(2H)-dione for a mild synthesis of indeno[2,1-e]pyrazolo[3,4-b]pyridine-5(1H)-one derivatives in excellent yields (Scheme 33).[84]

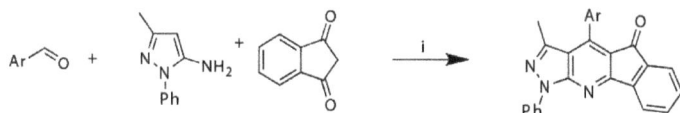

Scheme 34. Synthesise indeno[2, 1-e]pyrazolo[3, 4-b]pyridine-5(1H)-one: (i) [BMIM][Br], 95°C.

Zhang et al. reported the reaction of aldehydes, 5-amino-3-methyl-1-phenylpyrazole and malononitrile or ethyl cyanoacetate in [BMIM][BF$_4$] as a green route to pyrazolo[3,4-b]pyridines (Scheme 34).[84]

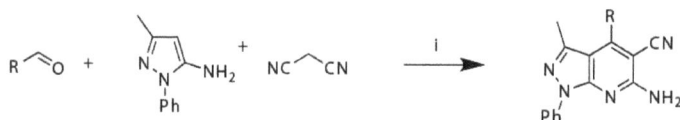

Scheme 35. Preparation of pyrazolo[3,4-b]pyridine derivatives: (i) [BMIM][BF$_4$], 80°C.

Yao et al. synthesised 4-aryl-3,4-dihydro-1H-pyrimido[1,2-a]benzimidazol-2-one via the reaction of aryl aldehyde, 1,3-dicarbonyl compounds and 1H-benzo[d]imidazol-2-amine in [BMIM][BF$_4$] (Scheme 35).[85] The reaction accomplished in [BMIM][BF$_4$] exhibited higher yield (75%) than other counterparts.

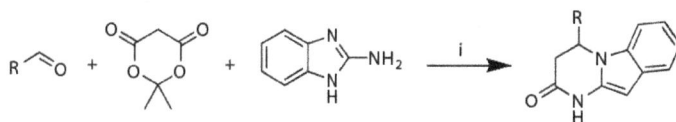

Scheme 36. Synthesis of 4-aryl-3,4-dihydro-1H-pyrimido[1,2-a]benzimidazol-2-one: (i) [BMIM][BF$_4$], 90°C.

Wang has a particular interest in the development novel methods for the preparation of various biologically important heterocyclic compounds by using ionic liquids. This group uses ILs as both novel reaction media and reaction promoters.[86],[87] Shi et al. also have similar interest and this led to the synthesis indeno[2,1:5,6]pyrido[2,3-d]pyrimidine and pyrimido[4,5-b]quinoline derivatives from aromatic aldehydes, 6-amino-3-substituted-1-methyl-pyrimidine-2,4(1H,3H)-diones and dimidone analogues derivatives in ionic liquid without any catalyst (Scheme 36).[88]

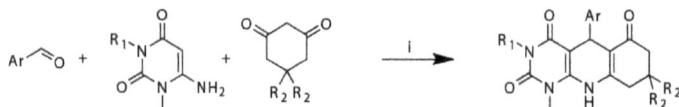

Scheme 37. Synthesis of 5-aryl-7,8,9,10-tetrahydropyrimido[4,5-b]quinoline-2,4,6-trione: [BMIM][Br], 95°C.

Pyrimidoquinolinedione derivatives are known to possess antitumor, anticancer, antihypertensive, antibacterial activity and are Kaposi's sarcoma-associated herpesvirus and topoisomerase inhibitors.[89],[90] Entry to this highly biologically active scaffolds can be obtained by the reaction of aldehydes, 1-naphthylamine and barbituric acid in [BMIM][BF$_4$] gave 7-aryl-11,12-dihydrobenzo[h]pyrimido-[4,5-b]quinoline-8,10(7H,9H)-diones (Scheme 37). While the reaction proceeded in traditional organic solvent, yield enhancements and shorter reaction tiomes were evident with the use of [BMIM][BF$_4$].[91]

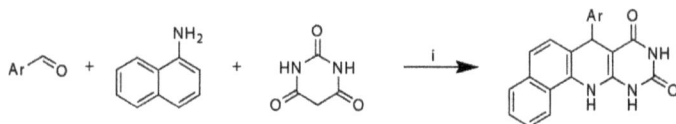

Scheme 38. The synthetic of 7-aryl-11,12-dihydrobenzo[h]pyrimido-[4,5-b]quinoline-8,10(7H,9H)-diones: (i) [BMIM][BF$_4$], 90°C.

2.1.4. 3CRs yielding heterocycles with >three ring nitrogens

The purine bioisosteres, [1,2,4]triazolo[1,5-a]pyrimidine, have been reported to possess antitumour activity.[92] Using both [BMIM][BF$_4$] and [Bpy][BF$_4$] as reaction solvents, Li synthesised 5-(trifluoromethyl)-4,7-dihydro-7-aryl-[1,2,4]-triazolo[1,5-a]pyrimidine derivatives from aldehydes, 3-amino-1,2,4-triazole and ethyl 4,4,4-trifluoro-3-oxobutanoate or 4,4,4-trifluoro-1-phenylbutane-1,3-dione yielding 4-, 5- and 7- substituted derivatives (Scheme 38). The 5- and 7- positions are known to be important for retention of antitumour activity.[93]

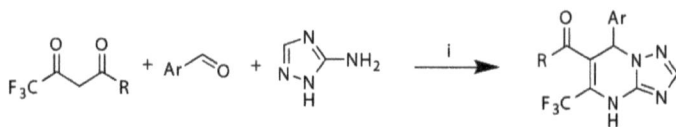

Scheme 39. Synthesis of 5-(trifluoromethyl)-4,7-dihydro-7-aryl-[1,2,4]triazolo[1,5-a]pyrimidine derivatives: (i) BMIM][BF$_4$] or [byp][BF$_4$].

2.1.5. 3CRs yielding heterocycles with one ring oxygen

The synthesis of heterocyclic compounds with oxygen in the ring is slightly more complex, only due to the reduced numbers of suitable oxygen nucleophiles to affect the final ring-closing step. The 4H-pyran core is found in a wide range of natural products and it has thus attraced a considerable degree of attention.[94]-[96] The high reactivity of 4H-pyran derivatives has led to their use as synthons in the synthesis of more complex species. Access to highly substituted 4H-pyrans is easily accomplished by the 1,1,3,3-tetramethylguanidine catalysed addition of aromatic aldehydes, malononitriles, and β-dicarbonyl in [BMIM][BF₄] (Scheme 39).[97]

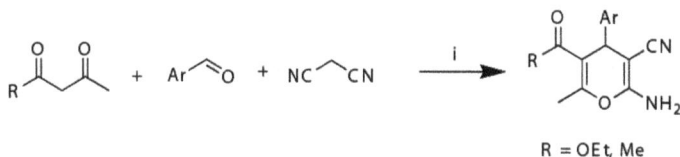

R = OEt, Me

Scheme 40. Synthesis of 4H-pyran derivatives: (i) TMG, [BMIM][BF₄], 80°C.

With dimedone as the 1,3-dicarbonyl source the corresponding 5-oxo-5,6,7,8-tetrahydro-4H-benzo[b]pyran derivatives were accessed in [BMIM][BF₄], [HMIM][BF₄], [OMIM][BF₄], [OMIM][PF₆] and [DMIM][PF₆]. In this instance no additional catalyst was required and the reactions were complete in 2-6 h with yields ranging from 52% to 98%.[98] Fang *et al.* reported a subtle variation leading to the syntheiss of more highly substituted 5-oxo-5,6,7,8-tetrahydro-4H-benzo[b]pyrans by condensation of aromatic aldehyde, malononitrile (or ethyl cyanoacetate), and dimedone (or 1,3-cyclo-hexanedione) in water catalyzed by acidic ionic liquids such as [TEBSA][HSO₄], [TBPSA][HSO₄], [EDPSA][HSO₄] (Scheme 40).[99] The reactions gave the products in good yields between 86 to 94%.

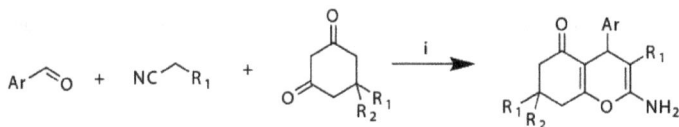

Scheme 41. Synthesis of 5-oxo-5,6,7,8-tetrahydro-4H-benzo[b]pyrans: (i) Ionic liquid chosen from [TEBSA][HSO₄], [TBPSA][HSO₄] and [EDPSA][HSO₄].

Interest in oxygen containing heterocycles is not limited to those with biological actiity. A number of analogues, such as the 2-amino-2-chromenes are natural products that have found utility in cosmetics and pigments. They also have a role as biodegradable agrochemicals.[100]-[102] Traditional approaches to this scaffold required the reaction of aldehydes, active methylene containing compounds and activated phenols. Stoichiometric quantities of organic base (piperidine) in volatile organic solvents are also required.[103],[104] By replacing the organic solvent with [BMIM][OH] the reaction proceeded with aromatic aldehydes, malononitrile with α- or β-naphthol in the absence of additional catalyst (Scheme 41).[105] After five reuses of the [BMIM][OH] the isolated product yield had dropped from 91% to 85%, which may be due to [BMIM][OH] degredation.

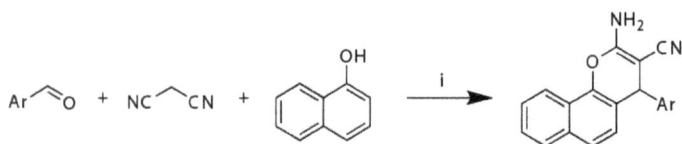

Scheme 42. Synthesis of 2-amino-2-chromenes: (i) [BMIM][OH], H₂O, reflux.

The basic 4H-pyran scaffold can be increased in complexity by modification of the basic building blocks described above, e.g. in Schemes 39 and 41. Both Khurana and Magoo, and Zakeri reported the synthesis of a series of 12-aryl-8,9,10,12-tetrahydrobenzo[a]xanthen-11-ones by the reaction of β-naphthol, aromatic aldehydes, and dimedone derivatives (Scheme 42).[106],[107] Smooth conversion was accomplished through the use of catalytic p-TSA in [BMIM][BF₄] at 80°C for 35-45 min. The [BMIM][BF₄] could be recyled without a reduction in product yield.

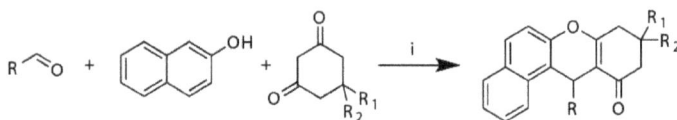

Scheme 43. Synthesis of a series of 12-aryl-8,9,10,12-tetrahydrobenzo[a]xanthen-11-ones: (i) [BMIM][BF₄], p-TSA or BAIL, 120°C.

Zheng and Li accessed the tetrahydrobenzo[b]pyran and pyrano[c]chrome scaffoled via a series of novel Lewis basic task-specific ionic liquids. These novel Ils were used to catalyse the addition of aromatic aldehydes, dimedone and malononitrile can also be used as catalysts in multicomponent reaction accession during the mixture of tetrahydrobenzo[b]pyran and pyrano[c]chromene derivatives (Scheme 43).[108]

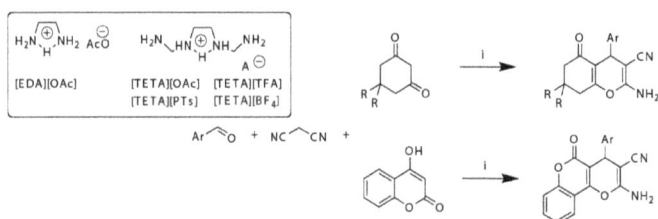

Scheme 44. Synthesis of tetrahydrobenzo[b]pyran and pyrano[c]chromene derivatives: (i) [TETA][TFA] (5%), H₂O-EtOH (1:1), reflux.

2.1.6. 3CRs yielding heterocycles with two ring oxygen atoms

Dihydropyrano[3,2-c]chromene-3-carbonitriles are important heterocycles with a wide range of biological properties.[109]-[111] A number of 2-amino-4Hpyrans are reported to be useful photoactive materials.[112] The task specific ionic liquid, hydroxyethanolammonium acetate [HEAA], was used to initiate a domino cascade of 4-hydroxycoumarin, aldehydes, and malononitrile at room temperature ultimately yielding 2-amino-5-oxo-4,5-dihydropyrano[3,2-c]chromene-3-carbonitrile derivatives (Scheme 44).[113]

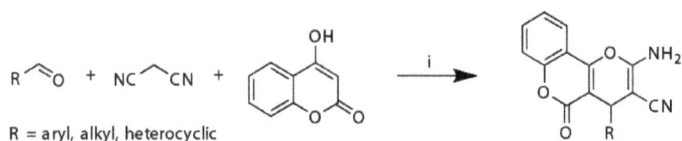

R = aryl, alkyl, heterocyclic

Scheme 45. Synthesis of 2-amino-5-oxo-4,5-dihydropyrano[3,2-c]chromene-3-carbonitrile derivatives: (i) [HEAA], pulverise, rt.

Gong *et al.* reported facile method for the synthesis of 4*H*-pyrans in the presence of basic ionic liquid [BMIM][OH] as catalyst in aqueous medium (Scheme 45).[114] The synthesis of 2-amino-4-aryl-5-oxo-4*H*,5H-pyrano-[3,2-c]chromence-3-carbonitrile was achieved by the three-component condensation of an aromatic aldehyde, malononitrile with 4-hydroxycoumarin in the presence of 10 mol% [BMIM][OH] at 100ºC.

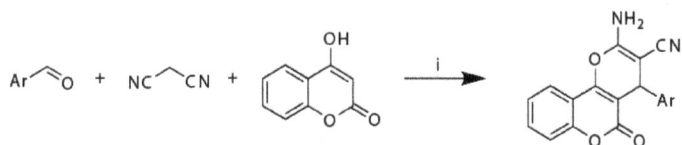

Scheme 46. Synthesis of 2-amino-4-aryl-3-cyano-5-oxo-4*H*,5*H*-pyrano[3,2-c]chromenes: (i) [BMIM][OH], H_2O, 100ºC.

The 3,4-dihydro-2*H*-furo[3,2-c]coumarin core is present in a number of natural products ranging including Novobiocin and warfarin. Earlier methods to this core include the oxidative cyclisation of the Michael adduct from the reaction of cyclic 1,3-diketones and chalcones using a phase transfer catalyst; from 1,3-dicarbonyl compounds and (*E*)-β-bromo-β-nitrostyrenes in the presence of tert-butylammonium bromide (TBAB) (20 mol %);[115]-[117] and the manganese acetate promoted radical cyclization of 4-hydroxycoumarin and 2-hydroxy-1,4-naphthoquinone with electron-rich alkenes.[118] Rajesh *et al.* reported a much simpler and greener approach for regio- and diastereo- selective synthesis of furocoumarins (Scheme 46).[119] These reactions proceeded chemo-, regio- and stereoselectively and furnished compounds in good to excellent yields (81-92%).

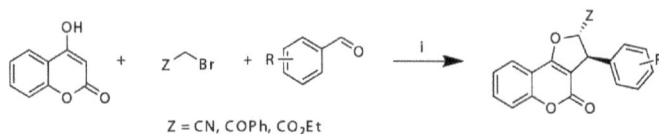

Scheme 47. Synthesis of furo[3,2-c]coumarins: (i) [BMIM][OH],pyridine, 80-90°C.

Coumarin derivatives have received considerable attention because they possess several types of pharmacological properties, such as antibacterial, anticancer.[120] The coumarins have attracted the attention of a number of research groups interested not only in their biological activity, but also in developing more activity, but also in debeloping more environmentally friendly approaches to their synthesis. Gong *et al.* reported the condensation of 4-hydroxycoumarin, aldehydes, and Meldrum's acids or malononitrile or α-cyanocinnamonitriles in the presence of [BMIM][OH].[114],[121] While Chen *et al.* used 4-hydroxycoumarin, benzaldehyde and 1,3-dicarbonyl by use 1,3-dimethyl-2-oxoimidazolidine-1,3-diium cation [DMDBSI][2HSO₄] were employed as the model reactions in the presence of different catalysts (Scheme 47).[122] Both approaches offered ease of access and considerable improvements over the traditional approaches to this class of compounds.[123]-[127]

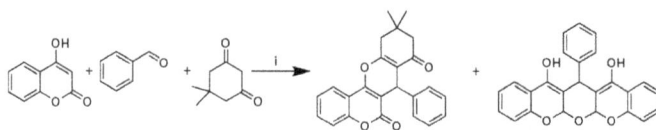

Scheme 48. Synthesis of 4-hydroxycoumarin: (i) [DMDBSI][2H₂SO₄], H₂O, reflux.

2.1.7. 3CRs yielding heterocycles with one ring sulfur atom

Thiophenes, dihydrothiophenes and tetrahydrothiophenes are known important constituents of a range of pharmacologically active compounds.[128]-[130] While these compounds are of significant interest to medicinal and synthetic chemists, the synthetic routes to highly functionalised sulfur heterocycles are not well developed. Notwithstanding this, Zhang *et al.* have reported the synthesis of thiopyrans from aldehydes, malononitrile and cyanothioacetamide in an ionic liquid [BMIM][BF₄] as a recyclable solvent and promoter without the need of a catalyst (Scheme 48).[131]

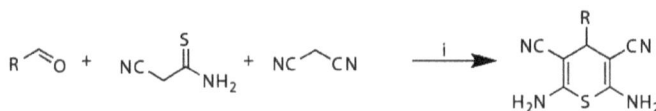

Scheme 49. Synthesis of thiopyrans: (i) [BMIM][BF₄], 80°C.

Given that most of the synthetic procedures towards sulfur heterocycles suffer from some drawbacks such as low yields, long reaction times, the requirement for harsh reaction conditions, it is not surprising that a number of groups have risen to the challenge and examined the use of ionic liquids as a potential method for enhancing the reaction outcomes whilst increasing the efficiency of the synthesis.[132]-[134] Kumar *et al.* have developed a series of novel amino acid derived functional ionic liquids that facilitated the synthesis of dihydro-thiophene and tarcine derivatives in good yield under mild conditions from 2-arylidenema-lononitrile, 1,3-thiazolidinedione, aliphatic or aromatic amines were added with ionic liquid [Bz-His(n-propyl)₂-OMe][Br] and water (Scheme 49). While the products were shown as single diastereoisomers, no details of the level of diastereoselectivity were provided.[135]

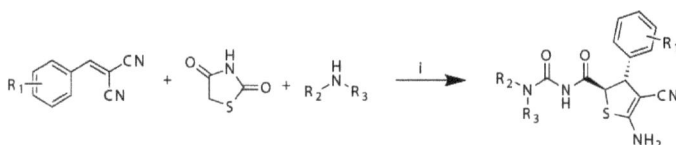

Scheme 50. Synthesis of dihydrothiophenes: (i) [Bz-His(n-propyl)₂-OMe][Br], 70°C.

2.1.8. 3CRs yielding heterocycles with ring oxygen and nitrogen atoms

The synthesis of simple heterocycles with a single type of heteroatom is important, but a considerable number of biologically active compounds have different types of heteroatom within a single structure. The benzo[*b*][1,4]oxazin scaffold as a privileged structure for the generation of drug-like libraries in drug-discovery programs has been amply demonstrated. Benzo[*b*][1,4]oxazin derivatives have been used as the basic framework for substances of interest in numerous therapeutic areas, such as anti-Candina albicans agents,[136] antifungals, [137] and kinase inhibitors.[138] Ebrahim *et al.* used [BMIM][Br] as both the solvent and reaction promotor for the room temperature three-component condensation of 2-aminophe-nole, an aldehyde and isocyanide to prepare benzo[*b*][1,4]oxazines (Scheme 50).[139]

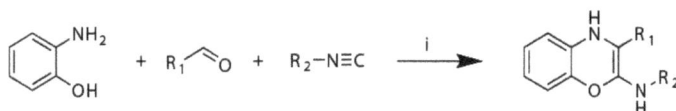

Scheme 51. Synthesis of benzo[*b*][1,4]oxazines: (i) [BMIM][Br], rt.

Asri *et al.* described use of ionic liquids as complementary new media for multicomponent reactions leading to the 2,6-diazabicyclo[2.2.2]octane core (Scheme 51).[140] Interestingly both hydrophobic and hydrophillic Ils [BMIM][BF₄] and [BMIM][NTf₂] gave acceptable yields, as the original synthesis required the use of toluene and a significant excess of molecular sives to remove water and drive the reaction forward. The original synthesis of these

2,6-diazabicyclo[2.2.2]octanes required 6g of molecular seives per 200 mg reagent. Thus the use of Ils represents a significant greening of this synthesis.[141]

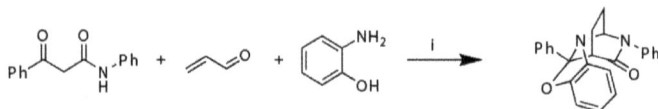

Scheme 52. Synthesis 2,6-diazabicyclo[2.2.2]octane core: (i) [BMIM][BF₄] or [BMIM][NTf₂], 110ºC, 4 Å sieves.

The synthesis of 6-amino-4-aryl-5-cyano-3-methyl-1-phenyl-1,4-dihydropyrano[2,3-c]pyrazoles was first reported by Otto in 1974.[142] These molecules have since been shown to possess interesting biological activity.[143] Balaskar *et al.* simplified the synthesis of this important class of compounds in a triethylammonium acetate [TEAA] ionic liquid catalyzed reaction of aromatic aldehydes, malononitrile and 3-methyl-1-phenyl-2-pyrazolin-5-one at room temperature (Scheme 52). TEAA plays dual role as reaction media and catalyst. These reactions are rapid, complete in 25 min, and typically high yielding (>90%).[144]

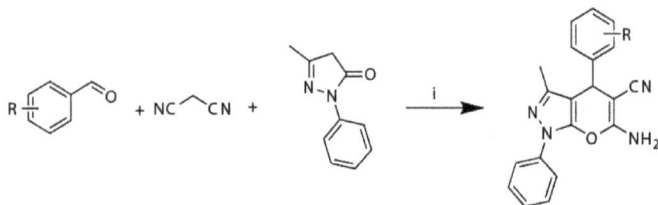

Scheme 53. Synthesis of 6-amino-4-aryl-5-cyano-3-methyl-1-phenyl-1,4 dihydropyrano[2,3-c]pyrazoles: (i) [TEAA], rt.

The furopyridine core is another privileged scaffold in medicinal chemistry.[145],[146] Shi *et al.* rectified what they perceived as an oversight in this area with their synthesis of the furo[3,4-b]pyridine motifs by reaction of an aldehyde, 5-amino-3-methyl-1-phenylpyrazole and tetronic acid. They explored the use of [BMIM][Br], [BMIM][BF₄], [PMIM][Br], water, glacial acetic acid, acetone, and ethanol as potential solvents for the synthesis of furo[3,4-e]pyrazolo[3,4-b]pyridine-5(7H)-one derivatives (Scheme 53).[147] Across the range of aromatic aldehydes examined, the ILS [BMIM][Br], [BMIM][BF₄], and [PMIM][Br] consistently gave the highest product yields and the shortest reaction times.

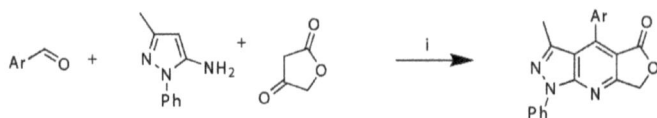

Scheme 54. Synthesise the furo[3,4-e]pyrazolo[3,4-b]pyridine-5(7H)-one derivatives: (i) ILs, 95ºC.

Moghaddam reported the synthesis of novel spiro[chromeno[2,3-d]pyrimidine-5,3'-indoline]tetraone derivatives by the combination of isatin, barbituric acid, and cyclohexane-1,3-dione derivatives in the presence of alum (KAl(SO$_4$)$_2$•12H$_2$O) as a catalyst for 15 min and [BMIM][PF$_6$] (Scheme 54).[148] The unique structural array and the highly pronounced pharmacological activity displayed by the class of spirooxindole compounds have made them attractive synthetic targets.[149]

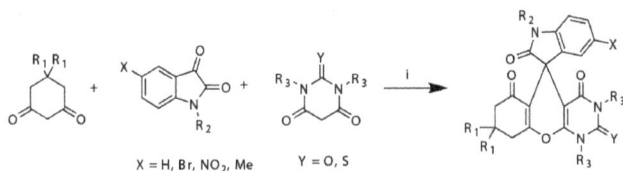

X = H, Br, NO$_2$, Me   Y = O, S

Scheme 55. Synthesis of spiro[chromeno[2,3-d]pyrimidine-5,3'-indoline]tetraone derivatives: (i) [BMIM][PF$_6$], alum, 100°C.

The synthesis of furopyrimidines and 2-aminofurans have received little attention with only a few procedures reported. Among these, the furo[2,3-d]pyrimidines have been shown to possess sedative, antihistamine, diuretic, muscle relaxant, and antiulcer properties.[150]-[156] The condensation of an aldehyde, N,N-dimethylbarbituric acid and alkyl or aryl isocyanide in [BMIM][Br] gave furo[2,3-d]pyrimidine-2,4(1H,3H)-diones in high yields at room temperature within 20 minutes (Scheme 55).[157]

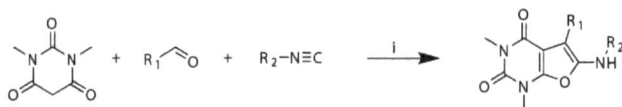

Scheme 56. Synthesis of furo[2,3-d]pyrimidine-2,4(1H,3H)-diones: (i) [BMIM][Br], rt, 15-20 min.

The potential antitumour pharmacophore, benzopyrano[2,3-d]pyrimidine,[158] was accessed by Gupta et al. by the condensation of the salicyladehyde, malononitrile, and dimethylamine at room temperature in [BMIM][BF$_4$] at room temperature (Scheme 56). Howver, this approach was limited to the use of dimethlamine, with the diethylamine resulted in no reaction.[159]

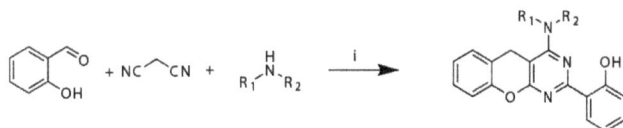

Scheme 57. Synthesis of benzopyranopyrimidines: (i) [BMIM][BF$_4$], rt.

2.1.9. 3CRs yielding heterocycles with ring sulfur and nitrogen atoms

4-Thiazolidinones have been exploited as potential bactericidal, antifungal, anticonvulsant, anti-HIV, and antituberculotic agents.[160],[161] While there have been multiple synthestic approaches, there is still considerable scope to develop a more environmentally friendy and efficient approach to this scaffold.[162],[163] Lingampalle et al. have developed a rapid entry to 4-thiazolidinones via the N-methylpyridinium tosylate [NMP][Ts] cyclocondensation of amines, aromatic ketones, and mercaptoacetic acid.[164], [165] The reaction proceeds via imine formation, followed by rapid cyclocondensation at 120 °C (Scheme 57).

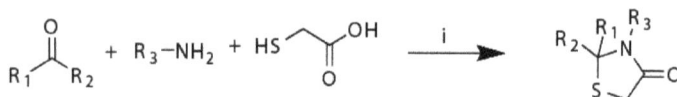

Scheme 58. Synthesis of 4-thiazolidinones: (i) [NMP][Ts], 120°C, 3h.

The (1H)-quinolones, thiazolo[3,2-a]pyrimidines and pyrimido[2,1-b]benzothiazoles display considerable bioactivity and are important lead compounds in the development of anti-viral, anti-platelet, anti-cancer and anti-malarial agents.[166]-[168] While there are many reported synthesis of pyrimido[2,1-b]benzothiazoles, arguably Yadav et al. 's 1-methoxyethyl-3-methylimidazolium trifluoroacetate [MOEMim][TFA] mediated MCR is the most direct and efficient reported thus far (Scheme 58).[169]

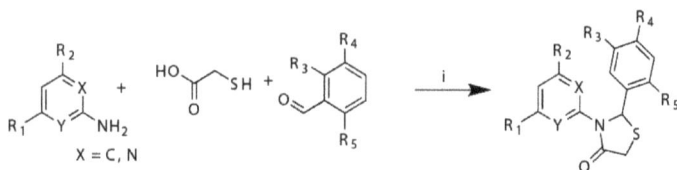

Scheme 59. Synthesis of 2,3-diaryl/2-aryl-3-heteroaryl-1,3-thiazolidin-4-ones: (i) [MOEMim][TFA].

2.1.10. 3CRs yielding heterocycles with ring oxygen, sulfur and nitrogen atoms

Application of a tandem Knoevenagel, Michael and ring transformation reactions involving 3-arylrhodanines, aromatic aldehydes and a mercaptoacetyl transfer agent, 2-methyl-2-phenyl-1,3-oxathiolan-5-one, in a chiral ionic liquid L-prolinium sulfate [Pro$_2$SO$_4$], gave 6-mercaptopyranothiazoles with diastereoselectivities of 88-95%ee (Scheme 59).[170] The reactions were conducted at room temperature 25-30 h, followed by isolation to yield a single diastereomer in 76-90% yields.

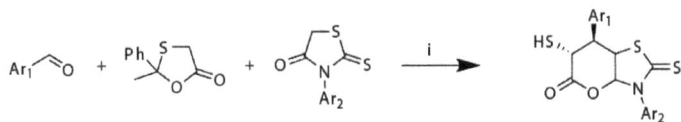

Scheme 60. Synthesis of 2-methyl-2-phenyl-1,3-oxathiolan-5-one: (i) [Pro₂SO₄], rt.

The biologically important 4*H*-benzo[*b*]pyrans can be smoothly accessed as shown in Scheme 60 in excellent yields (77-95%) after stirring for 30-50 min at ethanol reflux. In this instance [BMIM][OH] was used as a catalyst.[121]

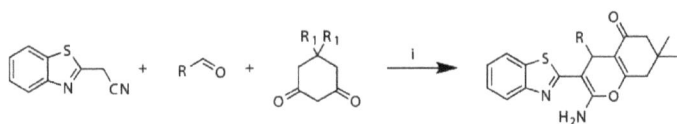

Scheme 61. Benzo[*b*]pyrans: (i) [BMIM][OH], EtOH, reflux.

2.2. 4CRs yielding heterocycles

2.2.1. 4CRs yielding heterocycles with one nitrogen in the ring

Increasing the number of components in MCRs from three to four offers the potential to increase substituent diversity, atom and step economy. This increased structural complexity allows for a facile access to highly decorated scaffolds, but interestingly in the four component IL mediated MCR, this has been limited to the synthesis of heterocyclic compounds. For example, rapid access to both alkyl and aromatic substituted 1,4-dihydropyridine derivatives can be accomplished via the reaction of an aldehyde, a 1,3-dicarbonyl compound, Meldrum's acid and ammonium acetate as the nitrogen source in [BMIM][BF₄] (Scheme 61).[172],[173]

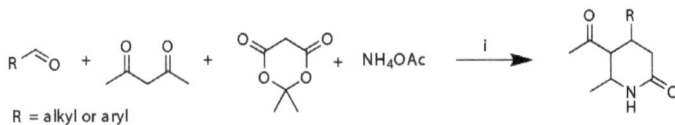

R = alkyl or aryl

Scheme 62. Synthesis of 1, 4-dihydropyridine derivatives: (i) [BMIM][BF₄].

2.2.2. 4CRs yielding heterocycles with ring oxygen and nitrogen atoms

Kanakarajuwe *et al.* have exploited the four-component MCR for the synthesis of novel antibacterial chromeno[2,3-*d*]pyrimidin-8-amines in [BMIM][BF₄] (Scheme 62). Simple stirring of a mixture of α-naphthol, malononitrile, aryl aldehydes and NH₄Cl in [BMIM] [BF₄] and with trace triethylamine (TEA) in DMF allowed direct isolation of the desired analogue, bypassing the more traditional route which involved isolation of corresponding iminochromenes.[174]

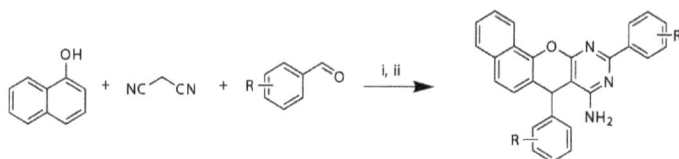

Scheme 63. General synthetic route of chromeno[2,3-d]pyrimidine-8-amine derivatives: (i) [BMIM][BF₄], TEA/DMF; (ii) NH₄Cl, 100ºC.

2.2.3. 4CRs yielding heterocycles with two ring nitrogens

The biologiocal roles of substituted imidazoles are well documented and numerous biologically active analoges have been reported.[175] Shaterian *et al.* have developed both a three and four component MCR route from benzil (or benzoin), substituted benzaldehydes and ammonium acetate (3-component), or with the addition of phenyl hydrazine (4-component) for the synthesis of 2,4,5-trisubstitutedimidazoles and 1,2,4,5-tetrasubstituted imidazoles respectively.[176] N-Methyl-2-pyrrolidonium hydrogen sulfate [NMP][HSO₄] at 100ºC was found to be superior to all previous reports which used a wide variety of catalys to conduct the same transformations. Recyling of the [NMP][HSO₄] saw a gradual diminution of the product yield from 98% to 82% over seven cycles. While this team examined the four component route (addition of phenylhydazine) giving 1,2,4,5-substituted imidazoles in high yields using Brønsted acidic ionic liquid, [NMP][HSO₄] (Scheme 63).

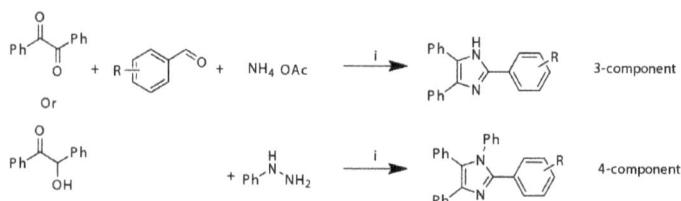

Scheme 64. Synthesis of 2,4,5- and 1,2,4,5- substituted imidazoles by a three or four component MCR: (i) [NMP][HSO₄].

Pyrano[2,3-c]pyrazoles represent an important scaffold in medicinal chemistry with multiple synthetic approaches developed. These approaches include synthesis in water, ethanol reflux, microwave assisted and solvent free aporaches. Each approach comes replete with its own set of advantages and disadvantages from excess solvent requirements, long reaction times and poor yields.[177]-[180] Khurana *et al.* synthesis of 4H-pyrano[2,3-c]pyrazoles avoids most of these disadvantages by providing for a high yielding (typically >85%), short duration cyclocondensation of hydrazine monohydrate or phenyl hydrazine, ethyl acetoacetate, aldehydes, and malononitrile in [BMIM][BF$_4$] with catalytic quantities of L-proline (10 mol%) (Scheme 64).[181]

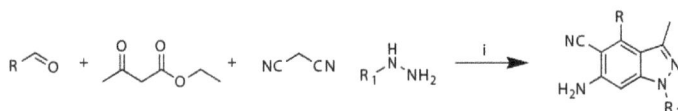

Scheme 65. Synthesis of pyrano[2,3-c]pyrazoles: (i) [BMIM][BF$_4$], L-proline (10 mol%), 50°C.

Xiao *et al.* have described a novel, efficient, and green procedure for the synthesis of 3-(5-amino-3-methyl-1*H*-pyrazol-4-yl)-3-arylpropanoic acid derivatives through the four-component reaction in [BMIM][BF$_4$] (Scheme 65).[182] Reactions were rapid (5 min) and the product isolated by pouring onto water and recrystallisation from EtOH / H$_2$O to afford pure product.

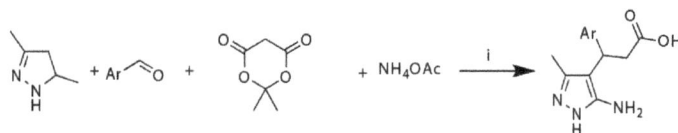

Scheme 66. Synthesis of 3-(5-amino-3-methyl-1*H*-pyrazol-4-yl)-3-arylpropanoic acid derivatives: (i) [BMIM][BF$_4$].

2.2.4. 4CRs yielding heterocycles with >three ring nitrogens

Ghahremanzadeh *et al.* reported the green synthesis of 1*H*-indolo[2,3-b]pyrazolo[4,3-e]pyridines from indolin-2-one, 3-oxo-3-phenylpropanenitrile, phenylhydrazine and benzaldehyde under a variety of conditions (Scheme 63).[183] The best reaction outcome was observed with the use of p-TSA in [BMIM][Br] at 140°C (Scheme 66).

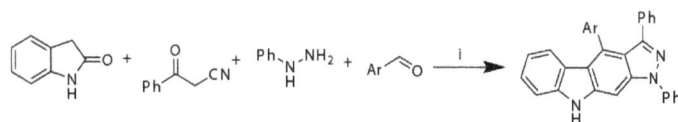

Scheme 67. Synthesis of 1*H*-indolo[2,3-b]pyrazolo[4,3-e]pyridines: (i) [BMIM][Br], p-TSA, 140°C.

Building on their earlier report on the synthesis of spiro[indolinepyrazolo[4',3':5,6]pyrido[2,3-d]pyrimidine]triones from barbituric acid, phenylhydrazine, 3-oxo-3-phenylpropanenitrile and isatin, Ghahremanzadeh *et al.* noted that the use of mixture of alum

(KAl$(SO_4)_2.12H_2O$) and [BMIM][PF$_6$] was a green approach to the same class of compounds (Scheme 67).[184],[185]

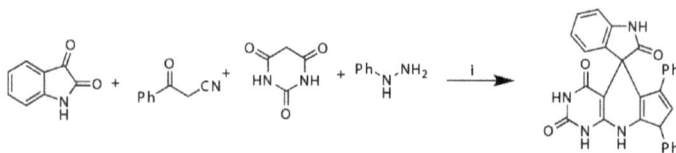

Scheme 68. Synthesis of spiro[indolinepyrazolo[4',3':5,6]pyrido[2,3-d]pyrimidine]triones: (i) [BMIM][PF$_6$], alum.

3. Conclusions

In this brief review we have demonstrated the considerable utility of room temperature ionic liquids in multicomponent reactions. Almost universally, the addition of an ionic liquid increases the speed of reaction and reaction yields. In many cases the ionic liquid was used as both the solvent and the reaction promotor. It was possible to add catalytic quantities of ionic liquids in conventional solvent and still achieve a much greener reaction outcome.

While the linear variant of the four-component MCR in ionic liquids is currently poorly described, there is little doubt that room temperature ionic liquids will aid in the synthesis of such species. Overall the IL-MCR approach is an extremely useful one, especially for the rapid entry to highly functionalised heterocyclic molecules of potentials use in medicianl chemistry.

Author details

Ahmed Al Otaibi and Adam McCluskey

Chemistry, School of Environmental & Life Sciences, The University of Newcastle, University Drive, Callaghan NSW, Australia

References

[1] Hill, T.A.; Odell, L.R.; Quan, A.; Ferguson, G.; Robinson, P.J.; McCluskey, A. Long Chain Amines and Long Chain Ammonium Salts as Novel Inhibitors of Dynamin GTPase Activity. Bioorganic & Medicinal Chemistry Letters 2004;14 3275-3278.

[2] Hill, T.; Odell, L. R.; Edwards, J. K.; Graham, M. E.; McGeachie, A. B.; Rusak, J.; Quan, A.; Abagyan, R.; Scott, J. L.; Robinson, P. J.; McCluskey, A. Small Molecule In-

hibitors of Dynamin I GTPase Activity: Development of Dimeric Tyrphostins. Journal of Medicinal Chemistry 2005;48 7781-7788.

[3] Hill, T. A.; Stewart, S. G.; Gordon, C. P.; Ackland, S. P.; Gilbert, J.; Sauer, B.; Sakoff, J. A.; McCluskey, A. Norcantharidin Analogues: Synthesis, Anticancer Activity and Protein Phosphatase 1 and 2A Inhibition. Chemistry Medicinal Chemistry 2008;3 1878-1892.

[4] Tarleton, M.; Robertson, M. J.; Gilbert, J.; Sakoff, J. A.; McCluskey, A. Library Synthesis and Cytotoxicity of A Family of 2-Phenylacrylonitriles and Discovery of An Estrogen Dependent Breast Cancer Lead Compound. Medicinal Chemistry Communications 2011;2 31-37.

[5] Odell, L. R.; Howan, D.; Gordon, C. P.; Robertson, M. J.; Chau, N.; Mariana, A.; Whiting, A. E.; Abagyan, R.; Daniel, J. A.; Gorgani, N. N.; Robinson, P. J.; McCluskey, A. The Pthaladyns: GTP Competitive Inhibitors of Dynamin I and II GTPase Derived from Virtual Screening. Journal of Medicinal Chemistry 2010;53 5267-5280.

[6] Hill, T. A.; Stewart, S. G.; Sauer, B.; Gilbert, J.; Ackland, S. P.; Sakoff, J. A.; McCluskey, A. Heterocyclic Substituted Cantharidin and Norcantharidin Analogues—Synthesis, Protein Phosphatase (1 and 2A) Inhibition, and Anti-cancer Activity. Bioorganic & Medicinal Chemistry Letters 2007;17 3392-3397.

[7] Hill, T. A.; Stewart, S. G.; Ackland, S. P.; Gilbert, J.; Sauer, B.; Sakoff, J. A.; McCluskey, A. Norcantharimides, Synthesis and Anticancer Activity: Synthesis of New Norcantharidin Analogues and their Anticancer Evaluation. Bioorganic & Medicinal Chemistry. 2007;15 6126-6134.

[8] Quan, A.; McGeachie, A. B.; Keating, D. J.; van Dam, E. M.; Rusak, J.; Chau, N.; Malladi, C. S.; Chen, C.; McCluskey, A.; Cousin, M. A.; Robinson, P. J. Myristyl Trimethyl Ammonium Bromide and Octadecyl Trimethyl Ammonium Bromide Are Surface-Active Small Molecule Dynamin Inhibitors that Block Endocytosis Mediated by Dynamin I or Dynamin II. Molecular Pharmacology 2007;72 1425-1439.

[9] Stewart, S. G.; Hill, T. A.; Gilbert, J.; Ackland, S. P.; Sakoff, J. A.; McCluskey, A. Synthesis and Biological Evaluation of Norcantharidin Analogues: Towards PP1 Selectivity. Bioorganic & Medicinal Chemistry 2007;15 7301-7310.

[10] Hill, T. A.; Gordon, C. P.; McGeachie, A. B.; Venn-Brown, B.; Odell, L. R.; Chau, N.; Quan, A.; Mariana, A.; Sakoff, J. A.; Chircop, M.; Robinson, P. J.; McCluskey, A. Inhibition of Dynamin Mediated Endocytosis by the Dynoles Synthesis and Functional Activity of a Family of Indoles. Journal of Medicinal Chemistry 2009;52 3762-3773.

[11] Strecker, A. Ueber die Künstliche Bildung der Mitchsaüre und einen Neuen, dem Glycocoll Homologen Körper. Liebigs Annalen der Chemie 1850;75 27.

[12] Hallett, J. P.; Welton, T. Room-Temperature Ionic Liquids: Solvents for Synthesis and Catalysis. 2. Chemical Reviews 2011;111 3508-3576.

[13] Dupont, J.; de Souza, R.; Suarez, P. A. Z. Ionic Liquid (Molten Salt) Phase Organome-
      tallic Catalysis. Chemical Reviews 2002;102 3667-3692.

[14] Fisher, T.; Sethi, A.; Welton, T.; Woolf, J. Diels-Alder Reactions in Room-temperature
      Ionic Liquids. Tetrahedron Letters 1999;40 793-796.

[15] Earle, M. J.; McCormac, P. B.; Seddon, K. R. Regioselective alkylation in ionic liquids.
      Chemical Communications 1998; 2245-2246.

[16] Boon, J. A.; Levisky, J. A.; Pflug, J. L.; Wilkes, J. S. Friedel-Crafts reactions in ambient-
      temperature molten salts. Journal of Organic Chemistry 1986;51 480-483.

[17] Ellis, B.; Keim, W.; Wasserscheid, P. Linear dimerisation of but-1-ene in biphasic
      mode using buffered chloroaluminate ionic liquid solvents. Chemical Communica-
      tions 1999; 337-340.

[18] Earle, M. J.; McCormac, P. B.; Seddon, K. R. Diels–Alder Reactions in Ionic Liquids. A
      Safe Recyclable Alternative to Lithium Perchlorate–diethyl Ether Mixtures. Green
      Chemistry 1999;1 23-25.

[19] Corey, E. J.; Zhang, F. Y. Highly Enantioselective Michael Reactions Catalyzed by a
      Chiral Quaternary Ammonium Salt. Illustration by Asymmetric Syntheses of (S)-Or-
      nithine and Chiral 2-Cyclohexenones. Organic Letters 2000;2 1097-1100.

[20] Schoefer, S. H.; Kaftzik, N.; Kragl, U.; Wasserscheid, P. Enzyme Catalysis in Ionic
      Liquids: Lipase Catalysed Kinetic Resolution of 1-Phenylethanol with Improved
      Enantioselectivity. Chemical Communications 2001; 425-426.

[21] Bertozzi, F.; Gustafsson, M.; Olsson, R. A Novel Metal Iodide Promoted Three-Com-
      ponent Synthesis of Substituted Pyrrolidines. Organic Letters 2002;4 3147-3150.

[22] Zolfigol, M. A.; Khazaei, A.; Moosavi-Zare, A. R.; Zare, A.; Khakyzadeh, V. Rapid
      Synthesis of 1-amidoalkyl-2-naphthols over Sulfonic Acid Functionalized Imidazoli-
      um Salts. Applied Catalysis A: General 2011;400 70-81.

[23] Hajipour, A. R.; Rajaei, A.; Ruoho, A. E. A Mild and Efficient Method for Preparation
      of Azides from Alcohols Using Acidic Ionic Liquid [H-NMP]HSO₄. Tetrahedron Let-
      ters 2009;50 708-711.

[24] Heravi, M. M.; Tavakoli-Hoseini, N.; Bamoharram, F. F. Brønsted Acidic Ionic Liq-
      uids as Efficient Catalysts for the Synthesis of Amidoalkyl Naphthols. Synthetic
      Communications 2011;41 298-306.

[25] Fang, D.; Shi, Q. R.; Cheng, J.; Gong, K.; Liu, Z. L. Regioselective mononitration of
      aromatic compounds using Brønsted acidic ionic liquids as recoverable catalysts. Ap-
      plied Catalysis A: General 2008;345 158-163.

[26] Gui, J.; Cong, X.; Liu, D.; Zhang, X.; Hu, Z.; Sun, Z. Novel Brønsted Acidic Ionic Liq-
      uid as Efficient and Reusable Catalyst System for Esterification. Catalysis Communi-
      cations 2004;5 473-477.

[27] Hajipour, A. R.; Ghayeb, Y.; Sheikhan, N.; Ruoho, A. E. Brønsted Acidic Ionic Liquid as An Efficient and Reusable Catalyst for one-pot Synthesis of 1-amidoalkyl 2-Naphthols Under Solvent-free Conditions. Tetrahedron Letters 2009;50 5649-5651.

[28] Zhang, Q.; Luo, J.; Wei, Y. A Silica Gel Supported Dual Acidic Ionic Liquid: An Efficient and Recyclable Heterogeneous Catalyst for the One-pot Synthesis of Amidoalkyl Naphthols. Green Chemistry 2010;12 2246-2254.

[29] Kotadia, D. A.; Soni, S. S. Silica Gel Supported–SO₃H Functionalised Benzimidazolium Based Ionic Liquid as A Mild and Effective Catalyst for Rapid Synthesis of 1-Amidoalkyl Naphthols. Journal of Molecular Catalysis A: Chemical 2012;353-354 44-49.

[30] Luo, J.; Zhang, Q. A One-Pot Multicomponent Reaction for Synthesis of 1-Amidoalkyl-2-Naphthols Catalyzed by PEG-based Dicationic Acidic Ionic Liquids Under Solvent-free conditions. Monatshefte für Chemie 2011;142 923-930.

[31] Yadav, L. D. S.; Rai, A. The first Ionic Liquid-promoted Three-component Coupling Strategy for An Expeditious Synthesis of β-nitrocarbonitriles/thiocyanates. Tetrahedron Letters 2009;50 640-643.

[32] Yan, S.; Gao, Y.; Xing, R.; Shen, Y.; Liu, Y.; Wu, P.; Wu, H. An Efficient Synthesis of (E)-nitroalkenes Catalyzed by Recoverable Diamino-Functionalized Mesostructured Polymers. Tetrahedron 2008;64 6294-6299.

[33] Zhao, X. L.; Liu, L.; Chen, Y. J.; Wang, D. Three-Component Barbier Allylation-Friedel-Crafts Alkylation Promoted by BuPyCl/SnCl₂•2H₂O: Application to The Synthesis of 4-(Substituted Phenyl)Chromans. Chinese Journal of Chemistry 2007;25 1312-1322.

[34] Davoodnia, A.; Heravi, M. M.; Rezaei-Daghigh, L.; Tavakoli-Hoseini, N. A Modified and Green Procedure for the Synthesis of β-Amido Ketones Using a Brønsted-Acidic Ionic Liquid as Novel and Reusable Catalyst. Chinese Journal of Chemistry 2010;28 429-433.

[35] Fang, D.; Gong, K.; Zhang, D. Z.; Liu, Z. L. One-pot, Three-component Mannich-type Reaction Catalyzed by Functionalized Ionic Liquid. Monatshefte für Chemie 2009;140 1325-1329.

[36] Gong, K.; Fang, D.; Wang, H.L.; and; Liu, Z.L. Basic Functionalized Ionic Liquid Catalyzed One-pot Mannich-type Reaction: Three Component Synthesis of ˙-Amino Carbonyl Compounds. Monatshefte fur Chemie 2007;138 1195-1198.

[37] Liu, B.; Xu, D.; Dong, J.; Yang, H.; Zhao, D.; Luo, S.; Xu, Z. Highly Efficient AILs/L-Proline Synergistic Catalyzed Three-Component Asymmetric Mannich Reaction. Synthetic Communications 2007;37 3003-3010.

[38] Akbari, J.; Heydari, A. A Sulfonic Acid Functionalized Iionic Liquid as A homogeneous and recyclable catalyst for the one-pot synthesis of α-Aminophosphonates. Tetrahedron Letters 2009;50 4236-4238.

[39] Reddy, M. V.; Dindulkar, S. D.; Jeong, Y. T. BF$_3$SiO$_2$-Catalyzed One-pot Synthesis of α-Aminophosphonates in Ionic Liquid and Neat Conditions. Tetrahedron Letters 2011;52 4764-4767.

[40] Gregory, R. J. H. Cyanohydrins In Nature and the Laboratory: Biology, Preparations, and Synthetic Applications. Chemical Reviews 1999;99 3649-3682.

[41] Brunel, J. M.; Holmes, I. P. Chemically Catalyzed Asymmetric Cyanohydrin Syntheses. Angewandte Chemie International Edition 2004;43 2752-2778.

[42] North, M. Synthesis and Applications of Non-racemic Cyanohydrins. Tetrahedron Asymmetry 2003;14 147-176.

[43] Chen, F. X.; Feng, X. M. Synthesis of Racemic Tertiary Cyanohydrins. Synthesis Letters 2005; 892-899.

[44] Shen, Z. L.; Ji, S. J. Ionic Liquid [bmim]BF4 as An Efficient and Recyclable Reaction Medium for the Synthesis of O-Acetyl Cyanohydrin via One-Pot Condensation of Aldehyde, TMSCN, and Ac2O. Synthetic Communications 2009;39 808-818.

[45] Davoodnia, A.; Bakavoli, M.; Moloudi, R.; Tavakoli-Hoseini, N.; Khashi, M. Highly Efficient, One-pot, Solvent-free Synthesis of 2,4,6-Triarylpyridines using A Brønsted-Acidic Ionic Liquid
as Reusable Catalyst. Monatshefte für Chemie 2010;141 867-870.

[46] Heravi, M. R. P.; Fakhr, F. Ultrasound-Promoted Synthesis of 2-Amino-6-(Arylthio)-4-Arylpyridine-3,5-Dicarbonitriles using ZrOCl2 8H2O/NaNH2 as the Catalyst in the Ionic Liquid [bmim]BF4 at Room Temperature. Tetrahedron Letters 2011;5 6779-6782.

[47] Ranu, B. C.; Jana, R.; Sowmiah, S. An Improved Procedure for the Three-Component Synthesis of Highly Substituted Pyridines Using Ionic Liquid. Journal of Organic Chemistry 2007;72 3152-3154.

[48] Wu, H.; Wan, Y.; Lu, L. L.; Shen, Y.; Ye, L.; Zhang, F. R. Catalyst-Free One-Pot Synthesis
of 2,4,6-Triaryl-1,4-dihydropyridines in Ionic Liquid and Their Catalyzed Activity on Two Simple Diels–Alder Reactions. Synthetic Communications 2008;38 666-673.

[49] Wu, X. Y. Facile and Green Synthesis of 1,4-Dihydropyridine Derivatives in n-Butyl Pyridinium Tetrafluoroborate. Synthetic Communications. 2012;[42] 454-459.

[50] Llinas-Brunet, M.; Bailey, M. D.; Ghiro, E.; Gorys, V.; Halmos, T.; Poirier, M.; Rancourt, J.; Goureare, N. A Systematic Approach to the Optimization of Substrate-Based Inhibitors of the Hepatitis C Virus NS3 Protease: Discovery of Potent and Specific Tripeptide Inhibitors. Journal of Medicinal Chemistry 2004;47 6584-6594.

[51] Frutos, R. P.; Haddad, N.; Houpis, I. N.; Johnson, M.; Smith-Keenan, L. L.; Fuchs, V.; Yee, N. K.; Farina, V.; Faucher, A. M.; Brochu, C.; Hache, B.; Duceepe, J. -S.; Beaulieu, P. Synthesis 2006; 256-XX.

[52] Huang, L. J.; Hsich, M. C.; Teng, C. M.; Lee, K. H.; Kno, S. C. Synthesis and Antiplate-let Activity of Phenyl Quinolones. Bioorganic & Medicinal Chemistry 1998;6 1657-1662.

[53] Gasparotto, V.; Castagliuolo, I.; Chiarelotto, G.; Pezzi, V.; Montanaro, D.; Brun, P.; Palu, G.; Viola, G.; Ferlin, M. G. Synthesis and Biological Activity of 7-Phenyl-6,9-di-hydro-3H-pyrrolo[3,2-f]quinolin-9-ones: A New Class of Antimitotic Agents Devoid of Aromatase Activity. Journal of Medicinal Chemistry 2006;49 1910-1915.

[54] Ruchelman, A. L.; Singh, S. K.; Ray, A.; Wu, X. H.; Yang, J.M.; Li, T.K.; Liu, A.; Liu, L. F.; LaVoie, E. J. 5H-Dibenzo[c,h]1,6-naphthyridin-6-ones: Novel Topoisomerase I-Tar-geting Anticancer Agents With Potent Cytotoxic Activity. Bioorganic & Medicinal Chemistry 2003;11 2061-2073.

[55] Theeraladanon, C.; Arisawa, M.; Nishidi, A.; Nakagawa, M. A Novel Synthesis of Substituted Quinolines Using Ring-closing Metathesis (RCM): its Application to the Synthesis of Key Intermediates for Anti-malarial Agents. Tetrahedron 2004;60 3017-3035.

[56] Yadav, A. K.; Sharma, G. R.; Dhakad, P.; Yadav, T. A Novel Ionic Liquid Mediated Synthesis of 4(1H)-Quinolones, 5H-thiazolo[3,2-a]pyrimidin-5-one and 4H-pyrimi-do[2,1-b]benzothiazol-4-ones. Tetrahedron Letters 2012;53 859-862.

[57] Wang, X. S.; Wu, J. R.; Li, Q.; Zhang, M. M. A Novel and Green Method for the Syn-thesis of Highly Substituted Isoquinoline Derivatives in Ionic Liquid. Journal of Het-erocyclic Chemistry 2009;46 1355-1363.

[58] Martin, N.; Quinteiro, M.; Seoane, C.; Soto, J. L.; Mora, A.; Suárez, M.; Ochoa, E.; Mo-rales, A.; Del Bosque, J. R. Synthesis and Conformational Study of Acridine Deriva-tives Related to 1,4-dihydropyridines. Journal of Heterocyclic Chemistry 1995;51 235-238.

[59] Li, Y.; Xu, X.; Shi, D.; Ji, S. One-pot Synthesis of 14-Aryl-1,6,7,14-tetrahydrodibenzo-[a,i]acridine-1,6-dione in Ionic Liquids. Chinese Journal of Chemistry 2009;27 1510-1514.

[60] Shi, D.; Ni, S.; Yang, F.; Ji, S. An Efficient and Green Synthesis of 3,3'-Benzylidene-bis(4-hydroxy-6-methylpyridin-2(1H)-one) Derivatives through Multi-Component Reaction in Ionic Liquid. Journal of Heterocyclic Chemistry 2008;45 1275-1280.

[61] Agarwal, A.; Srinivas, K.; Puri, S. K.; Chauhan, P. M. S. Synthesis of 2,4,6-trisubstitut-ed pyrimidines as antimalarial agents. Bioorganic & Medicinal Chemistry. 2005;13 4645-4650.

[62] Gholap, A. R.; Venkatesan, K.; Daniel, T.; Lahoti, R. J.; Srinivasan, K. V. Ionic liquid Promoted Novel and Efficient Onepot Synthesis of 3,4-dihydropyrimidin-2-(1H)-ones at Ambient Temperature Under Ultrasound Irradiation. Green Chemistry 2004;6 147-150.

[63] Gui, J.; Liu, D.; Wang, C.; Lu, F.; Lian, J.; Jiang, H.; Sun, Z. TITLES Synthetic Commu-
      nications. 2009;[39] 3436-3443.

[64] Peng, J.; Deng, Y. Ionic liquid Catalyzed Biginelli Reaction Under Solvent-free Condi-
      tions. Tetrahedron Letters 2001;42 5917-19.

[65] Zolfigol, M. A.; Khazaei, A.; Moosavi-Zare, A. R.; Zare, A. 3-Methyl-1-Sulfonic Acid
      Imidazolium Chloride as A New, Efficient and Recyclable Catalyst and Solvent for
      the Preparation of N-sulfonyl Imines at Room Temperature. Journal of the Iranian
      Chemical Society 2010;7 646-651.

[66] Cole, A. C.; Jensen, J. L.; Ntai, I.; Tran, K. L. T.; Weaver, K. J.; Forbes, D. C.; Davis, J.;
      H. J. Novel Brønsted Acidic Ionic Liquids and Their Use as Dual Solvent–Catalysts.
      Journal of the American Chemical Society 2002;124 5962-5963.

[67] Hajipour, A. R.; Seddighi, M. Pyridinium-Based Brønsted Acidic Ionic Liquid as a
      Highly Efficient Catalyst for One-Pot Synthesis of Dihydropyrimidinones. Synthetic
      Communications 2012; 42 227-235.

[68] Fang, D.; Zhang, D. Z.; Liu, Z. L. One-Pot Three-Component Biginelli-Type Reaction
      Catalyzed by Ionic Liquids in Aqueous Media. Monatshefte fur Chemie 2010;141
      419-423.

[69] Mirzai, M.; Valizadeh, H. Microwave-Promoted Synthesis of 3,4-Dihydropyrimi-
      din-2(1H)-(thio)ones Using IL-ONO as Recyclable Base Catalyst Under Solvent-Free
      Conditions. Synthetic Communications 2012;42 1268-1277.

[70] Valizadeh, H.; Heravi, M. M.; Amiri, M. Unexpected Synthesis of N-Methylben-
      zo[d]isoxazolium Hydroxides Under Microwave Irradiation Conditions Molecular
      Diversity JOURNAL 2010;14 575-579.

[71] Shaabani, A.; Rahmati, A. 1,1,3,3-N,N,N,N-Tetramethylguanidinium Trifluoroacetate
      Ionic Liquid–Promoted Efficient One-Pot Synthesis of Trisubstituted Imidazoles.
      Synthetic Communications 2006;36 65-70.

[72] Khurana, J. M.; Kumar, S. Ionic liquid: An Efficient and Recyclable Medium for the
      Synthesis of Octahydroquinazolinone and Biscoumarin Derivatives. Monatshefte fur
      Chemie 2010;141 561-564.

[73] Pawar, O. B.; Chavan, F. R.; Sakate, S. S.; Shinde, N. D. Ultrasound Promoted and
      Ionic Liquid Catalyzed Cyclocon densation Reaction for the Synthesis of 4(3H)-Qui-
      nazolinones. Chinese Journal of Chemistry 2010;28 69-71.

[74] Dabiri, M.; Salehi, P.; Bahramnejad, M. Ecofriendly and Efficient One-Pot Procedure
      for The Synthesis of Quinazoline Derivatives Catalyzed by an Acidic Ionic Liquid
      Under Aerobic Oxidation Conditions. Synthetic Communications 2010;40 3214-3225.

[75] Shaabani, A.; Maleki, A. Ionic Liquid Promoted One-Pot Three-Component Reaction:
      Synthesis of Annulated Imidazo[1,2-a]azines Using Trimethylsilylcyanide. Monat-
      shefte fur Chemie 2007;138 51-56.

[76] Shaabani, A.; Soleimani, E.; Maleki, A. Ionic liquid Promoted One-Pot Synthesis of 3-Aminoimidazo[1,2-*a*]Pyridines. Tetrahedron Letters 2006;47 3031-3034.

[77] Hajipour, A. R.; Ghayeb, Y.; Sheikhan, N.; Ruoho, A. E. Brønsted Acidic Ionic Liquid as an Efficient and Reusable Catalyst for One-Pot, Three-Component Synthesis of Pyrimidinone Derivatives via Biginelli-Type Reaction Under Solvent-Free Conditions. Synthetic Communications 2011;41 2226-2233.

[78] Raghuvanshi, D. S.; Singh, K. N. A Highly Efficient Green Synthesis of 1*H*-pyrazolo[1,2-*b*]Phthalazine-5,10-dione Derivatives and their Photophysical Studies. Tetrahedron Letters 2011;52 5702-5705.

[79] Khurana, J. M.; Magoo, D. Efficient One-Pot Syntheses of 2H-Indazolo[2,1-*b*]Phthalazine-triones by Catalytic H2SO4 in Water–ethanol or Ionic Liquid. Tetrahedron Letters 2009;50 7300-7303.

[80] Wang, H. X.; JI, S. J.; GU, D. G. Synthesis of N-(*α*-Alkoxyalkyl)benzotriazoles Catalyzed by Acidic Ionic Liquid at Room Temperature. Chinese Journal of Chemistry 2007;25 1041-1043.

[81] Ehlert, F. J.; Ragan, P.; Chen, A.; Roeske, W. R.; Yamamura, H. I. Modulation of Benzodiazepine Receptor binding: Insight into Pharmacological Efficacy. European Journal of Pharmacology 1982;78 249-253.

[82] Sanghvi, Y. S.; Larson, S. B.; Willis, R. C.; Robins, R. K.; Revankar, G. R. Synthesis and Biological Evaluation of Certain C-4 Substituted Pyrazolo[3,4-*b*]Pyridine Nucleosides. Journal of Medicinal Chemistry 1989;32 945-951.

[83] Zhang, X.; Li, D.; Fan, X.; Wang, X.; Li, X.; Qu, G.; Wang, J. Ionic Liquid-Promoted mMlti-component Reaction: Novel
and Efficient Preparation of Pyrazolo[3,4-*b*]Pyridinone, Pyrazolo[3,4-*b*]Quinolinone and their Hybrids with Pyrimidine Nucleoside. Molecular Diversity 2010;14 159-167.

[84] Zhang, X. Y.; Li, X. Y.; Fan, X. S.; Wang, X.; Wang, J. J.; Qu, G. R. A Novel Synthesis of Pyrazolo[3,4-*b*]Pyridine Derivatives through Multi-component Reaction in Ionic Liquids. Chinese Chemical Letters 2008;19 153-156.

[85] Yao, C.; Lei, S.; Wang, C.; Li, T.; Yu, C.; Wang, X.; Tua, S. Three-Component Synthesis of 4-Aryl-1*H*-Pyrimido[1,2-*a*] Benzimidazole Derivatives in Ionic Liquid. Journal of Heterocyclic Chemistry 2010;26 47-32.

[86] Zhang, X.; Fan, X.; Niu, H.; Wang, J. An Ionic Liquid as A Recyclable Medium for the Green Preparation of • • • -bis (substituted benzylidene)Cycloalkanones Catalyzed by FeCl3 6H2O. Green Chemistry 2003;5 267-269.

[87] Fan, X.; Hu, X.; Zhang, X.; Wang, J. Ionic Liquid Promoted Knoevenagel and Michael Reactions. Australian Journal of Chemistry 2004;57 1067-1071.

[88] Shi, D. Q.; Ni, S. N.; Yang, F.; Shi, J. W.; Dou, G. L.; Li, X. Y.; Wang, X. S.; Ji, S. J. An Efficient Synthesis of Pyrimido[4,5-*b*]Quinoline and Indeno[2',1':5,6]pyrido[2,3-*d*]Pyr-

imidine Derivatives via Multicomponent Reactions in Ionic Liquid. Journal of Heterocyclic Chemistry 2008;45 963-702.

[89] Kimachi, T.; Yoneda, F.; Sasaki, T. New Synthesis of 5-Amino-5-Deazaflavin Derivatives by Direct Coupling of 5-Deazaflavins and Amines. Journal of Heterocyclic Chemistry 1992;29 763-765.

[90] Bond, A.; Reichert, Z.; Stivers, J. T. Novel and Specific Inhibitors of a Poxvirus Type I Topoisomerase. Molecular Pharmacology 2006;69 547-557.

[91] Guo, H. Y.; Yu, Y. One-pot Synthesis of 7-Aryl-11,12-Dihydrobenzo[h]Pyrimido-[4,5-b]Quinoline-8,10(7H,9H)-diones via Three-Component Reaction in Ionic Liquid. Chinese Chemical Letters 2010;21 1435-1438.

[92] Zhang, N.; Ayral-Kaloustian, S.; Nguyen, T.; Afragola, J.; Hernandez, R.; Lucas, J. Synthesis and SAR of [1,2,4]Triazolo[1,5-a]pyrimidines, a Class of Anticancer Agents with a Unique Mechanism of Tubulin Inhibition. Journal of Medicinal Chemistry 2007; 50 319-327.

[93] Li, T.; Yao, C.; Lei, S.; Yu, C.; Tu, S. A Facile One-Pot Three-Component Synthesis of 5-(Trifluoromethyl)-4,7-dihydro-[1,2,4]-Triazolo[1,5-a]Pyrimidine Derivatives in Ionic Liquid. Chinese Journal of Chemistry 2011;29 2427-2432.

[94] Kuthan, J. New Developments in the Chemistry of Pyrans. Advances in Heterocyclic Chemistry 1995;62, 19-135.

[95] Hatakeyama, S.; Ochi, N.; Numata, H.; Takano, S. A New Route to Substituted 3-methoxycarbonyldihydropyrans: Enantioselective Synthesis of (–)-Methyl Elenolate. Chemical Communications 1988; 1202-1204.

[96] Cingolant, G. M.; Pigini, M. Research in the Field of Antiviral Compounds. Mannich Bases of 3-Hydroxycoumarin. Journal of Medicinal Chemistry 1969;12 531-532.

[97] Peng, Y.; Song, G.; Huang, F. Tetramethylguanidine-[bmim][BF4].
An Efficient and Recyclable Catalytic System for One-Pot Synthesis of 4H-Pyrans. Monatshefte fur Chemie 2005;[136] 727-731.

[98] Jiang, Z. Q.; JI, S. J.; Lu, J.; Yang, J. M. A Mild and Efficient Synthesis of 5-Oxo-5,6,7,8-tetrahydro-4H-benzo[b]pyran Derivatives in Room Temperature Ionic Liquids. Chinese Journal of Chemistry 2005;23 1085-1089.

[99] Fang, D.; Zhang, H. B.; Liu, Z. L. Synthesis of 4H-Benzopyrans Catalyzed by Acyclic Acidic Ionic Liquids in Aqueous Media. Journal of Heterocyclic Chemistry 2010;47 63-67.

[100] Hafez, E. A. A.; Elnagdi, M. H. Nitriles in Heterocyclic Synthesis: Novel Synthesis of Benzo[c]coumarin and of Benzo[c]pyrano[3,2-c]quinoline Derivatives. Heterocycles 1987;26 903-907.

[101] Kidwai, M.; Saxena, S. Aqua mediated synthesis of substituted 2-amino-4H-chromenes and in vitro study as antibacterial agents. Bioorganic & Medicinal Chemistry Letters 2005;15 4295-4298.

[102] Shestopalov, A. M.; Niazimbetova, Z. I.; Evans, D. H. Synthesis of 2-Amino-4-aryl-3-cyano-6-methyl-5-ethoxycarbonyl-4H-pyrans. Heterocycles 1999;51 1101-1107.

[103] Bloxham, J.; Dell, C. P.; Smith, C. W. Preparation of Some New Benzylidenemalononitriles by an SNAr Reaction: Application to Naphtho[1,2-b]pyran Synthesis. Heterocycles 1994;38 399-408.

[104] Zhuang, Q. Y.; Rong, L. C.; Shi, D. Q. Synthesis and Crystal Structure of Substituted Naphthopyran. Chinese Journal of Organic Chemistry 2003;23 671-673.

[105] Gong, K.; Wang, H. L.; Fang, D.; Liu, Z. L. Basic Ionic Liquid as Catalyst for the Rapid and Green Synthesis of Substituted 2-amino-2-chromenes in Aqueous Media. Catalysis Communications 2008;9 650-653.

[106] Khurana, J. M.; Magoo, D. pTSA-Catalyzed One-Oot synthesis of 12-aryl-8,9,10,12-Tetrahydrobenzo[a]xanthen-11-Ones in Ionic Liquid and Neat Conditions. Tetrahedron Letters 2009;50 4777-4780.

[107] Zakeri, M.; Heravi, M. M.; Saeedi, M.; Karimi, N.; Oskooie, H. A.; Tavakoli-Hoseini, N. One-pot Green Procedure for Synthesis of Tetrahydrobenzo[a]- xanthene-11-One Catalyzed by Brønsted Ionic Liquids under Solvent-free Conditions. Chinese Journal of Chemistry 2011;29 1441-1445.

[108] Zheng, J.; Li, Y. Basic ionic liquid-catalyzed multicomponent synthesis of tetrahydrobenzo[b]pyrans and pyrano[c]chromenes. Mendeleev Communications 2011;21 280-281.

[109] Burgard, A.; Lang, H.J.; Gerlach, U. Asymmetric synthesis of 4-amino-3,4-dihydro-2,2-dimethyl-2H-1-benzopyrans. Tetrahedron 1999;55 7555-7562.

[110] Evans, J. M.; Fake, C. S.; Hamilton, T. C.; Poyser, R. H.; Showell, G. A. Synthesis and Antihypertensive Activity of 6,7-disubstituted trans-4-amino-3,4-dihydro-2,2-dimethyl-2H-1-benzopyran-3-ols. Journal of Medicinal Chemistry 1984;27 1127-1131.

[111] Evans, J. M.; Fake, C. S.; Hamilton, T. C.; Poyser, R. H.; Watts, E. A. Synthesis and antihypertensive activity of substituted trans-4-amino-3,4-dihydro-2,2-dimethyl-2H-1-benzopyran-3-ols. Journal of Medicinal Chemistry 1983;26 1582-1589.

[112] D. Arnesto, D.; Horspool, W. M.; Martin, N.; Ramos, A.; Seaone, C. Synthesis of Cyclobutenes by the Novel Photochemical Ring Contraction of 4-substituted 2-amino-3,5-dicyano-6-phenyl-4H-pyrans. The Journal of Organic Chemistry 1989;54 3069-3072.

[113] Shaterian, H. R.; Honarmand, M. Task-Specific Ionic Liquid as the Recyclable Catalyst for the Rapid and Green Synthesis of Dihydropyrano[3,2-c]chromene Derivatives. Synthetic Communications 2011;41 3573-3581.

[114] Gong, K.; Wang, H. L.; Luo, J.; Liu, Z. L. One-Pot Synthesis of Polyfunctionalized Pyrans Catalyzed by 1145 Basic Ionic Liquid in Aqueous Media. Journal of Heterocyclic Chemistry 2009;46 1145-1150.

[115] Xie, J.W.; Li, P.; Wang, T.; Zhou, F.T. Efficient and Mild Synthesis of Functionalized 2,3-dihydrofuran Derivatives via Domino Reaction in Water. Tetrahedron Letters 2011;52 2379-2382.

[116] Fan, L.P.; Li, P.; Li, X.S.; Xu, D.C.; Ge, M.M.; Zhu, W.D.; Xie, J.W. Facile Domino Access to Chiral Mono-, Bi-, and Tricyclic 2,3-Dihydrofurans. The Journal of Organic Chemistry 2010;75 8716-8719.

[117] Rueping, M.; Parra, A.; Uria, U.; Bessselièvre, F.; Merino, E. Catalytic Asymmetric Domino Michael Addition–Alkylation Reaction: Enantioselective Synthesis of Dihydrofurans. Organic Letters 2010;12 5680-5683.

[118] Yılmaz, M.; Yakut, M.; Pekel, A. T. Synthesis of 2,3-Dihydro-4H-furo[3,2-c]chromen-4-ones and 2,3-Dihydronaphtho[2,3-b]furan-4,9-diones by the Radical Cyclizations of Hydroxyenones with Electron-Rich Alkenes using Manganese(III) Acetate. Synthetic Communications 2008;38 914-927.

[119] Rajesh, S. M.; Perumal, S.; Menéndez, J. C.; Pandian, S.; Murugesan, R. Facile Ionic Liquid-mediated, Three-Component Sequential Reactions for the Green, Regio- and Diastereo-selective Synthesis of Furocoumarins. Tetrahedron 2012;68 5631-5636.

[120] Manolov, I.; Maichle-Moessmer, C.; Danchev, N. D. Synthesis, structure, toxicological and pharmacological investigations of 4-hydroxycoumarin derivatives. European Journal of Medicinal Chemistry. 2006;[41] 882-890.

[121] Gong, K.; Wang, H. L.; Luo, J.; Liu, Z. L. One-Pot Synthesis of Polyfunctionalized Pyrans Catalyzed by 1145 Basic Ionic Liquid in Aqueous Media. Journal of Heterocyclic Chemistry. 2009;[46] 1145-1150.

[122] Chen, Z.; Zhu, Q.; Su, W. A Novel Sulfonic Acid Functionalized Ionic Liquid Catalyzed Multicomponent Synthesis of 10,11-dihydrochromeno[4,3-b]chromene-6,8(7H, 9H)-dione Derivatives in Water. Tetrahedron Letters 2011;52 2601-2604.

[123] Bloxham, J.; Dell, C. P.; Smith, C. W. Preparation of Some New Benzylidenemalononitriles by an SNAr Reaction: Application to Naphtho[1,2-b]pyran Synthesis. Heterocycles 1994;38 399-408.

[124] Wang, X. S.; Shi, D. Q.; Yu, H. Z. Synthesis of 2-Aminochromene Derivatives Catalyzed by KF/Al2O3. Synthetic Communications 2004;34 509-514.

[125] Jin, T. J.; Xiao, J. C.; Wang, S. J.; Li, T. S. An Efficient and Convenient Approach to the Synthesis of Benzopyrans by a Three-Component Coupling of One-Pot Reaction. Synthesis Letters 2003;13 2001-2004.

[126] Wang, X. S.; Shi, D. Q.; Tu, S. J. A Convenient Synthesis of 5-Oxo-5,6,7,8-tetrahy-dro-4H-benzo[b]pyran Derivatives Catalyzed by KF-Alumina. Synthetic Communications 2003;33 119-126.

[127] Jin, T. S.; Wang, A. Q.; Wang, X. A Clean One-pot Synthesis of Tetrahydroben-zo[b]pyran Derivatives Catalyzed by Hexadecyltrimethyl Ammonium Bromide in Aqueous Media. Synthesis Letters 2004;5 871-873.

[128] M. J. de Groot, M. J.; Alex, A. A.; Jones, B. C. Development of a Combined Protein and Pharmacophore Model for Cytochrome P450 2C9. Journal of Medicinal Chemistry 2002;45 1983-1993.

[129] M. Lee, M.; D. Hesek, D.; and S. Mobashery, S. A Practical Synthesis of Nitrocefin. The Journal of Organic Chemistry 2005;70 367-369.

[130] Flynn, B. L.; Flynn, G. P.; Hamel, E.; Jung, M. K. The Synthesis and Tubulin Binding Activity of Thiophene-Based Analogues of Combretastatin A-4. Bioorganic & Medicinal Chemistry Letters 2001;11 2341-2343.

[131] Zhang, X.; Li, X.; Fan, X.; Wang, X.; Li, D.; Qu, G.; Wang, J. Ionic liquid promoted preparation of 4H-thiopyran
and pyrimidine nucleoside-thiopyran hybrids through one-pot multi-component reaction of thioamide. Molecular Diversity 2009;13 5761-5765.

[132] Shvekhgeimer, M.G. A. Dihydrothiophenes. Synthesis and properties (review). Chemistry of Heterocyclic Compounds 1998;34 1101-1122.

[133] McIntosh, J. M.; Goodbrand, H. B.; Masse, G. M. Dihydrothiophenes. II. Preparation and properties of some alkylated 2,5-dihydrothiophenes. Journal of Organic Chemistry 1974;39 202-206.

[134] Leusen, A. M.; Berg, K. J. Formation and Reactions of 2,3-dimethylene-2,3-dihydro-thiophene. Tetrahedron Letters 1988;29 2689-2692.

[135] Kumar, A.; Gupta, G.; Srivastava, S. Functional Ionic Liquid Mediated Synthesis (FILMS) of dihydrothiophenes and Tacrine Derivatives. Green Chemistry 2011;13 2459-24963.

[136] Fringuelli, R.; Pietrella, D.; Schiaffella, F.; Guarrac, I. A.; Perito, S.; Bistoni, F.; Vec-chiarelli, A. Anti-Candida albicans Properties of Novel Benzoxazine Analogues. Bioorganic & Medicinal Chemistry 2002;10 1681-1686.

[137] Macchiarulo, A.; Costantino, G.; Fringuelli, F.; Vecchiarelli, A.; Schiaffella, F.; Frin-guelli, R. 1,4-Benzothiazine and 1,4-Benzoxazine Imidazole Derivatives with Anti-fungal Activity: A Docking Study. Bioorganic & Medicinal Chemistry 2002;10 3415-3423.

[138] Adams, N. D.; Darcy, M. G.; Dhanak, D.; Duffy, K. J.; Fitch, D. M.; Knight, S. D.; Newlander, K. A.; Shaw, A. N. WO Int. Patent 2006113432, 2006.

[139] Soleimani, E.; Khodaei, M. M.; Koshvandi, A. T. K. Three-Component, One-Pot Synthesis of Benzo[b][1,4]oxazines in Ionic Liquid 1-Butyl-3-methylimidazolium Bromide. Synthetic Communications 2012;42 1367-1371.

[140] Asri, Z. E.; Nisson, Y. G.; Guillen, F. D. R.; Basle, O.; Isambert, N.; Duque, M. D. M. S.; Ladeira, S.; Rodriguez, J.; Constantieux, T.; Plaquevent, J. C. Multicomponent Reactions in Ionic Liquids: Convenient and Ecocompatible Access to the 2,6-DABCO Core. Green Chemistry 2011;13 2549-2552.

[141] F. Liéby-Muller, F.; Constantieux, T.; Rodriguez, J. Multicomponent Domino Reaction from β-Ketoamides: Highly Efficient Access to Original Polyfunctionalized 2,6-Diazabicyclo[2.2.2]octane Cores. Journal of the American Chemical Society 2005;127 17176-17177.

[142] Otto, H.H. Synthesis of some 4H-pyrano[2,3-c]pyrazoles. Archiv der Pharmazie 1974; [307] 444-447.

[143] Lehmann, F.; Holm, M.; Laufer, S. Three-Component Combinatorial Synthesis of Novel Dihydropyrano[2,3-c]pyrazoles. Journal of Combinatorial Chemistry 2008;10 364-367.

[144] Balaskar, R. S.; Gavade, S. N.; Mane, M. S.; Shingate, B. B.; Shingare, M. S.; Mane, D. V. Greener Approach Towards the Facile Synthesis of 1,4-dihydropyrano[2,3-c]pyrazol-5-yl Cyanide Derivatives at Room Temperature. Chinese Chemical Letters 2010;21 1175-1179.

[145] New, J. S.; Christopher, W. L.; Yevich, J. P.; Butler, R.; Schlemmer, R. F., Jr; Vander-Maelen, C. P.; Cipollina, J. A. The Thieno[3,2-c]pyridine and Furo[3,2-c]pyridine Rings: New Pharmacophores with Potential Antipsychotic Activity. Journal of Medicinal Chemistryy 1989;32 1147-1156.

[146] Bukoski, R. D.; Bo, J.; Xue, H.; Bian, K. Antiproliferative and Endothelium-Dependent Vasodilator Properties of 1,3-Dihydro-3-p-chlorophenyl-7-hydroxy-6-methyl-furo[(3,4c]pyridine hydrochloride (Cicletanine). Journal of Pharmacology and Experimental Therapeutics 1993;265 30-35.

[147] Shi, D. Q.; Yang, F.; Ni, S. N. A Facile Synthesis of Furo[3,4-e]pyrazolo[3,4-b]pyridine-5(7H)one Derivatives via Three-Component Reaction in Ionic Liquid without Any Catalyst. Journal of Heterocyclic. Chemistry 2009;46 469-476.

[148] Moghaddam, M. M.; Bazgir, A.; Mehdi, A. M.; Ghahremanzadeh, R. Alum (KAl(SO4)2•12H2O) Catalyzed Multicomponent Transformation: Simple, Efficient, and Green Route to Synthesis of Functionalized Spiro[chromeno[2,3-d]pyrimidine-5,3'-indoline]tetraones in Ionic Liquid Media. Chinese Journal of Chemistry 2012;30 709-714.

[149] Alper, P. B.; Meyers, C.; Lerchner, A.; Siegel, D. R.; Carreira, E. M. Facile, Novel Methodology for the Synthesis of Spiro[pyrrolidin-3,3'-oxindoles]: Catalyzed Ring

Expansion Reactions of Cyclopropanes by Aldimines. Angewandte Chemie International Edition 1999;38 3186-3189.

[150] Figueroa-Villar, J. D.; Carneiro, C. L.; Cruz, E. R. Synthesis of 6-Phenylaminofuro[2,3-d]pyrimidine-2,4(1H,3H)diones from Barbiturylbenzylidenes and Isonitriles. Heterocycles 1992;34 891-894.

[151] Kobayashi, K.; Tanaka, H.; Tanaka, K.; Yoneda, K.; Morikawa, O.; Konishi, H. One-Step Synthesis of Furo[2,3-d]Pyrimidine-2,4(1H,3H)-Diones Using the CAN-Mediated Furan Ring Formation. Synthetic Communications 2000;30 4277-4291.

[152] Vilsmaier, E.; Baumheier, R.; Lemmert, M. Trapping of Diacceptor-Substituted Methylenecyclopropanes with Isocyanides A Further Application of the Principle of the Two-fold Nucleophilic Substitution at a Cyclopropane. Synthesis 1990; 995-998.

[153] Kawahara, N.; Nakajima T.; Itoh T.; Ogura, H. Simple Syntheses of Pyrrolo- and Furopyrimidine Derivatives. Heterocycles 1984;22 2217-2220.

[154] Qian, C. Y.; Nishino, H.; Kurosawa, K.; Korp, J. D. Manganese(II) Acetate-Mediated Double 2-hydroperoxyalkylations of Barbituric Acid and its Derivatives. Journal of Organic Chemistry. 1993;58 4448-4451.

[155] Nair, V.; Vinod, A. U.; Abhilash, N.; Menon, R. S.; Santhi, V.; Varma, R. L.; Viji, S.; Mathewa, S.; Srinivasb, R. Multicomponent reactions involving zwitterionic intermediates for the construction of heterocyclic systems: one pot synthesis of aminofurans and iminolactones. Tetrahedron 2003;59 10279-10286.

[156] Yadav, J. S.; Subba, Reddy, B. V.; Shubashree, S.; Sadashiv, K.; Naidu, J. J. Ionic Liquids-Promoted Multi-Component Reaction: Green Approach for Highly Substituted 2-Aminofuran Derivatives. Synthesis 2004;2376-2380.

[157] Shaabani, A.; Soleimani, E.; Darvishi, M. Ionic Liquid Promoted One-Pot Synthesis of Furo[2,3-d]pyrimidine-2,4(1H,3H)diones. Monatshefte fur Chemi 2007;138 43-46.

[158] Hadfield, J. A.; Pavlidis, V. H.; Perry, P. J.; McGown, A. T. Synthesis and anticancer activities of 4-oxobenzopyrano[2,3-d]pyrimidines. Anti-Cancer Drugs 1999;10 591-595.

[159] Gupta, A. K.; Kumari, K.; Singh, N.; Singh, D.; Raghuvanshi; Singh, K. N. An Eco-safe Approach to Benzopyranopyrimidines and 4H-chromenes in Ionic Liquid at Room Temperature. Tetrahedron Letters. 2012;53 650-653.

[160] Rawal, R. K.; Tripathi, R.; Katti, S. B.; Pannecouque, C.; Clercq, E. D. Design, synthesis, and evaluation of 2-aryl-3-heteroaryl-1,3-thiazolidin-4-ones as anti-HIV agents. Bioorganic & Medicinal Chemistry 2007;15 1725-1731.

[161] Srivastava, T.; Gaikwad, A. K.; Haq, W.; Sinha, S.; Katti, S. B. Synthesis and biological evaluation of 4-thiazolidinone derivatives as potential antimycobacterial agents (NA-1265FP). Arkivoc 2005;2 120-130.

[162] Veeresena, G.; Vie, N.; James, T. D.; Duane, D. M. Efficient Microwave Enhanced Synthesis of 4-Thiazolidinones. Synthesis Letters 2004;13 2357-2358.

[163] Fraga-Dubreuil, J.; Bazureau, J. P. Efficient combination of task-specific ionic liquid and microwave dielectric heating applied to one-pot three component synthesis of a small library of 4-thiazolidinones. Tetrahedron 2003;59 6121-6130.

[164] Lingampalle, D.; Jawale, D.; Waghmare, R.; Mane, R. Ionic Liquid–Mediated, One-Pot Synthesis for 4-Thiazolidinones. Synthetic Communications 2010;40 2397-2401.

[165] Kroutil, J.; Budesınsky, M. Preparation of diamino pseudodisaccharide derivatives from 1,6-anhydro-β-d-hexopyranoses via aziridine-ring cleavage. Carbohydrate Research 2007;342 147-153.

[166] Llinas-Brunet, M.; Bailey, M. D.; Ghiro, E.; Gorys, V.; Halmos, T.; Poirier, M.; Rancourt, J.; Goudreau, N. A Systematic Approach to the Optimization of Substrate-Based Inhibitors of the Hepatitis C Virus NS3 Protease: Discovery of Potent and Specific Tripeptide Inhibitors. Journal of Medicinal Chemistry 2004;47 6584.

[167] Huang, L. J.; Hsich, M. C.; Teng, C. M.; Lee, K. H.; Kno, S. C. Synthesis and antiplatelet activity of phenyl quinolones. Bioorganic & Medicinal Chemistry 1998;6 1657-1662.

[168] Theeraladanon, C.; Arisawa, M.; Nishidi, A.; Nakagawa, M. A novel synthesis of substituted quinolines using ring-closing metathesis (RCM): its application to the synthesis of key intermediates for anti-malarial agents. Tetrahedron 2004;60 3017-3035.

[169] Yadav, A. K.; Kumar, M.; Yadav, T.; Jain, R. An Ionic Liquid Mediated One-Pot Synthesis of Substituted Thiazolidinones and Benzimidazoles. Tetrahedron Letters 2009;50 5031-5034.

[170] Yadav, L. D. S.; Yadav, B. S.; Rai, V. K. Multicomponent Reactions in Chiral Ionic Liquids: A Stereocontrolled Route to Mercaptopyranothiazoles. Journal of Heterocyclic. Chemistry 2008;45 1315-1319.

[171] Wen, L. R.; Xie, H. Y.; Li, M. A Basic Ionic Liquid Catalyzed Reaction of Benzothiazole, Aldehydes, and 5,5-Dimethyl-1,3-cyclohexanedione: Efficient Synthesis of Tetrahydrobenzo[b]pyrans. Journal of Heterocyclic Chemistry 2009;46 954-959.

[172] Zhang, X. Y.; Li, Y. Z.; Fan, X. S.; Qu, G. R.; Hu, X. Y.; Wang, J. Multicomponent Reaction in Ionic Liquid: A Novel and Green Synthesis of 1, 4-Dihydropyridine Derivatives. Chinese Chemical Letters 2006;17 150-152.

[173] Tu, S.; Zhu, X.; Zhang, J.; Xu, J.; Zhang, Y.; Wang, Q.; Jia, R.; Jiang, B.; Zhang, J.; Yao, C. New potential biologically active compounds: Design and an efficient synthesis of N-substituted 4-aryl-4,6,7,8-tetrahydroquinoline-2,5(1H,3H)diones under microwave irradiation. Bioorganic & Medicinal Chemistry Letters 2006;16 2925-2928.

[174] Kanakaraju, S.; Prasanna, B.; Basavoju, S.; Chandramouli, G. V. P. Ionic Liquid Catalyzed One-Pot Multi-Component Synthesis, Characterization

and Antibacterial Activity of Novel Chromeno[2,3-*d*]pyrimidin-8-amine derivatives. Journal of Molecular Structure 2012;1017 60-64.

[175] R. Breslow, R. Biomimetic Chemistry and Artificial Enzymes: Catalysis by Design. Accounts of Chemical Research 1995;28 146-153.

[176] Shaterian, H. R.; Ranjbar, M. An Environmental Friendly Approach for The Synthesis of Highly Substituted Imidazoles Using Brønsted Acidic Ionic Liquid, N-methyl-2-pyrrolidonium hydrogen sulfate, as reusable catalyst. Journal of Molecular Liquids 2011;160 40-49.

[177] Lehmann, F.; Holm, M.; Laufer, S. T. Three-Component Combinatorial Synthesis of Novel Dihydropyrano[2,3-*c*]pyrazoles. Journal of Combinatorial Chemistry 2008;10 364-367.

[178] Zhou, J. F.; Tu, S. J.; Zhu, H. Q.; Zhi, S. J. A Facile One-Pot Synthesis Of Pyrano[2,3-*c*]Pyrazole Derivatives Under Microwave Irradiation. Synthetic Communications 2002;32 3363.

[179] Guo, S. B.; Wang, S. X.; Li, J. T. D,L-Proline-Catalyzed One-Pot Synthesis of Pyrans and Pyrano[2,3-*c*]pyrazole Derivatives by a Grinding Method under Solvent-Free Conditions. Synthetic Communications 2007;37 2111-2120.

[180] Ren, Z.; Cao, W.; Tong, W.; Jin, Z. Solvent-Free, One-Pot Synthesis of Pyrano[2,3-*c*]pyrazole Derivatives in the Presence of KF 2H$_2$O by Grinding. Synthetic Communications 2005;35 2509-2513.

[181] Khurana, J. M.; Nand, B.; Kumar, S. Rapid Synthesis of Polyfunctionalized Pyrano[2,3-*c*]pyrazoles via Multicomponent Condensation in Room- Temperature Ionic Liquids. Synthetic Communications 2011;41 41, 405-410.

[182] Xiao, Z.; Lei, M.; Hu, L. An Unexpected Multi-Component reaction to synthesis of 3-(5-amino-3-methyl-1*H*-pyrazol-4-yl)-3-arylpropanoic acids in ionic liquid. Tetrahedron Letters 2011;52 7099-7102.

[183] Ghahremanzadeh, R.; Ahadi, S.; Bazgir, A. A one-pot, four-component synthesis of α-carboline derivatives. Tetrahedron Letters 2009;50 7379-7381.

[184] Ghahremanzadeh, R.; Sayyafi, M.; Ahadi, S.; Bazgir, A. Novel One-Pot, Three-Component Synthesis of Spiro[Indoline-pyrazolo[4′,3′:5,6]pyrido[2,3-*d*]pyrimidine]trione Library. Journal of Combinatorial Chemistry 2009;11 393-396.

[185] Ghahremanzadeh, R.; Moghaddam, M. M.; Ayoob Bazgir, A.; Akhondi, M. M. An Efficient Four-component Synthesis of Spiro[indolinepyrazolo[4′,3′:5,6]pyrido[2,3-*d*]pyrimidine]triones. Chinese Journal of Chemistry 2012;30 321-326.

Chapter 5

Ionic Liquids as Doping Agents in Microwave Assisted Reactions

Marcos A. P. Martins, Jefferson Trindade Filho,
Guilherme S. Caleffi, Lilian Buriol and
Clarissa P. Frizzo

Additional information is available at the end of the chapter

1. Introduction

The use of microwave (MW) irradiation as a tool for organic synthesis has been a fast growth area [1-8]. Several examples have shown that the application of MW irradiation re‐ duces the reaction time, increases the product yield and sometimes results in a different product distribution compared to conventional thermal heating method [1-6,9-20]. The rate acceleration observed in organic reactions using MW irradiation is due to material-wave in‐ teractions leading to thermal and nonthermal effects. The thermal effects result from a more efficient energy transfer to the reaction mixture, which is known as dielectric heating. This process relies on the ability of a substance (solvent or reactant) to absorb MW and convert them into heat. The reaction mixture is heated from the inside since the MW energy is trans‐ ferred directly to the molecules (solvent, reactants, and catalysts). This process is known as 'volumetric core heating' and results in a temperature gradient that is reversed compared to the one resulting from conventional thermal heating [1,9-14]. Nonthermal effects result in differences in product distributions, yields, and reaction times. They may result from the orientation effects of polar species in the electromagnetic field that makes a new reaction path with lower activation energy [9-14, 21-23]. It has been suggested [24] that MW activa‐ tion could originate from hot spots generated by dielectric relaxation on a molecular scale. Currently, thermal and nonthermal effects are being extensively studied mainly to verify the existence or not of nonthermal effects [21-23].

Several studies have reported the application of ionic liquids (ILs) in different areas and, in particular, their use in organic reactions [25-33]. ILs are generally defined as liquid electrolytes composed entirely of ions. Occasionally, a melting point criterion has been proposed to distinguish between molten salts and ILs (mp < 100 °C). However, both molten salts and ILs are better described as liquid compounds that display ionic-covalent crystalline structures [34-35]. Suitably selected, many combinations of cations and anions allow the design of ILs that meets all the requirements for the chemical reaction under study; based on this, they are also known as 'designer solvents' [36]. Properties such as solubility, density, refractive index, and viscosity can be adjusted to suit requirements simply by making changes to the structure of the anion, the cation, or both [37-43].

The junction of the use of MW irradiation with the use of ILs provides a method of high interest in organic synthesis. ILs interact very efficiently with MW irradiation through the ionic conduction mechanism [7-8] and are rapidly heated at rates easily exceeding 10 °C per second [44-50]. Despite few reports on the exact measurement of their dielectric properties and loss tangent values, the experimentally attained heating rates of ILs applying MW irradiation attest to their extremely high MW absorptivity [46,51]. This ability allows that small amounts of ILs can be employed as additives in order to increase the dielectric constant of nonpolar solvents characterizing them as doping agents [52-57]. In particular, ILs can be used as support in the synthesis of organic compounds which are carried out using MW irradiation and less polar solvents. Research groups have used ILs as doping agents for MW heating of otherwise nonpolar solvents such as hexane, toluene, tetrahydrofuran, and dioxane [52-57]. Thus, in view of the good relation between MW and IL, the following topics will be discussed in this chapter: (*i*) behavior of the solvents under MW environment with emphasis in the heating effects of adding a small quantity of ILs in solvents with different loss tangent, such as *N,N*-dimethyl formamide (DMF), acetonitrile (ACN), hexane (HEX), toluene (TOL), tetrahydrofurane (THF); (*ii*) ILs as doping agents in MW assisted reactions especially in *N*-alkylation reaction of pyrazole with alkyl halides (Figure 1).

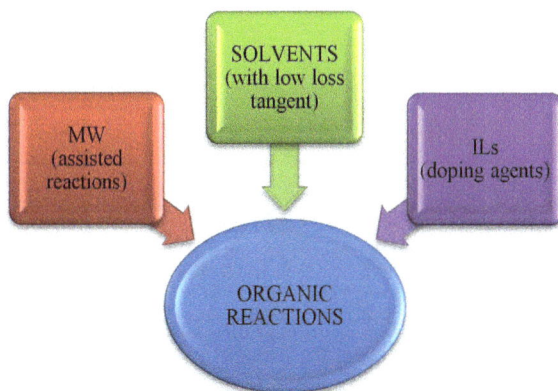

Figure 1. Ionic Liquids as doping agents in microwave assisted reactions.

2. Behavior of the Solvents Under Microwave Environment

Several organic solvents are used in various types of organic reactions under MW irradiation. The particular ability of the solvents to convert electromagnetic energy into thermal energy is directly related to their dielectric properties. The magnitude of the heating efficiency, in the specific temperature and frequency, is determined by the so-called 'loss tangent' (tan δ), whose formula is represented by Eq. 1 [52].

$$\tan \delta = \frac{\varepsilon''}{\varepsilon'} \tag{1}$$

In Eq. 1, ε'' is the dielectric loss and ε' is the dielectric constant. A reaction medium with a high tan δ at the standard operating frequency of a MW synthesis reactor (2.45 GHz) is required for good absorption and, consequently, for efficient heating. Solvents used for MW synthesis can be classified as high with tan δ > 0.5, medium with tan δ 0.1 – 0.5, and low MW absorbing with tan δ < 0.1 (Table 1) [14,58]. In general, the reactions which used solvents with a high tan δ have a good absorption of MW irradiation and, accordingly, an efficient heating [8,59,60]. Solvents such as DMSO and DMF are essential to reactions performed in MW. While these are great solvents for performing the reaction, the subsequent workup procedure is difficult to remove them due to their high boiling point and miscibility with the product [53]. Thus, in certain situations, it is convenient to use solvents which are less polar such as THF, TOL and HEX [14,58,59-60]. However, it is necessary to use a heating agent for the reactions carried out in solvents with low absorption in the MW irradiation. ILs, for instance, can be added to the reaction medium to increase the absorbance level of the MW irradiation [51,59]. Therefore, the use of ILs appears as a support to increase the temperature of the reactions carried out in a MW transparent solvents [53].

Solvent	tan δ	Solvent	tan δ
Ethylene glycol	1.350	1,2-Dichloroethane	0.127
Ethanol	0.941	Water	0.123
Dimethyl sulphoxide	0.825	Chloroform	0.091
Methanol	0.659	Acetonitrile (ACN)	0.062
1,2-Dichlorobenzene	0.280	Tetrahydrofurane (THF)	0.047
Methylpyrrolidone	0.275	Dichloromethane	0.042
Acetic acid	0.174	Toluene (TOL)	0.040
N,N-Dimethylformamide (DMF)	0.161	Hexane (HEX)	0.020

Table 1. Loss tangent of several solvents [59-60].

3. Heating Effects of Adding a Small Quantity of Ionic Liquid in Solvents

Systematic studies on temperature profiles and the thermal stability of IL under MW irradiation conditions were studied [52]. In these studies it was found that even the addition of a small amount of an IL resulted in dramatic changes in the heating profiles due to changes in the overall dielectric properties of the reaction medium.

Leadbeater and Torenius [53] studied the heating and contamination effects of several ILs in less polar solvents, such as HEX, TOL, THF and dioxane (DIO) (Figure 2) under MW irradiation. These authors have shown that all solvents used can be heated way above their boiling point in sealed vessels using a small quantity of an IL, thereby allowing them to be used as media for MW assisted chemistry. Table 2 shows the temperatures reached for pure solvents and for doped solvents with ILs using 200 W of power under MW irradiation.

The effects of varying the quantity of IL used to the solvent heating were investigated. The authors found that the best condition used was 2 mL of solvent and 0.2 mmol of IL, resulting in rapid heating. In these studies the contamination, if any, of the parent solvent with the IL or any decomposition products formed as they are heated were also studied [53].

Results showed that both [BMIM][PF$_6$] and [BMIM][BF$_4$] proved to be useful in MW heating of solvents, with [BMIM][PF$_6$] being more effective (Table 3). There was no contamination of the solvent when using [BMIM][PF$_6$] with any of the solvents screened or when [BMIM][BF$_4$] was used with HEX. There was contamination due to the decomposition of [BMIM][BF$_4$] when used with TOL or DIO; the extent was much less in the case of the latter. The [BMIM][BF$_4$] was slightly soluble in THF thus in this case the only source of contamination at the end of the heating experiments was a trace of the parent IL rather than any decomposition. To the experiments, 100 W of the power was used [53].

Leadbeater et al. [54] also investigated the decomposition of some ILs and found out that when the IL was heated above 200 °C, decomposition occurred to give an alkyl halide and alkyl imidazole as shown in Scheme 1. Halide ion (X$^-$) acts as a nucleophile in attaching the cation with the subsequent elimination of alkyl-X. This decomposition was verified for elevated temperatures, which was not totally unexpected.

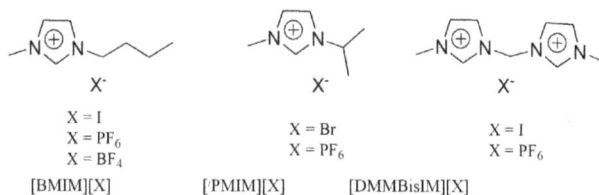

Figure 2. Ionic liquids used as doping agent.

Solvent	IL Added	T(attained) (°C)	Time (taken) (s)	T (without IL) (°C)	Solvent Boiling Point (°C)
HEX	[BMIM][I]	217	10	46	69
	[iPMIM][Br]	228	15		
TOL	[BMIM][I]	195	150	109	111
	[iPMIM][Br]	234	130		
THF	[BMIM][I]	268	70	112	66
	[iPMIM][Br]	242	60		
DIO	[BMIM][I]	264	90	101	101
	[iPMIM][Br]	246	90		

Table 2. The Microwave Irradiation Effects of Adding a Small Quantity of ILs in Less Polar Solvents [53].

Solvent	IL Added	T. attained (°C)	Time Taken (s)	Level of Contamination
HEX	[BMIM][PF$_6$]	279	20	None
	[iPMIM][PF$_6$]	90	300	None
	[BMIM][BF$_4$]	192	60	None
TOL	[BMIM][PF$_6$]	280	60	None
	[iPMIM][PF$_6$]	79	120	None
	[BMIM][BF$_4$]	165	90	Contaminated
THF	[BMIM][PF$_6$]	231	60	None
	[BMIM][BF$_4$]	95	50	contaminated[b]
DIO	[BMIM][PF$_6$]	149	100	None
	[BMIM][BF$_4$]	184	120	Contaminated

[b][BMIM][BF$_4$] is slightly soluble in THF and so cannot totally be removed; thus contamination is due to [BMIM][BF$_4$] rather than decomposition.

Table 3. Microwave irradiation effects in the presence of a small quantity of ILs in less polar solvents [53].

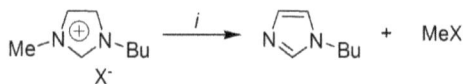

X = Cl, Br, I
i: MW

Scheme 1.

Hoffmann et al. [61] showed that ILs of the 1,3-dialkylimidazolium-type revealed (Figure 3) great potential for the application of MW for organic synthesis. These authors verified that the increase of MW power resulted in a drastic decrease in heating time.

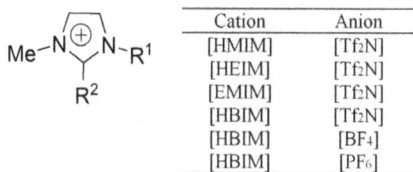

	Cation	Anion
	[HMIM]	[Tf$_2$N]
	[HEIM]	[Tf$_2$N]
	[EMIM]	[Tf$_2$N]
	[HBIM]	[Tf$_2$N]
	[HBIM]	[BF$_4$]
	[HBIM]	[PF$_6$]

Figure 3. Ionic liquids used in the study [61].

A supplementary investigation covered the heating behavior of ILs as doping agent when mixed with solvents less polar in MW irradiation such as TOL and cyclohexane (100 mL of solvent in 1 mL, 3 mL and 5 mL of IL) (Table 4). Therefore, the authors concluded that small amounts of ILs are necessary to significantly reduce the heating time of TOL or cyclohexane under MW conditions. An increase in the MW power generates a reduction in heating time (Table 4). Also in this case, heating time approaches a limiting value even with an increase of the MW power. This was also true for the addition of ILs to both non-polar solvents (TOL and cyclohexane).

IL	Power (W)	Ht$_{(35-105°C/s)}$ (TOL : IL (mL))		
		100 : 1	100 : 3	100 : 5
[HMIM][Tf$_2$N]	300	318	90	70
[HMIM][Tf$_2$N]	400	167	66	53
[HMIM][Tf$_2$N]	500	112	54	39
[HBIM][PF$_6$]	300	548	168	143
[HBIM][PF$_6$]	400	319	126	92
[HBIM][PF$_6$]	500	229	88	86

Table 4. Heating times (Ht) of toluene/ionic liquid-mixtures.

Following the direction of these studies, we also performed some experiments using ILs as doping agents with several solvents under MW irradiation. The objective of this study was to check if the data of our MW equipment are in accordance with the data already published. Thus, we performed investigations of power profiles in different solvents with distinguished loss tangent values as DMF, ACN, THF, TOL and HEX in the presence of small quantities of [BMIM][BF$_4$] as doping agent. The solvents doped were submitted under MW irradiation in an attempt to reach a temperature of 150°C (temperature which may be used in organic reactions) [62]. For this, we used various concentrations of IL in different solvents, as shown in Table 5. After reaching the desired temperature, the doped solvents were irradiated for 5 min and we verified that lower concentrations of IL required higher power for all solvents tested (Table 5). During the 5 min of MW irradiation the power remained substantially constant. Solvents with the low loss tangent such as HEX and TOL achieved only 99 and 108 °C, respectively, even though 300 W of power was applied.

Entry	Solvent	[BMIM][BF$_4$] (mmol.mL^{-1})	Power (W)
1	DMF	0.057	20.402
2	DMF	0.107	18.679
3	DMF	0.196	16.887
4	DMF	0.397	15.895
5	ACN	0.048	38.967
6	ACN	0.104	31.229
7	ACN	0.205	30.478
8	ACN	0.407	29.402
9	THF	0.063	240.079
10	THF	0.103	185.834
11	THF	0.218	112.582
12	THF	0.401	69.415
13	TOL	0.045	-^{b}
14	TOL	0.102	119.429
15	TOL	0.197	67.805
16	TOL	0.403	47.317
17	HEX	0.049	-^{c}
18	HEX	0.105	130.718
19	HEX	0.210	68.858
20	HEX	0.398	60.301

^{a}In a sealed vessel, under simultaneous cooling, 150 °C for 5 min, temperature was measured with fiber-optic probe. ^{b}Achieved 108 °C, 300 W, 20 min. ^{c}Achieved 99 °C, 300 W, 20 min.

Table 5. Power dependence of IL concentration in some solvents^a.

Figure 4 illustrates the dependence between the concentrations of [BMIM][BF$_4$] in the solvents and the power irradiated by MW equipment. At low concentrations of [BMIM][BF$_4$] (~ 0.05 mmol.mL^{-1}) a significant increase in the power is required to maintain the temperature of 150 °C. Another point is that, to maintain the temperature of 150 °C, solvents such as

DMF and ACN did not require substantial variation of power as that found to HEX, TOL and THF when the concentration of IL ranged from ~ 0.05 mmol.mL^{-1} to ~ 0.4 mmol.mL^{-1}. These data corroborate previous studies reported [53,61] and highlight the efficiency of [BMIM][BF$_4$] as doping agent of poorly MW absorbing solvents.

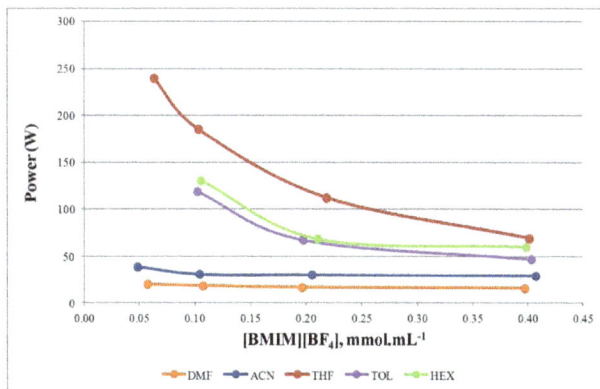

Figure 4. Power profiles of solvents with different concentrations of [BMIM][BF$_4$].

4. Ionic Liquids as Doping Agents (ILDA) in Microwave Assisted Reactions

The efficient use of ILs as a doping agent in reaction under MW irradiation was firstly introduced by Ley *et al.* [55]. The authors described the synthesis of thioamides from the secondary or tertiary amides (Scheme 2) and nitriles from primary amides (Scheme 3) in presence of thiophosphorylated amine resin using small quantity of IL [EMIM][PF$_6$] (120 mg) in TOL (2.5 mL).

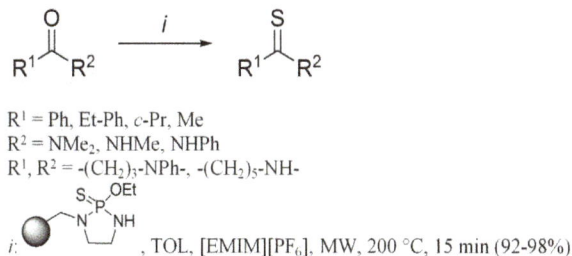

R^1 = Ph, Et-Ph, *c*-Pr, Me
R^2 = NMe$_2$, NHMe, NHPh
R^1, R^2 = -(CH$_2$)$_3$-NPh-, -(CH$_2$)$_5$-NH-

i: , TOL, [EMIM][PF$_6$], MW, 200 °C, 15 min (92-98%)

Scheme 2.

Protocols used the reactants thiophosphorylated amine resin and secondary or tertiary amides in a molar ratio of 1:3-20, respectively, to obtain thioamides, and used the reactants thiophosphorylated amine resin and primary amides in a molar ratio of 1:3.5, respectively to furnish nitriles. Reactions were carried out under both MW irradiation at 200 °C for 15 min to obtain the thioamides in 92-98% (Scheme 2) and nitriles in 95 - < 99% yields (Scheme 3). Acetonitrile was also investigated as an alternative MW absorbent and proved to be effective, in spite of being less efficient than the IL.

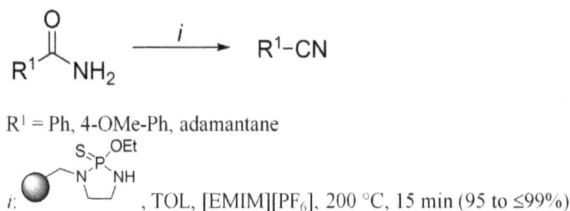

R^1 = Ph, 4-OMe-Ph, adamantane

i: [structure], TOL, [EMIM][PF$_6$], 200 °C, 15 min (95 to ≤99%)

Scheme 3.

Eycken *et al.* [56] initially investigated the intramolecular hetero-Diels-Alder reaction in a series of 2(1*H*)-pyrazinones to obtain the chloro-bicycles and dione-bicycles, as showed in Scheme 4. In their initial experiments they used pyrazinone (R = Bu, n = 2) as a model substrate involving DCE as solvent to obtain the chloro-bicycles (R = Bu, n = 2). Using a preselected maximum temperature of 190 °C (300 W maximum power), neat DCE could be heated to ca. 170 °C within 10 min under sealed vessel conditions. Prolonged time heating is needed to reach higher temperatures. In an effort to promote the enhance of the maximum attainable reaction temperature, the solvent (DCE) was doped with different amounts of [BMIM][PF$_6$]. Adding 0.035 mmol of IL to the neat solvent (2 mL of DCE), the preselected temperature of 190 °C could be reached in 3 min upon MW heating. These results clearly demonstrated that even small amounts of IL were able to change the dielectric properties of a less polar solvent. These changes are sufficiently significant to heat more rapidly the reaction medium and to reach higher reaction temperatures. Increasing the amount of IL to 0.075 mmol led to a more rapid heating of the reaction mixture, as expected. When 0.150 mmol concentration was used, it provided a profile that allowed heating the DCE doped with IL to 190 °C in 1 min. To minimize the risk of potential contaminations or side reactions caused by the IL, all the following cycloaddition studies were carried out using this set of conditions (0.150 mmol IL for 2 mL of DCE) in 100 mg of 2(1*H*)-pyrazinones to obtain the chloro-bicycles. After, the hydrolysis reaction was carried out to obtain the dione-bicycles with yields of 57-77%.

The same authors [56] reported the synthesis of the chloro-pyridine and pyridonefrom the cycloaddition reaction of 2(1*H*)-pyrazinone with dimethylacetylenedicarboxylate (DMAD) under the MW/IL conditions (Scheme 5). The reactants 2(1*H*)-pyrazinone and DMAD were used in a molar ratio of 1:1. The reaction conditions used were the same reported previously, 190 °C, DCE/[BMIM][PF$_6$] (0.150 mmol IL for 2 mL of DCE) in 5 min to furnish yields of

82% of chloro-pyridine and 2% of pyridine. Another cycloaddition reaction used hetero-dienes with ethene, leading to the bicyclic cycloadducts was investigated by these authors. However, using IL as a doping agent in the DCE was not successful because this reaction was not suitable for MW irradiation.

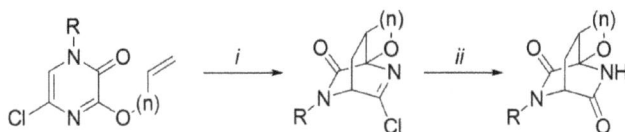

R = Bn, Ph; n = 2, 3
i : DCE, [BMIM][PF₆], MW, 190 °C, 8-15 min
ii : H₂O, MW, 190 °C, 5 min (57-77%)

Scheme 4.

Leadbeater and Torenius [53] described the Diels-Alder reaction from equimolar amounts of 2,3-dimethylbutadieneand methyl acrylate to furnish the [4 + 2] adduct cyclohex-3-ene using a mixture of TOL (2 mL) and [^{i}PrMIM][PF₆] (55 mg) under MW irradiation (Scheme 6). The mixture was irradiated at 200 °C for 5 min and led to the cyclohex-3-ene in 80% yield. The power used during the reaction performed under MW irradiation was 100 W. In a control experiment, the reaction was repeated in the absence of [^{i}PrMIM][PF₆], and it was found that after the same time (5 min at 100 W power) there was no product formed.

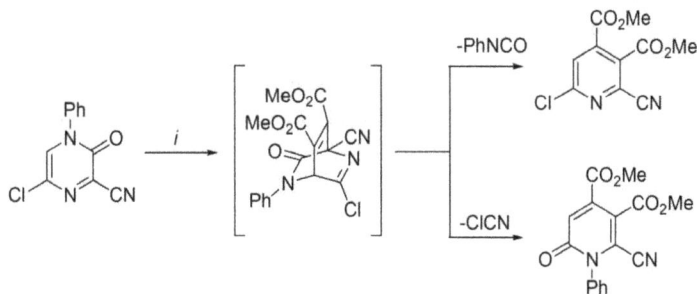

i : DMAD, DCE, [BMIM][PF₆], MW, 190 °C, 5 min (2-82%)

Scheme 5.

The same authors studied [53] the reaction of Michael addition from equimolar amounts of imidazole and methyl acrylate to furnish the methyl 3-(imidazol-1-yl) propionate (Scheme 7). The mixture of TOL (2 mL) and [^{i}PrMIM][PF₆] (55 mg) was irradiated for 2 min (200 °C, 100 W) and led to the methyl 3-(imidazol-1-yl)propionatein 75% yield. The reaction was re-

peated firstly in the absence of IL and TOL and secondly in the absence of TOL; in both cases after the same time and power (2 min at 100 W) there was no product formed.

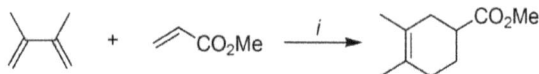

i: TOL, [ᴵPMIM][PF₆], MW, 100 W, 200 °C, 5 min (80%)

Scheme 6.

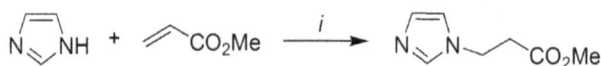

i:TOL, [ᴵPMIM][PF₆], Et₃N, MW, 100 W, 200 °C, 2 min (75%)

Scheme 7.

Garbacia *et al.* [57] described the ring-closing metathesis reactions (RCM) using diene substrates to furnish rings of five-, six-, or seven membered carbo- or heterocycles under MW irradiation (Scheme 8). The mixture of dienes (X = NTs, m,n = 1) and 0.5 mol% Grubbs' catalyst in
the presence of DCM/[BMIM][PF₆] (0.04 M of IL) was irradiated in MW for 15 s, furnishing the
desired product in > 98% yields. When neat DCM was used after the same time period only
57% conversion was observed. The authors mentioned that this was not a surprise since the reaction temperature during the full irradiation event (0-15 s) was significantly lower for the neat
solvent. On the other hand, it was not possible to use the cationic ruthenium allenylidene catalyst in conjunction with an IL-doped solvent. With both [BMIM][PF₆] and [BMIM][BF₄] (0.04 M
in DCM), conversions were below 30%, presumably due to catalyst deactivation.

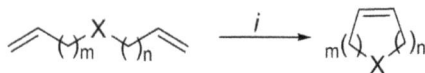

X = NTs, O, CHOTs, C(CO₂Et)₂
m, n = 1,2

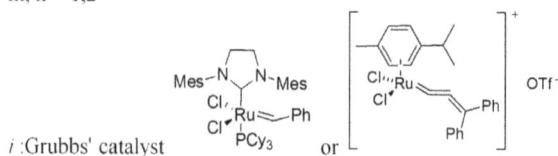

i :Grubbs' catalyst

Scheme 8.

Leadbeater *et al.* [54] also reported the conversion of alcohols to alkyl halides using IL. Initially, they screened a range of reaction conditions mediated by MW irradiation using 100 W of power. Focusing on 1-octanol, they varied the MW irradiation time (0.5-10 min), the ILs ([PMIM][I], [ʲPMIM][Br], [BMIM][Cl]) and the acid (PTSA, H_2SO_4). The reaction was performed from equimolar amounts of alcohol, IL and acid. The authors also investigated the effects of the addition of TOL as co-solvent (2 mL). When these reactions were carried out with neat IL, they reached 200 °C in a few seconds (≤ 15 s). On the other hand, using TOL as co-solvent it took a little longer to heat up but still reached 200 °C within a matter of 30–40 s (Table 6). Results showed that PTSA was more efficient than H_2SO_4 in he reactions involving the iodo, bromo and chloro ILs. Reaction times were in an increasing order: iodo < bromo < chloro substitutions with 0.5, 3 and 10 min, respectively. Most of the reactions using neat ILs presented higher product yields. The use of 2 mL of TOL as a co-solvent decreased the yield of the product formed.

IL/(Nucleofile)	Time (min)	Acid	Product	Yield[a] (%) Without co-solvent	Yield[a] (%) With co-solvent
[PMIM][I]	0.5	PTSA	$CH_3(CH_2)_7$-I	81	56
[PMIM][I]	1	PTSA	$CH_3(CH_2)_7$-I	53	38
[PMIM][I]	0.5	H_2SO_4	$CH_3(CH_2)_7$-I	3	55
[PMIM][I]	1	H_2SO_4	$CH_3(CH_2)_7$-I	38	15
[ʲPMIM][Br]	0.5	PTSA	$CH_3(CH_2)_7$-Br	68	42
[ʲPMIM][Br]	3	PTSA	$CH_3(CH_2)_7$-Br	95	32
[ʲPMIM][Br]	0.5	H_2SO_4	$CH_3(CH_2)_7$-Br	73	59
[ʲPMIM][Br]	1	H_2SO_4	$CH_3(CH_2)_7$-Br	42	40
[BMIM][Cl]	3	PTSA	$CH_3(CH_2)_7$-Cl	32	0
[BMIM][Cl]	3	H_2SO_4	$CH_3(CH_2)_7$-Cl	49	8
[BMIM][Cl]	10	PTSA	$CH_3(CH_2)_7$-Cl	42	35

[a]Yield of isolated product.

Table 6. Reaction conditions of 1-octanol with IL/(Nucleofile) [54].

Having found suitable conditions, the reaction was performed to a range of different alcohols. Further optimization of the reaction showed that the best reaction conditions for obtaining the 1-octanol were when IL was used in reaction medium. On the other hand, some dihalogenate 1,8-octanediol have furnished the best results when the co-solvent method was used as showed in Table 7. When using geraniol, not unexpectedly, geranyl iodide could not be isolated, but bromide and chloride could be obtained (Table 7). When using benzyl alcohol, it was possible to obtain the iodide in moderate yield (46%), the bromide in good yield

(68%) but only the chloride in low yield (17%). The authors believe that the co-solvent method is better because the organic product is more soluble in the organic solvent than in the IL and that once formed it moves to the organic layer and is protected from decomposition which can occur in the higher-temperature, acid IL environment.

$$R\diagup OH \xrightarrow{\ i\ } R^1\diagup X$$

R = HO(CH$_2$)$_7$, CH=C(Me)-(CH$_2$)$_2$-CH=CMe$_2$, Ph
R^1 = X-(CH$_2$)$_7$, CH=C(Me)-(CH$_2$)$_2$-CH=CMe$_2$, Ph
i : TOL, ILs ([PMIM][I], [^{i}PMIM][Br],[BMIM][Cl]) Et$_3$N, MW, 100 W, 200 °C, 1-10 min (17-86%)

Alcohol	IL	Product	Time (min)	Yield[a] (%)
1,8-Octanediol[b]	[PMIM][I]	I-(CH$_2$)$_8$-I	3	53
1,8-Octanediol[b]	[^{i}PMIM][Br]	Br-(CH$_2$)$_8$-Br	3	86
1,8-Octanediol[b]	[BMIM][Cl]	Cl-(CH$_2$)$_8$-Cl	10	50
Geraniol	[PMIM][I]	I-CH=C(Me)-(CH$_2$)$_2$-CH=CMe$_2$	0.5	Dec[c]
Geraniol	[^{i}PMIM][Br]	Br-CH=C(Me)-(CH$_2$)$_2$-CH=CMe$_2$	3	47
Geraniol	[BMIM][Cl]	Cl-CH=C(Me)-(CH$_2$)$_2$-CH=CMe$_2$	10	30
Benzyl alcohol	[PMIM][I]	PhCH$_2$-I	1	46 (72)[d]
Benzyl alcohol	[^{i}PMIM][Br]	PhCH$_2$-Br	3	68
Benzyl alcohol	[BMIM][Cl]	PhCH$_2$-Cl	10	17

[a]Yield of isolated product. [b]0.5 mmol alcohol. [c]Dec = decomposition observed. [d]3 min.

Table 7. Conversion of alcohols to alkyl halides using co-solvent method [54].

Silva *et al.* [63] used the MW irradiation technique in the Diels–Alder reaction of tetrakis(pentafluorophenyl)porphyrin with pentacene and naphthacene. One of the synthetic methods used for the synthesis of these compounds was the use of IL-doped under MW irradiation. In order to increase the product yields, the authors used NMP and DCB as solvent systems with higher loss tangents, doped with an [BMIM][PF$_6$]. Unfortunately, none of these experiments gave better results.

5. Ionic Liquids as Doping Agents in Microwave Assisted *N*-Alkylation Reactions

Reactions of *N*-alkylation of pyrazoles using IL as doping agent under MW irradiation have been little explored. Leadbeater and Torenius [53] studied the reaction of alkylation of pyrazoles used 1H-pyrazole and alkyl halides to furnish 1-alkylpyrazoles under MW irradiation (Scheme 9). The authors found that to this reaction the product was not obtained using 2 mL of TOL and 55 mg of [^{i}PrMIM][PF$_6$], which were reaction conditions previously established for other reactions (Diels-Alder and Michael addition). Although the authors did not manage to characterize the reaction products, they affirm that "it is clear to see that all the IL is destroyed since the biphasic starting mixture (solvent and IL) becomes a monophasic mixture after just a few seconds of MW irradiation. This shows the limitations of our protocol; it not being possible to undertake reactions which use or generate nucleophiles such as halide ions".

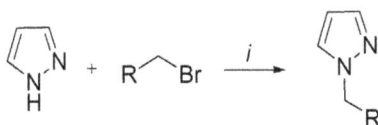

R = aryl
i : TOL, [^{i}PMIM][PF$_6$], Et$_3$N, MW, 100 W, 200 °C

Scheme 9.

Taking into account the results found by Leadbeater and Torenius [53], Kresmsner et al. [51] described the use of passive heating elements (PHEs) in N-alkylation of pyrazoles using NH-pyrazole and 1-(2-bromoethyl) benzene to obtain 1-phenethyl-1H-pyrazole. PHEs are materials which allow the compounds with low absorption of MW irradiation or poorly absorbing solvents such as HEX, carbon tetrachloride, THF, DIO, or TOL to be effectively heated to temperatures far above their boiling points (200-250 °C) under sealed vessel MW conditions. Thus, the authors used cylinders of sintered silicon carbide (SiC), PHE, which are chemically inert and strongly MW absorbing materials in the reactions of alkylation of pyrazoles.

Based on the studies mentioned above, we decided to explore the doping capacity of IL under MW irradiation in the N-alkylation of pyrazoles. This is a fundamental reaction of broad synthetic utility that often requires basic catalysis and thermodynamic reaction conditions. In addition, N-alkylation reaction of this heterocycle is a synthetic approach useful in the preparation of building blocks for the synthesis of important active compounds like pharmaceuticals [64] and agrochemicals [65]. In this way, it is clear the importance to develop a new methodology regarding this reaction. Our research group has previously reported the N-alkylation of pyrazoles using IL as solvent in oil bath [31]. Thus, we focused the use of efficient MW irradiation to perform the N-alkylation of pyrazoles in less polar solvents. Since these molecular solvents poorly absorb MW irradiation due to their lower loss tangent, the use of IL as doping agents becomes essential to achieve high temperatures. A symmetrical pyrazole and two reactive alkyl halides were chosen to perform these tests. [BMIM][BF$_4$] was selected due to its successful results in our previous work of N-alkylation in oil bath [31]. The amount of IL employed was ~ 0.1 mmol.mL^{-1}, which represents the minimum quantity required to achieve 150 °C in the four solvents under

study – HEX, TOL, THF and DIO (Figure 3) [66]. We also selected a base, KOH, to investigate the influence of basic catalysis on this reaction [31]. Initially, the reaction between
butyl bromine and 3,5-dimethylpyrazole was performed in absence of basic catalysis.
Based on data presented in Table 8, we could see that the reaction in HEX achieved the
highest conversion followed by TOL, THF and DIO. In a basic medium, the conversion
was increased only for TOL and DIO. The maintenance of moderate conversions could be
explained by the low solubility of KOH in the solvents employed (Table 8).

Thus, we decided to investigate if a change in the alkylant agent reactivity could lead to
higher conversions. Since iodine is a better leaving group than bromine, ethyl iodine was
chosen to react with 3,5-dimethylpyrazole. Higher conversions were achieved for all tested
solvents when compared with the results mentioned previously (Table 9). These results suggest that the nature of the leaving group would have greater influence than the basic catalysis on the product conversion.

Contrary to the results of Leadbeater and Torenius [53], we chose substrates for the reaction
that showed moderate to good conversions. Thus, the IL is shown as an alternative to passive heating elements PHE [13].

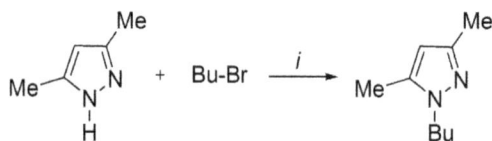

i = [BMIM][BF$_4$], MW, 150 °C, 30 min

Entry	Solvent[a]	Base	[BMIM][BF$_4$] (mmol.mL^{-1})	Conversion (%)[b]
1	HEX	-	0.117	59
2	HEX	KOH	0.123	50
3	TOL	-	0.123	17
4	TOL	KOH	0.127	37
5	THF	-	0.113	17
6	THF	KOH	0.124	17
7	DIO	-	0.108	9
8	DIO	KOH	0.118	41

[a]3mL. [b]Determined by ^{1}H NMR.

Table 8. Conversion of 1*H*-pyrazole in 1-butylpyrazolein low polar solvents in presence of [BMIM][BF$_4$].

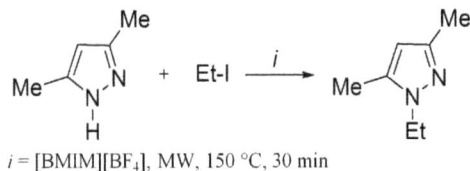

i = [BMIM][BF$_4$], MW, 150 °C, 30 min

Entry	Solvent[a]	[BMIM][BF$_4$] (mmol.mL^{-1})	Conversion (%)[b]
1	HEX	0.123	75
2	TOL	0.114	43
3	THF	0.120	71
4	DIO	0.118	43

[a]3mL.[b]Determined by ^{1}H NMR.

Table 9. Conversion of 1H-pyrazole in 1-ethylpyrazolein low polar solvents in presence of [BMIM][BF$_4$].

6. Conclusions

After analysis of the literature and results previously obtained by us about ILs as doping agents under MW irradiation, it is possible to conclude that: (i) the use of a small amount of IL in less polar solvents such as THF, TOL, and HEX promotes efficient heating under MW irradiation in sealed vessels; (ii) solvents with low tan δ when doped with small amounts of ILs are generally ideal reaction media as they allow a very rapid heating by MW irradiation in sealed vessels; (iii) an important limitation in the use of ILs as a doping agent is the chance of IL decomposition at temperatures higher than its thermal stability.

The examples of ILs as doping agents reviewed in this chapter showed that their applications are little explored and they have the potential to become an area of greater interest in the organic synthesis.

7. List of Abbreviations

ACN	Acetonitrile
[BMIM][BF4]	1-Butyl-3-methylimidazolium tetrafluoroborate
[BMIM][Br]	1-Butyl-3-methylimidazolium bromide

[BMIM][I]	1-Butyl-3-methylimidazolium iodide
[BMIM][PF6]	1-Butyl-3-methylimidazolium hexafluorophosphate
DCE	1,2-Dichloroethane
DCM	Dichloromethane
DIO	Dioxane
DMF	N,N-Dimethylformamide
DMAD	Dimethylacetalenedicarboxylate
[DMMBisIM][I]	(Bis(1-methylimidazol-3-yl))methane iodide
[DMMBisIM][PF6]	(Bis(1-methylimidazol-3-yl))methane hexafluorophosphate
[EMIM][Tf2N]	1-Ethyl-3-methylimidazolium bis(trifluoromethylsulfonyl)amide
Grubbs' catalyst	Grubbs' Catalysts (are a series of transition metal carbene complexes used as catalysts for olefin metathesis)
[HBIM][BF4]	1-Buthylimidazolium tetrafluoroborate
[HBIM][PF6]	1-Buthylimidazolium hexafluorophosphate
[HBIM][Tf2N]	1-Buthylimidazolium bis(trifluoromethylsulfonyl)amide
HEX	Hexane
[HEIM][Tf2N]	1-Ethylimidazolium bis(trifluoromethylsulfonyl)amide
[HMIM][Tf2N]	1-Methylimidazolium bis(trifluoromethylsulfonyl)amide
ILDA	Ionic Liquids as Doping Agents
['PMIM][Br]	1-*iso*-Propyl-3-methylimidazolium bromide
['PMIM][PF6]	1-*iso*-Propyl-3-methylimidazolium hexafluorophosphate
[PMIM][I]	1-Propyl-3-methylimidazolium iodide
PTSA	*p*-Toluenesulfonicacid
PHEs	Passive Heating Elements
RCM	Ring-Closing Metathesis
TOL	Toluene
THF	Tetrahydrofurane

Acknowledgements

The authors are grateful to Conselho Nacional de Desenvolvimento Científico e Tecnológico (CNPq Procs. No. 578426/2008-0; 471519/2009-0), Fundação de Amparo à Pesquisa do Estado do Rio Grande do Sul (FAPERGS/CNPq-PRONEX Edital No. 008/2009, Proc. No. 10/0037-8) and Coordenação de Aperfeiçoamento de Pessoal de Nível Superior (CAPES/PROEX) for financial support. The fellowships from CNPq (M.A.P.M., J.T.F.), and CAPES (G.S.C., C.P.F., L.B.) are also acknowledged.

Author details

Marcos A. P. Martins*, Jefferson Trindade Filho, Guilherme S. Caleffi, Lilian Buriol and Clarissa P. Frizzo

*Address all correspondence to: mmartins@base.ufsm.br

Department of Chemistry, NUQUIMHE, Federal University of Santa Maria, Brazil

References

[1] Lindström, P., Tierney, J., Wathey, B., & Westman, J. (2001). Microwave assisted organic synthesis. A review. *Tetrahedron*, 57(45), 9225-9283.

[2] Perreux, L., & Loupy, A. (2001). A tentative rationalization of microwave effects in organic synthesis according to the reaction medium, and mechanistic considerations. *Tetrahedron*, 57(45), 9199-9223.

[3] Deshayes, S., Liagre, M., Loupy, A., Luche-L, J., & Petit, A. (1999). Microwave activation in phase transfer catalysis. *Tetrahedron*, 55(36), 10851-10870.

[4] Strauss, C. R. (1999). Invited Review. A Combinatorial Approach to the Development of Environmentally Benign Organic Chemical Preparations. *Australian Journal of Chemistry*, 52(2), 83-96.

[5] Galema, S. A. (1997). Microwave chemistry. *Chemical Society Reviews*, 26(3), 233-238.

[6] Bacsa, B., Horváti, K., Bõsze, S., Andreae, F., & Kappe, C. O. (2008). Solid-Phase Synthesis of Difficult Peptide Sequences at Elevated Temperatures: A Critical Comparison of Microwave and Conventional Heating Technologies. *The Journal of Organic Chemistry*, 73(19), 7532-7542.

[7] Gabriel, C., Gabriel, S., Grant, E. H., Halstead, B. S., & Mingos, D. M. P. (1998). Dielectric parameters relevant to microwave dielectric heating. *Chemical Society Reviews*, 27(3), 213-224.

[8] Mingos, D. M. P., & Baghurst, D. R. (1991). Tilden Lecture. Applications of microwave dielectric heating effects to synthetic problems in chemistry. *Chemical Society Reviews*, 20(1), 1-47.

[9] Polshettiwar, V., & Varma, R. (2008). Microwave-Assisted Organic Synthesis and Transformations using Benign Reaction Media. *Accounts of Chemical Research*, 41(5), 629-639.

[10] Varma, R. S. (1991). Solvent-free organic syntheses. using supported reagents and microwave irradiation. *Green Chemistry*, 1(1), 43-55.

[11] Loupy, A. (2004). Solvent-free microwave organic synthesis as an efficient procedure for green chemistry. *C. R. Chim.*, 7103-112.

[12] Varma, R. S., & Polshettiwar, V. (2008). Aqueous microwave chemistry: a clean and green synthetic tool for rapid drug discovery. *Chemical Society Reviews*, 37(8), 1546-1557.

[13] Varma, R. S. (1999). Solvent-free synthesis of heterocyclic compounds using microwaves. *Journal of Heterocyclic Chemistry*, 36(6), 1565-1571.

[14] Kappe, C. O. (2004). *Controlled Microwave Heating in Modern Organic Synthesis* (Angewandte Chemie International Edition), 43(46), 6250-6284.

[15] Vargas, P. S., Rosa, F. A., Buriol, L., Rotta, M., Moreira, D. N., Frizzo, Bonacorso. H. G., Zanatta, N., & Martins, M. A. P. (2012). Efficient microwave-assisted synthesis of 1-aryl-4-dimethylamino methyleno-pyrrolidine-2, 3, 5-triones. *Tetrahedron Letters*, 53(25), 3131-3134.

[16] Buriol, L., Frizzo, Moreira. D. N., Prola, L. D. T., Marzari, M. R. B., München, T. S., Zanatta, N., Bonacorso, H. G., & Martins, M. A. P. (2011). An E-factor minimized solvent-free protocol for the preparation of 4,5-dihydro-5-(trifluoromethyl)-1H-pyrazoles. *Monatshefte für Chemie*, 142(5), 515-520.

[17] Buriol, L., Frizzo, C. P., Prola, L. D. T., Moreira, D. N., Marzari, M. R. B., Scapin, E., Zanatta, N., Bonacorso, H. G., & Martins, M. A. P. (2011). Synergic Effects of Ionic Liquid and Microwave Irradiation in Promoting Trifluoromethylpyrazole Synthesis. *Catalysis Letters*, 141(8), 1130-1135.

[18] Buriol, L., Frizzo, C. P., Marzari, M. R. B., Moreira, D. N., Prola, L. D. T., Zanatta, N., Bonacorso, H. G., & Martins, M. A. P. (2010). Pyrazole synthesis under microwave irradiation and solvent-free conditions. *Journal of the Brazilian Chemical Society*, 21(6), 1037-1044.

[19] Martins, M. A. P., Beck, P., Moreira, D. N., Buriol, L., Frizzo, C. P., Zanatta, N., & Bonacorso, H. G. (2010). Straightforward microwave-assisted synthesis of 1-carboxymethyl-5-trifluoromethyl-5-hydroxy-4,5-dihydro-1H-pyrazoles under solvent-free conditions. *Journal of Heterocyclic Chemistry*, 47(2), 301-308.

[20] Martins, M. A. P., Frizzo, Moreira. D. N., Buriol, L., & Machado, P. (2009). Solvent-Free Heterocyclic Synthesis. *Chemical Reviews*, 109(9), 4140-4182.

[21] Katritzky, A. R., & Singh, S. K. (2003). Microwave-assisted heterocyclic synthesis. AR-KIVOC (13), 68-86.

[22] Hosseini, M., Stiasni, N., Barbieri, V., & Kappe, C. O. (2007). Microwave-Assisted Asymmetric Organocatalysis. A Probe for Nonthermal Microwave Effects and the Concept of Simultaneous Cooling. *The Journal of Organic Chemistry*, 72(4), 1417-1424.

[23] Herrero, M. A., Kremsner, J. M., & Kappe, C. O. (2008). Nonthermal Microwave Effects Revisited: On the Importance of Internal Temperature Monitoring and Agitation in Microwave Chemistry. *The Journal of Organic Chemistry*, 73(1), 36-47.

[24] Laurent, R., Laporterie, A., Dubac, J., Lefeuvre, S., & Audhuy, M. (1992). Specific activation by microwaves: myth or reality? *The Journal of Organic Chemistry*, 57(26), 7099-7102.

[25] Moreira, D. N., Frizzo, C. P., Longhi, K., Soares, A. B., Marzari, M. R. B., Buriol, L., Brondani, S., Zanatta, N., Bonacorso, H. G., & Martins, M. A. P. (2011). Ionic liquid and Lewis acid combination in the synthesis of novel (E)-1-(benzylideneamino)-3-cyano-6-(trifluoromethyl)-1H-2-pyridones. *Monatshefte für Chemie*, 142(12), 1265-1270.

[26] Guarda, E. A., Marzari, M. R. B., Frizzo, Guarda. P. M., Zanatta, N., Bonacorso, H. G., & Martins, M. A. P. (2012). Enol ethers and acetals: Acylation with dichloroacetyl, acetyl and benzoyl chloride in ionic liquid medium. *Tetrahedron Letters*, 53(2), 170-172.

[27] Moreira, D. N., Longhi, K., Frizzo, C. P., Bonacorso, H. G., Zanatta, N., & Martins, M. A. P. (2010). Ionic liquid promoted cyclocondensation reactions to the formation of isoxazoles, pyrazoles and pyrimidines. *Catalysis Communications*, 11(5), 476-479.

[28] Martins, M. A. P., Guarda, E. A., Frizzo, C. P., Moreira, D. N., Marzari, M. R. B., Zanatta, N., & Bonacorso, H. G. (2009). Ionic Liquids Promoted the C-Acylation of Acetals in Solvent-free Conditions. *Catalysis Letters*, 130(1-2), 93-99.

[29] Frizzo, C. P., Marzari, M. R. B., Buriol, L., Moreira, D. N., Rosa, F. A., Vargas, P. S., Zanatta, N., Bonacorso, H. G., & Martins, M. A. P. (2009). Ionic liquid effects on the reaction of beta-enaminones and tert-butylhydrazine and applications for the synthesis of pyrazoles. *Catalysis Communications*, 10(15), 1967-1970.

[30] Moreira, D. N., Longhi, K., Frizzo, Bonacorso. H. G., Zanatta, N., & Martins, M. A. P. (2009). Ionic liquid promoted cyclocondensation reactions to the formation of isoxazoles, pyrazoles and pyrimidines. *Catalysis Communications*, 11(5), 476-479.

[31] Frizzo, C. P., Moreira, D. N., Guarda, E. A., Fiss, G. F., Marzari, M. R. B., Zanatta, N., Bonacorso, H. G., & Martins, M. A. P. (2009). Ionic liquid as catalyst in the synthesis of N-alkyl trifluoromethylpyrazoles. *Catalysis Communications*, 10(8), 1153-1156.

[32] Moreira, D. N., Frizzo, C. P., Longhi, K., Zanatta, N., Bonacorso, H. G., & Martins, M. A. P. (2008). An efficient synthesis of 1-cyanoacetyl-5-halomethyl-4,5-dihydro-1H-pyrazoles in ionic liquid. *Monatshefte für Chemie*, 139(9), 1049-1054.

[33] Martins, M. A. P., Frizzo, C. P., Moreira, D. N., Zanatta, N., & Bonacorso, H. G. (2008). Ionic Liquids in Heterocyclic Synthesis. *Chemical Reviews*, 108(6), 2015-2050.

[34] Wasserscheid, P., & Keim, W. (2000). *Ionic Liquids- New "Solutions" for Transition Metal Catalysis* (Angewandte Chemie International Edition), 39(21), 3772-3789.

[35] Seddon, K. R. (1987). *In Molten Salt Chemistry; Mamantov G, Marassi R, Eds.*, Reidel Publishing Co., Dordrecht, The Netherlands.

[36] Fremantle, M. (1998). Designer solvents- Ionic liquids may boost clean technology development. *Chemical & Engineering News*, 7632-37.

[37] Wasserscheid, P., & Welton, T. (2002). *Ionic Liquids in Synthesis*, Wiley-VCH Verlag, Stuttgart, Germany.

[38] Gordon, C. M., Holbrey, J. D., Kennedy, A. R., & Seddon, K. R. (1998). Ionic liquid crystals: hexafluorophosphate salts. *Journal of Materials Chemistry*, 8(12), 2627-2636.

[39] Seddon, K. R., Stark, A., & Torres, M. J. (2000). Influence of chloride, water, and organic solvents on the physical properties of ionic liquids. *Pure and Applied Chemistry*, 72(12), 2275-2287.

[40] Rogers, R. D., & Seddon, K. R. (2005). *Ionic Liquids III A: Fundamentals, Progress, Challenges, and Opportunities Properties and Structure*, American Chemical Society, Washington.

[41] Wilkes, J. S. (2004). Properties of ionic liquid solvents for catalysis. *Journal of Molecular Catalysis A: Chemical*, 214(1), 11-17.

[42] Holbrey, J. D., & Seddon, K. R. (1999). Ionic Liquids. *Clean Technologies and Environmental Policy*, 1(4), 223-236.

[43] Hardacre, C. (2005). Application of exafs to molten salts and ionic liquid technology. *Annual Review of Materials Research*, 35, 29-49.

[44] Horikoshi, S., Hamamura, T., Kajitani, M., Yoshizawa-Fujitaand, M., & Serpone, N. (2008). Green Chemistry with a Novel 5.8-GHz Microwave Apparatus. Prompt One-Pot Solvent-Free Synthesis of a Major Ionic Liquid: The 1-Butyl-3-methylimidazolium Tetrafluoroborate System. *Organic Process Research & Development*, 12(6), 1089-1093.

[45] Dimitrakis, G., Villar-Garcia, I. J., Lester, E., Licence, P., & Kingman, S. (2008). Dielectric spectroscopy: a technique for the determination of water coordination within ionic liquids. *Physical Chemistry Chemical Physics*, 10(20), 2947-2951.

[46] Damm, M., & Kappe, C. O. (2009). Parallel microwave chemistry in silicon carbide reactor platforms: an in-depth investigation into heating characteristics. *Molecular Diversity*, 13(4), 529-543.

[47] Martinez-Palou, R. (2009). Microwave-assisted synthesis using ionic liquids. *Molecular Diversity*, 14(1), 3-25.

[48] Leadbeater, N. E., & Torenius, H. M. (2006). *In Microwaves in Organic Synthesis, ed. A. Loupy* (2nd edition), Wiley-VCH, Weinheim, 327-361.

[49] Habermann, J., Ponzi, S., & Ley, S. V. (2005). Organic Chemistry in Ionic Liquids Using Non-Thermal Energy-Transfer Processes. *Mini-Reviews in Organic Chemistry*, 2(2), 125-137.

[50] Leadbeater, N. E., Toreniusand, H. M., & Tye, H. (2004). Microwave-Promoted Organic Synthesis Using Ionic Liquids: A Mini Review. *Combinatorial Chemistry & High Throughput Screening*, 7(5), 511-528.

[51] Kremsner, J. M., & Kappe, C. O. (2006). Silicon Carbide Passive Heating Elements in Microwave-Assisted Organic Synthesis. *The Journal of Organic Chemistry*, 71(12), 4651-4658.

[52] Kappe, C. O., Dallinger, D., & Murphree, S. S. (2009). *Pratical Microwave Synthesis for Organic Chemists*, Germany, Wiley-VCH.

[53] Leadbeater, N. E., & Torenius, H. M. (2002). A Study of the Ionic Liquid Mediated Microwave Heating of Organic Solvents. *The Journal of Organic Chemistry*, 67(9), 3145-3148.

[54] Leadbeater, N. E., Torenius, H. M., & Tye, H. (2003). Ionic liquids as reagents and solvents in conjunction with microwave heating: rapid synthesis of alkyl halides from alcohols and nitriles from aryl halides. *Tetrahedron*, 59(13), 2253-2258.

[55] Ley, S. V., Leach, A. G., & Storer, R. I. (2001). A polymer-Supported Thionating Reagent. *Journal of the Chemical Society, Perkin Transactions*, 1(4), 358-361.

[56] Eycken, E. V., der Appukkuttan, P., De Borggraeve, W., Dehaen, W., Dallinger, D., & Kappe, C. O. (2002). High-Speed Microwave-Promoted Hetero-Diels-Alder Reactions of 2(1H)-Pyrazinones in Ionic Liquid Doped Solvents. *The Journal of Organic Chemistry*, 67(22), 7904-7907.

[57] Garbacia, S., Desai, B., Lavastre, O., & Kappe, C. O. (2003). Microwave-Assisted Ring-Closing Metathesis Revisited. On the Question of the Nonthermal Microwave Effect. *The Journal of Organic Chemistry*, 68(23), 9136-9139.

[58] Obermayer, D., & Kappe, C. O. (2010). On the importance of simultaneous infrared/ fiber-optic temperature monitoring in the microwave-assisted synthesis of ionic liquids. *Organic & Biomolecular Chemistry*, 8(1), 114-121.

[59] Kappe, C. O. (2008). Microwave dielectric heating in synthetic organic chemistry. *Chemical Society Reviews*, 37(6), 1127-1139.

[60] Loupy, A. (2006). *Microwaves in Organic Synthesis*, Wiley-VCH, Weinheim, 2nd edn, and references therein.

[61] Hoffmann, J., Nuchter, M., Ondruschka, B., & Wasserscheid, P. (2003). Ionic liquids and their heating behaviour during microwave irradiation- a state of the art report and challenge to assessment. *Green Chemistry*, 5(3), 296-299.

[62] The experiments were performed in a Discover CEM MW using the mode of operation: with simultaneous cooling and temperature sensor fiber optics. The power of

the equipment was established at 200 W (or 300 W when necessary). A microwave vessel (10 mL) equipped with a standard cap (vessel commercially furnished by Discover CEM) was filled with solvent (3 mL) and [BMIM][BF$_4$] (quantities indicated in Table 5). After the vessel was sealed, the sample was irradiated for 5 min at 150 °C, which was plotted in Synergies Version 3.5.9 software and a maximum level of internal vessel pressure of 250 psi. The irradiation powers are indicated in Table 5. The solvents doped were subsequently cooled to 50 °C by compressed air.

[63] Silva, A. M. G., Tomé, A. C., Neves, M. G. P. M. S., Cavaleiro, J. A. S., & Kappe, C. O. (2005). Porphyrins in Diels-Alder reactions. Improvements on the synthesis of barrelene-fused chlorins using microwave irradiation. *Tetrahedron Letters*, 46(28), 4723-4726.

[64] Nebel, K., Brunner-G, H., & Pissiotas, G. (1996). *Int. Appl. Pat.*, WO 96/01254.

[65] Matos, I., Pérez-Mayora, E., Soriano, E., Zukal, A., Martín-Aranda, R. M., López-Peinado, A. J., Fonseca, I., & Cejka, J. (2010). *Chemical Engineering Journal*, 161(3), 377-383.

[66] The experiments were performed in a Discover CEM MW using the mode of operation: with temperature sensor fiber optics; without simultaneous cooling. The power of the equipment was established at 200 W. A microwave vessel (10 mL) equipped with a standard cap (vessel commercially furnished by Discover CEM) was filled with 1 mmol of 3,5-dimethylpyrazole and 1.2 mmol of 1-bromobutaneor iodoethane besides the addition of ~ 0.1 mmol.mL of [BMIM][BF$_4$] (quantities indicated on Tables 8 and 9) and 3 mL of solvent (Tables 8 and 9). KOH was also added in equimolar amount to 3,5-dimethylpyrazole (1 mmol) at experiments indicated in Table 8. The vessel was sealed. The sample was irradiated for 30 min at 150 °C under high stirring and a maximum level of internal vessel pressure of 250 psi. The solvent of the resultant mixture was evaporated under reduced pressure. After this step, the conversion was determinate by ^{1}H NMR. The ^{1}H NMR spectra were recorded on a Bruker DPX 400 (^{1}H at 400.13 MHz) and in CDCl$_3$/TMS solutions at 298 K. The spectroscopy data for compounds 1-butyl-3,5-dimethyl-1H-pyrazole and 1-ethyl-3,5-dimethyl-1H-pyrazole are present in the references: [31] and Potapov A. S., Khlebnikov A. I., Ogorodnikov V. D. (2006). Synthesis of 1-Ethylpyrazole-4-carbaldehydes,1,1'-Methylenebis(3,5-dimethylpyrazole-4-carbaldehyde), and Schiff Bases Derived There from. *Russian Journal of Organic Chemistry*, 42(4), 550-554, respectively.

Chapter 6

Safer and Greener Catalysts – Design of High Performance, Biodegradable and Low Toxicity Ionic Liquids

Rohit kumar G. Gore and Nicholas Gathergood

Additional information is available at the end of the chapter

1. Introduction

Molten salts which are ionic (i.e. a mixture of cation and anion) in nature and have a melting point below 100 °C are termed as Ionic liquids (ILs). [1] Preferably salts which are liquid at room temperature are called room temperature ionic liquids (RTILs). ILs have received great attention in the last couple of decades due to their unique properties such as low vapour pressure, high thermal stability, recyclability, non-flammability, and control over the product distribution. [2,3] Due to the control over fugative emission; ILs can be a replacement for volatile organic compounds (VOCs) which are commonly used as solvents in organic processes. Since the first ionic liquid was reported, [4] there has been a large number of articles been published with different types of cations and anions. One can easily design 10^{18} possible structures of ILs by varying cations and anions. This makes them "designer" molecules. [5-9] These designed combinations have already been found useful in different fields of chemistry, such as organic chemistry, [10-14] electrochemistry, [15-19] analytical chemistry, [20-24] and biochemistry. [15,25]

There are five major classes of cations in ILs e.g. ammonium, pyridinium, imidazolium, phosphonium and sulfonium (Figure 1).

Along with these, there are a large number of commonly used anions such as halides (chloride, bromide, iodide), bis(trifluoromethanesulfonimide) (NTf_2), tetrafluoroborate (BF_4), hexafluorophosphate (PF_6), octyl sulfate ($OctOSO_3$), acetate (OAc) and dicynamide ($N(CN)_2$) to name a few. Change in the anionic component can drastically affect physical properties of an ionic liquid such as hydrophilicity, viscosity and melting point.

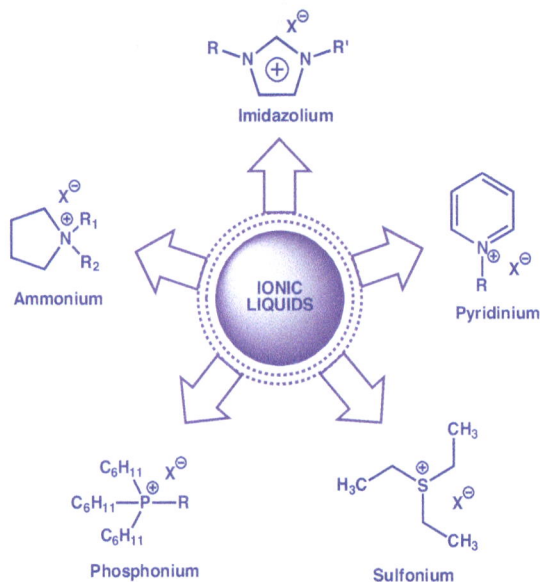

Figure 1. Major types of cations in ILs

2. Applications of ionic liquids in organic synthesis

Ionic liquids have been widely exploited in numerous organic reactions due to the versatility in the physical properties such as ease of product separation, [26,27] enhancement in rate of reaction, [28-30] catalyst immobilization, [31-33] and recyclability. [34-36] Modifications in cations and/or anions have facilitated their use in organic reactions while playing a role of reagent, solvent or catalyst. This can be reflected in a huge number of publications. Hence we are discussing, in our opinion only interesting representative examples here in this chapter.

In this chapter the aim is to demonstrate the versatility of ionic liquids in organic synthesis. We are also going to discuss the environmental fate of ionic liquids by addressing the importance of toxicity, eco(toxicity), biodegradation and green chemistry metrics. By exploring these parameters one can design and synthesise safer and greener catalyst/solvent.

2.1. Heck reaction

The palladium catalysed C-C bond forming reaction between aryl halide or vinyl halide (or triflate) and activated alkene in presence of base is known as the Heck reaction. [37] This reaction is named after Prof. Richard F. Heck, for which he was awarded Nobel Prize in Chemistry 2010, *"for palladium-catalyzed cross couplings in organic synthesis"* jointly with Prof. Ei-ichi Negishi and Prof. Akira Suzuki. This reaction is also known as Mizoroki-Heck reaction, as Tsutomu Mizoroki was the first to report this reaction. [38] A large variety of organic and inorganic bases can be used in this reaction. Phosphine ligands have been used to stabilize the catalytic system in molecular solvents. Although the reaction conditions are mild, the major drawback is that it is difficult to recycle the palladium catalyst in traditional solvents. Kaufmann and co-workers (1996) were first to demonstrate the use of tetraalkylammonium salts as an effective solvent in Heck reaction. [39] Since then a large number of publications have shown that different class of ILs can be used as solvent, catalyst or as a ligand in the Heck reaction. [40] L. Wang and co-workers have reported the Heck reaction of aryl halide and styrene (Scheme 1) in an ethanolamine-functionalized quaternary ammonium bromide which act as base, ligand and solvent. [41]

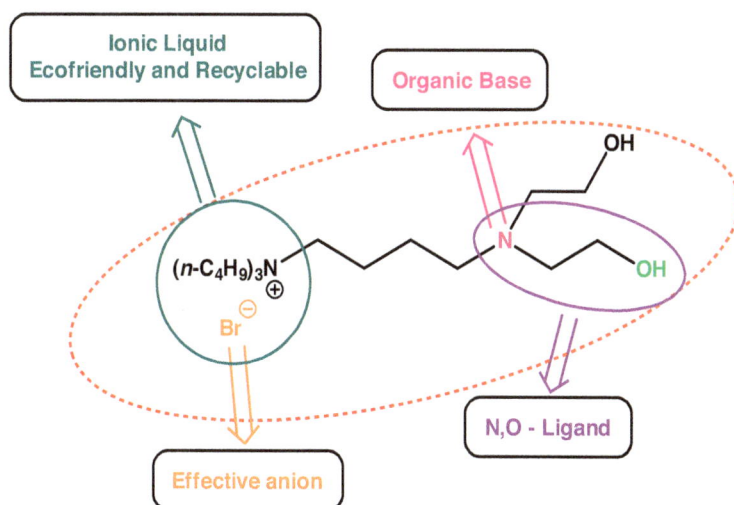

Figure 2. The functions of DHEABTBAB IL in the Heck reaction

The task specific ionic liquid i.e. 4-Di(hydroxyethyl)aminobutyl tributylammonium bromide (DHEABTBAB) (Figure 2) and palladium acetate served as an excellent catalytic system for the cross-coupling of a variety of olefins and aryl halides to give good to excellent results. The Heck reactions of styrene and iodobenzene/bromobenzene have shown excellent conversions and yield (>99%), whereas reaction of styrene and chlorobenzene have given only 66% yield. The reactions of activated and deactivated bromobenzenes and styrene/acrylates generated good to excellent yields (82 to 99%). This catalytic system was also successfully recycled and reused up to 6 times without significant loss of activity. Transmission electron microscopy (TEM) image of Pd-nanoparticles formed in (DHEABTBAB) showed even distribution due to ethanolamine moiety of the IL, which can either coordinate to the palladium or point away from the surface of the nanoparticle.

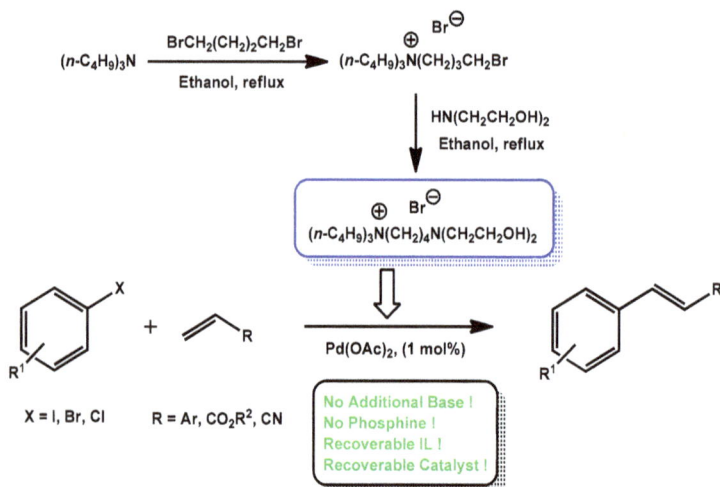

Scheme 1. Synthesis of ionic liquid and its application in the Heck reaction

In an attempt to eliminate the use of phosphine ligands, Xaio and co-workers demonstrated the *in situ* formation of a *N*-heterocyclic carbene complex with palladium when 1,3-dialkylimidazolium ILs were used as a solvent under basic conditions to generate carbene ligand. [42] They had successfully isolated the palladium carbene complex, by deprotonation of imidazolium-based ionic liquids in presence of base to form the catalytic precursor. Such participation of *N*-heterocyclic carbene as a ligand was predicted by Seddon. [43]

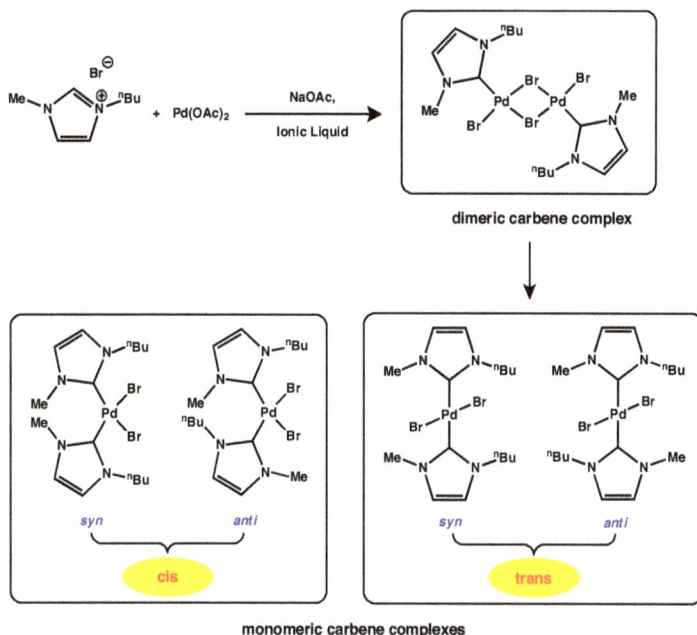

Scheme 2. Stepwise formation of *N*-hetrocyclic carbene complex of palladium

[BMIM] based ionic liquid and palladium acetate in presence of base such as sodium acetate first formed dimeric carbene complex, which eventually gave monomeric carbene complexes (Scheme 2). Existence of all four isomers of monomeric carbene complex was confirmed by ^{1}H-NMR. The Heck reaction of aryl halides with acrylates/styrene in ionic liquid under the reaction conditions have performed better than the isolated *trans* isomer of *N*-hetrocyclic carbene complex in ionic liquid. This might be due to the presence of other active palladium species formed in situ. Shrinivasan and co-workers have further supported such formation of Pd-carbene complex in [BMIM] based IL and accelerated the reaction under ultrasonic irradiation even at room temperature. [44]

Attempts and further efforts into increasing the recyclability of palladium catalyst and to reduce the use of solvent resulted in exploration of solid supported ionic liquids for use in the Heck reaction. [45,46] B. Han and co-workers have reported copolymerized ionic liquid supported palladium nanoparticles as an effective catalyst for the Heck reaction under solvent-free conditions. [47] The 1-aminoethyl-3-vinylimidazolium bromide i.e. [VAIM][Br] ionic liquid was grafted on cross-linked polymer polydivinylbenzene (PDVB). The palladium nanoparticles were anchored onto the polymer via the amino group in the ionic liquid (Scheme 3). Formation of the catalyst was confirmed by a number of analytical techniques such as X-ray photoelectron spectroscopy, transmission electron microscopy, Fourier transform infrared spectroscopy, etc.

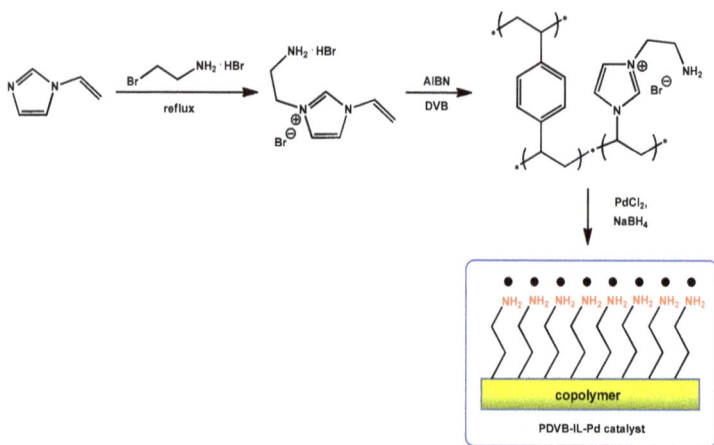

Scheme 3. Preparation of copolymerized ionic liquid supported palladium nanoparticles

The Heck reactions of a variety of iodobenzenes and acrylates have shown excellent conversions (above 93%) irrespective to the substitution on benzene ring. Triethylamine served as a good base under the reaction conditions. Due to the insoluble nature of cross-linked polymer and strong co-ordination between amino group and palladium nanoparticles, the catalyst was recovered very easily by filtration and washed with ethanol. The PDVB-IL-Pd catalyst was very active even after the 4th recycle and was confirmed by TEM image. The excellent stability of the catalyst was due to its insoluble nature in both reactants and product, and high thermal stability i.e. up to above 220 °C.

2.2. Sonogashira reaction

The palladium catalysed C-C coupling reaction of aryl halide and terminal acetylene is known as the Sonogashira reaction. Copper iodides have also been used as a co-catalyst in this reaction. [48] A stoichiometric amount of base is always used as acid (HX) scavenger. This is a widely used and efficient way to prepare substituted or unsubstituted acetylenes. Although Sonogashira coupling reactions works well under mild conditions, the drawback of this reaction is that copper catalysts used can promote side reactions, such as Glaster-type homocoupling of acetylenes. [49-53]

Ryu and co-workers have reported a palladium(II) catalysed efficient Sonogashira coupling in ionic liquid, without any copper co-catalyst. [54] The reactions with an aryl halides and alkyl/aryl acetylenes were carried out in [BMIM][PF$_6$] as a solvent and diisopropylamine or piperidine as a base. A number of palladium catalyst were screened in the Sonogashira reaction, where bis(triphenylphosphine)palladium(II) dichloride showed high catalytic activity in absence of copper co-catalyst. The Sonogashira reactions of aryl halides and alkyl/aryl acetylenes gave respective dialkyl/diaryl acetylenes in good yields (87-97%).

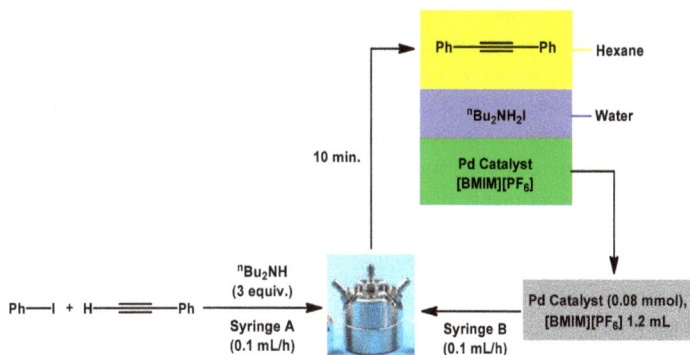

Scheme 4. Sonogashira reaction in a Microflow system

The group has successfully demonstrated the application of this reaction in a microflow reactor with IMM micromixer. Iodobenzene, phenylacetylene and base dibutylamine (syringe A) was added via one inlet to IMM's micromixer and Pd catalyst and [BMIM][PF$_6$] (syringe B) at the other inlet by using syringe pump (Scheme 4). After reacting in micromixer for 10 min., the product was easily isolated by Hexane/Water extraction, where Pd catalyst in ionic liquid was recycled and reused with slight loss of activity.

In an effort to develop an air stable copper free Sonogashira reaction, Wu, Liu and co-workers have reported palladium complex functionalized ionic liquid as a catalyst in Sonogashira reaction in [BMIM][PF$_6$] under aerobic and copper free conditions. [55]

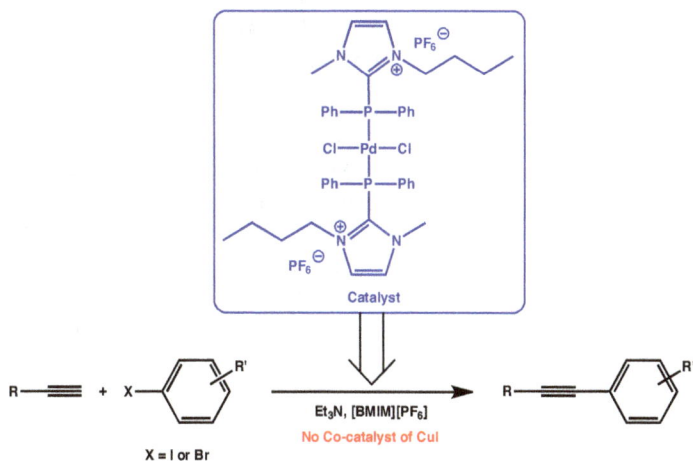

Scheme 5. The Sonogashira reactions in [BMIM][PF$_6$]

The functionalized ionic liquid i.e. di-(1-butyl-2-diphenylphosphino-3-methylimidazoli-um)-dichloridopalladium(II) hexafluorophosphate showed efficient catalytic activity and recyclability in coupling reactions (Scheme 5). A clear trend i.e. I>Br>Cl was observed in aryl halide and phenylacetylene couplings. Iodobenzene have shown excellent reactivity with variety of terminal acetylenes (90-99%) in Sonogashira coupling. The phosphine-ligated palladium complex functionalized ionic liquid was easily recycled and reused. Recyclability experiments displayed a gradual loss of activity of the catalyst in [BMIM][PF$_6$] after 6 recycles (100% to 68% yields), whereas rapid loss of activity in CH$_3$CN after 4 recycles (98% to 48% yields).

2.3. Suzuki coupling

The Suzuki reaction is a palladium catalysed coupling between aryl/vinyl boronic acid and aryl/vinyl halide in presence of base. This reaction is named after Prof. Akira Suzuki (Nobel Prize in Chemistry, 2010) and also referred to as Suzuki-Miyaura coupling. [56-58] It is one of the important C-C bond forming reaction in the synthesis of styrene and substituted biaryl compounds.

Wei and co-workers have developed a highly efficient silica supported ionic liquid with palladium incorporated anion catalyst for the Suzuki-Miyaura cross-coupling in water under reflux conditions. [59] The catalyst was prepared by an anion exchange reaction between silica-immobilized diimidazolium ionic liquid brushes with the sodium salt of Pd-EDTA (Scheme 6). This catalyst has shown great stability in air and excellent reactivity without any phosphine ligands. The Suzuki coupling of a large variety of aryl bromide and aryl iodides with phenylboric acid in water and PdEDTA-Ionic liquid brush as a catalyst gave very high yields ranging from 89% to 100%. This catalyst did not show loss of activity even after 10 recycles. Another advantage of SiO$_2$-BisIL-sOct[PdEDTA] catalyst was that it also act as a phase transfer catalyst in the reaction of water insoluble aryl halides.

Lombardo and co-workers have reported the triethylammonium ion-tagged diphenyl-phosphine palladium(II) complex for Suzuki-Miyaura reaction in pyrrolidinium ionic liquids under mild reaction conditions. [60] In an effort towards increasing the recyclability of palladium catalyst, triethylammonium ionic liquid supported diphenyl-phosphine ligand have been prepared. 1-Butyl-1-methyl-pyrrolidinium bis(trifluorome-thanesulfonimide) ([bmpy][NTf$_2$]) prove to be the best solvent along with water in presence of potassium phosphate as a base in the coupling of o-bromotoluene and phe-nylboronic acid. A Suzuki reaction of a number of electron donating and electron with-drawing groups on aryl halides and aryl boronic acids showed good to excellent results. 2-methylphenylboronic acid has given 99% yields when coupled with 1-naph-thylbromide and also with 4-bromobiphenyl. The coupling of p-anisylboronic acid and 4-bromobiphenyl gave 84% yield with triethylammonium ion-tagged diphenylphosphine palladium(II) complex.

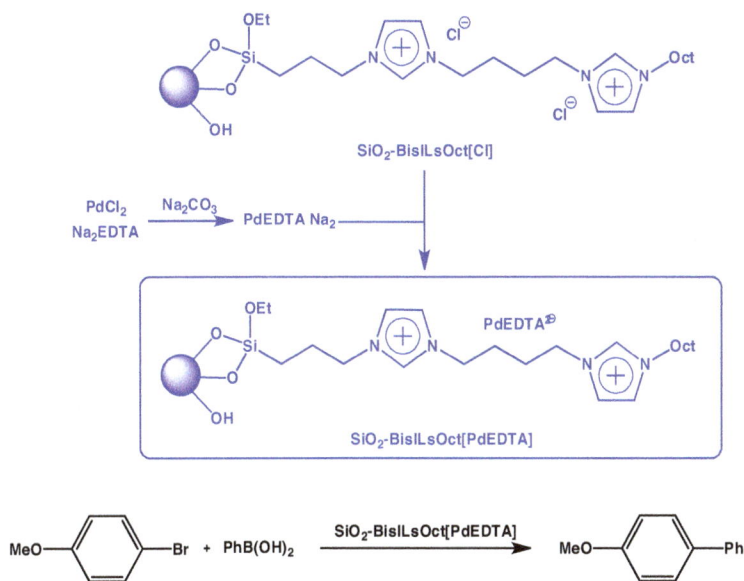

Scheme 6. Suzuki reaction with SiO$_2$-BisILsOct[PdEDTA] catalyst

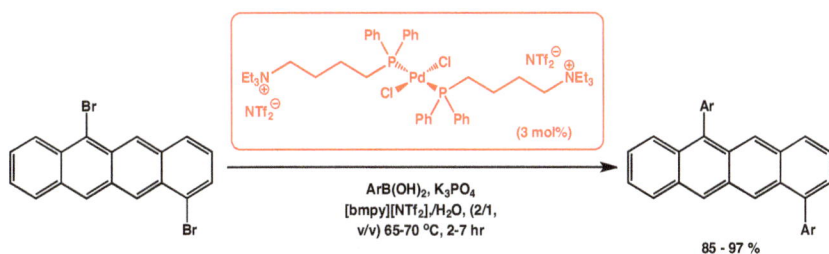

Scheme 7. Suzuki cross-coupling between 5,11-dibromotetracene with arylboronic acids

Miozzo and co-workers have demonstrated an excellent use of such ionic liquid ligated palladium complex in the challenging Suzuki cross-coupling between 5,11-dibromotetracene

with arylboronic acids under mild conditions (Scheme 7). [61] These couplings have given very high yields with phenyl and substituted phenylboronic acid (93-97%) even with 2-naphthylboronic acid (95% yield).

2.4. Stille coupling

The palladium catalyzed C-C bond formation reaction between organotin reagents and sp^2-hybridised organohalides are typically classed as Stille coupling reactions. [62,63] It is an important method of alkylation/arylation of vinyl/aryl halide. Organotin reagents used in this reaction are stable and easily stored in air. But the major drawbacks of Stille reaction are the toxicity of organotin reagents and recyclability of palladium catalyst.

To increase the recyclability of palladium catalyst and solvent, Handy and Zhang have reported the use of ionic liquid as a effective media for Stille coupling. [64] Stille coupling reactions were compared between NMP and [BMIM][BF$_4$] as a solvent with bis(benzonitrile)palladium(II) chloride as a catalyst and in the presence of triphenylarsine and Copper(I) iodide. These reactions demonstrated the compatibility of ionic liquid in Stille reactions.

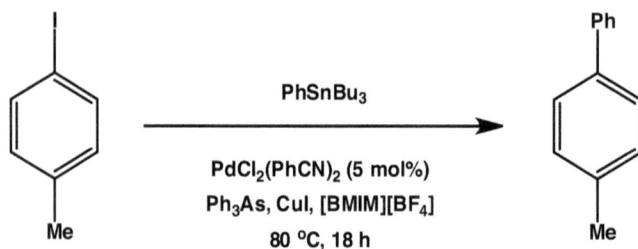

Scheme 8. Stille coupling of 4-iodotoluene and tributylphenyltin in [BMIM][BF$_4$]

Aryl coupling of a variety of aryl iodides and bromides and tributylphenyltin afforded the respective products with good yields (Scheme 8). The coupling of 4-iodotoluene and tributylphenyltin showed the highest conversion with 95% yield, whereas p-bromoanisole and tributylphenyltin gave only 15% yield towards the desired product. Ionic liquid and the catalyst bis(benzonitrile)palladium(II) chloride were recycled without loss of activity even after 5th cycle.

Due to the toxicity of organotin reagents and contamination of product by tin, organotin reagents have been boycotted by the pharmaceutical industry. Legoupy and co-workers reported an ionic liquid supported organotin reagent which can recycled and minimise contamination of product by organotin compound, without the use of solvent and additives. [65] Ionic liquids supported dibutylphenyltin was successfully synthesized and used as a reagent in palladium catalyzed Stille cross-coupling reactions involving brominated substrates under solvent-free conditions.

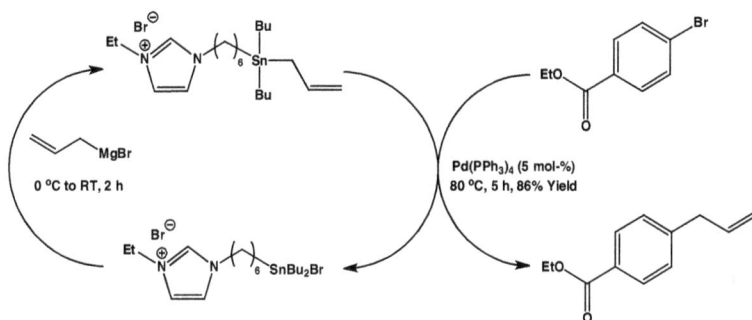

Scheme 9. Recyclable organotin reagent for Stille coupling

An effective use of different ionic liquid supported vinyl, allyl, aryl and heteroaryl organo-tin reagents with aryl bromides have seen use in Stille cross-coupling reaction. Such ionic liquid incorporated organotin reagent was recycled and reused 5 times with good yields and without loss of reactivity by using Grignard reaction (Scheme 9). It also helped to minimise tin contamination to less than 3 ppm.

2.5. Diels-Alder reaction

The cycloaddition reaction between the conjugated diene and dienophile/substituted alkene is known as Diels-Alder reaction. [66] Prof. Otto Paul Hermann Diels and Prof. Kurt Alder was awarded Nobel Prize in Chemistry in 1950 for "*for their discovery and development of the diene synthesis*". Diels-Alder reaction is an important tool in synthesis of huge and complex cyclic molecules such as cholesterol, reserpine, etc. Heterocyclic compounds can also be prepared with this reaction by using heteroatom (most of the times N and O) either as the diene or dienophile component. Diels-Alder reaction has immense importance due to the 100% atom economy in product formation. The reaction can be performed either by heating or by using Lewis/Brønsted acid catalysts such as $ZnCl_2$, HBF_4, $Sc(OTf)_3$ etc. in organic solvents.

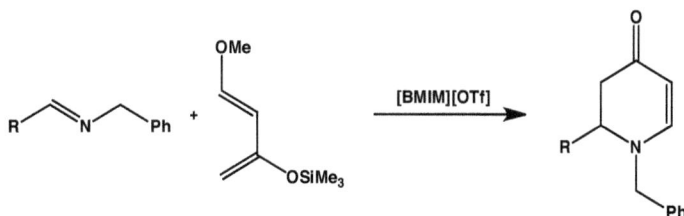

Scheme 10. Aza-Diels-Alder reaction of Danishefsky's diene with imines

Pégot and Vo-Thanh have reported aza-Diels-Alder reaction of Danishefsky's diene with imines in ionic liquids, at room temperature without any acid catalyst and organic solvents. [67] The reaction of N-benzylidenebenzylamine and Danishefsky's diene in [BMIM][OTf] showed high i.e. 94% conversion (91% yield) in 1 hour at room temperature (Scheme 10). Only half an equivalent amount of ionic liquid as used with respect to N-benzylidenebenzyl-amine. In the study of an effect of counter anion of [BMIM] cation, triflate (OTf) and bis(tri-fluoromethanesulfonimide) (NTf$_2$) has shown high yields i.e. 91% and 94% respectively in comparison with tetrafluoroborate (BF$_4$) and hexafluorophosphate (PF$_6$) i.e. 62% and 53% respectively. Reactions using pyridinium and ammonium cations with triflate anion gave good yields (91% and 89% respectively). These studies have shown that ionic liquids can be used as both polar solvent and as a catalyst in Aza-Diels-Alder reaction.

Zhou and co-workers reported C_2-symmetric ionic liquid-tagged bis(oxazoline) copper catalyst for Diels-Alder reaction in ionic liquid. [68] Bis(oxazoline)-copper(II) complexes have already been used as a Lewis acid catalyst in enantioselective Diels-Alder reactions. [69] In order to increase recyclability of the catalyst, the imidazolium-tagged bis(oxazoline) ligand copper catalyst was synthesized. (Scheme 11) The ionic liquid part of the ligand increased the insolubility of the copper catalyst in typical reaction solvents like diethyl ether, which makes workup procedure very simple. The product was separated from catalyst just by washing with diethyl ether.

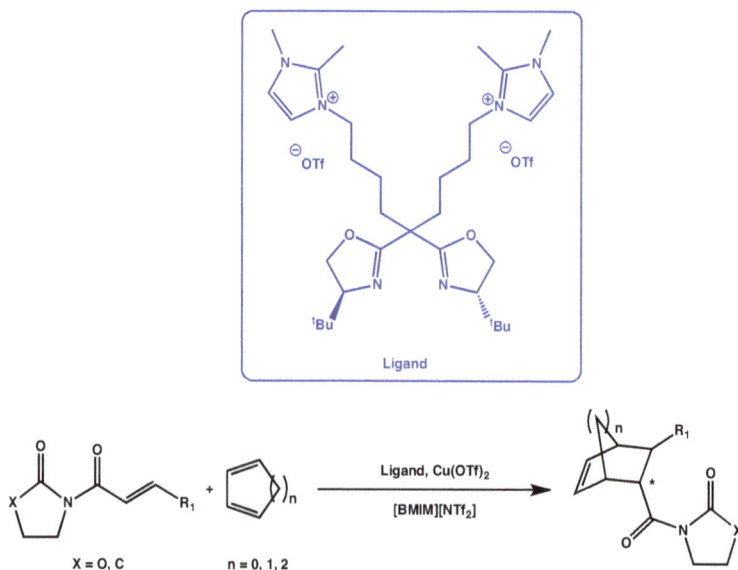

Scheme 11. Screening of the ligands in an asymmetric Diels-Alder reaction

The ionic liquid-tagged bis(oxazoline) copper catalyst did not show any conversion in the Diels-Alder reaction of N-acryloyloxazolidinone and cyclohexa-1,3-diene in DCM as a solvent. When the same reaction was carried out in [BMIM][NTf$_2$] has given required product with 98% conversion and 97% ee. N-Acryloyloxazolidinone was found to be more active than N-acryloylpyrrolidinone with cyclopentadine/cyclohexadiene in presence of C$_2$-symmetric ionic liquid-tagged (S,S)-t-Bu-box copper catalyst in [BMIM][NTf$_2$]. This efficient catalytic system (catalyst + IL) was recycled 20 times without loss of activity or enantioselectivity. This excellent recyclability was due to the ionic character of the ligand. The toxicity testing of the ligands synthesized were carried out on luminescent bacteria. The traditional t-Bu-box ligand has shown higher LC50 values (45 µg/mL) than most active ligand (11 µg/mL).

2.6. Acetalisation reactions

The acid catalysed nucleophilic addition of an alcohol to aldehyde or ketone to form respective acetal or ketal is termed as an acetalisation reaction. This is one of the important reactions in organic synthesis. As the carbonyl functionality is very reactive, it is important to protect against the attack of nucleophiles, acidic, basic or reducing agents. [70] There are several methods to protect aldehydes/ketones. Acetalisation i.e. formation of acetal has its own advantages, as it is stable to all nucleophilic and basic reagents. This reaction can be catalysed by traditional liquid acids such as HCl, H$_2$SO$_4$, etc. and also by solid acids i.e. Lewis/Brønsted acid catalysts such as ZnCl$_2$, FeCl$_3$, Zeolites, p-TSA, etc. [70-72] A water molecule is formed as a by-product in this reaction, which is important from a Green Chemistry perspective. The major drawback of this reaction is involvement of harmful liquid acids, which also involves handling hazards.

Forbes, Davis and co-workers reported Brønsted acidic ionic liquids with covalently bonded sulfonic acid functionality containing imidazolium and phosphonium cations. [73] These ILs has shown dual use as both catalyst and solvent in Fisher esterification and pinacol/benzopinacol rearrangement. Fang and co-workers further exploited such covalently bonded sulfonic acid functionality in ionic liquid and its dual use in acetalisation reaction. [74]

The Brønsted acidic ionic liquid N,N,N-trimethyl-N-propanesulfonic acid ammonium hydrogen sulfate ([TMPSA][HSO$_4$]) has been prepared economically and used as a catalyst and as a solvent in acetalisation reactions. A number of aldehydes and ketones were reacted with 1,2-diols and methanol to form acetal/ketal in [TMPSA][HSO$_4$] (Scheme 12). All reactions showed 100% selectivity with excellent conversions within 5-60 minutes. Most of the reactions gave quantitative yields, except the protection of acetophenone with methanol with 65% conversion. This Brønsted acidic ionic liquid was recycled 9 times without loss of catalytic activity and selectivity.

Du and Tian have also demonstrated the use of simple protonated 1-methylimidazolium ionic liquids as a Brønsted acid catalyst in the protection of aldehydes and ketones. [75] The IL catalyst was inexpensively prepared by protonation of 1-methylimidazole. The protection of various aldehydes and ketones with triethyl orthoformate in presence of 1-methylimidazolium tetrafluoroborate displayed very high conversions (84% - 93% yields) at room temperature (Scheme 13). The catalyst was easily recycled just by filtration and reused without any loss of activity.

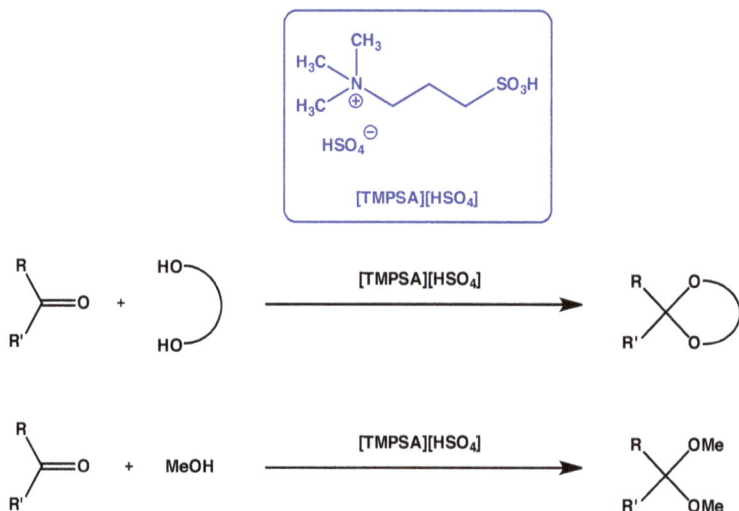

[TMPSA][HSO₄]

Scheme 12. Protection of aldehydes/ketones with alcohols in presence of [TMPSA][HSO₄]

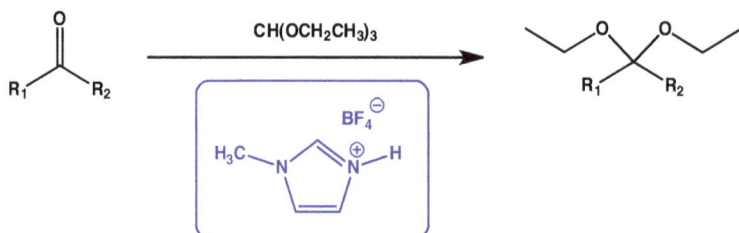

Scheme 13. Protection of aldehydes/ketones with triethyl orthoformate and IL catalyst

3. Environmental fate of ionic liquids

Due to the wide range of applications and versatility, ionic liquids are continually being used extensively in industry, [2] which has triggered an issue of waste management. Also, many of these are totally synthetic novel compounds. Hence it is important to study the environmental impact of such ionic liquids before releasing into the natural environment. Due to their low vapour pressure, ILs can reduce the possibility of air pollution. But bearing an ionic nature; ILs have a notably high solubility in water [7,76,77] (except NTf_2^- & PF_6^-) which is a viable and common means by which these ILs get released in nature. In order to check the biocompatibility of ILs, toxicity, eco-toxicity and biodegradation studies have to be carried out. ILs are usually referred to as "Green" alternatives to Volatile Organic Compounds (VOCs). Instead of the "Green" label, ILs can be categorized in the pattern of '*Traffic Signal Lights*' as discussed at the BATIL (Biodegradation And Toxicity of Ionic Liquids) meeting in DECHEMA, Frankfurt, 2009 (Fig. 1.3). [78] As we start classifying ILs in three colours (Red, Yellow and Green), we can find that most ILs are in the Red and Yellow regions, although this information was solely based on toxicity data. For an IL to be classified more accurately by a 'Traffic Signal Lights' pattern, detailed information about the toxicity, biodegradation and ease of synthesis etc. are required. Similar classification can be applied to commonly used organic solvents (Figure 3). [79]

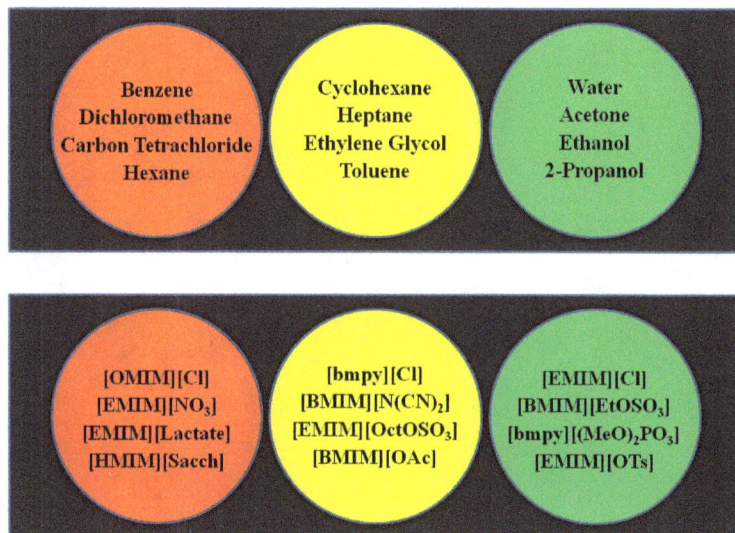

Figure 3. Recommendation for data representation of toxicity of ionic liquids and commonly used organic solvents [78,79]

3.1. Toxicity and eco(toxicity) of ionic liquids

As mentioned before, many ionic liquids are non-natural (synthetic) molecules. While a single toxicological test yields useful, albeit limited data, over the last decade, a large number of publications have demonstrated a wide variety of 'biological test systems' for toxicity testing of ionic liquids (Figure 4). [80,81] This includes fungi, bacteria, algae, enzymes, rat cell line, fish, etc. Only by assessing the toxicity of IL across a broad range of organisms can a 'true' understanding of how environmentally friendly the compound is?

Stock and co-workers reported the effect of ionic liquids on acetylcholinesterase. [82] Enzymes are a crucial part of the human nervous system. Acetylcholinesterase is known to catalyse the hydrolysis of the neurotransmitter acetylcholine, to acetate and choline. Inhibition of acetylcholinesterase results in muscular paralysis and other medically significant nervous problems. Organophosphates are a major class of acetylcholinesterase inhibitors. A range of commonly used imidazolium, pyridinium and phosphonium ionic liquids were tested in this assay. Imidazolium and pyridinium ionic liquids showed high toxicity to acetylcholinesterase at very low concentrations, whereas phosphonium ionic liquids were non-toxic within the test limits. This testing showed that toxicity of these ionic liquids lies in the cationic part and alkyl side chain and not in the anionic part.

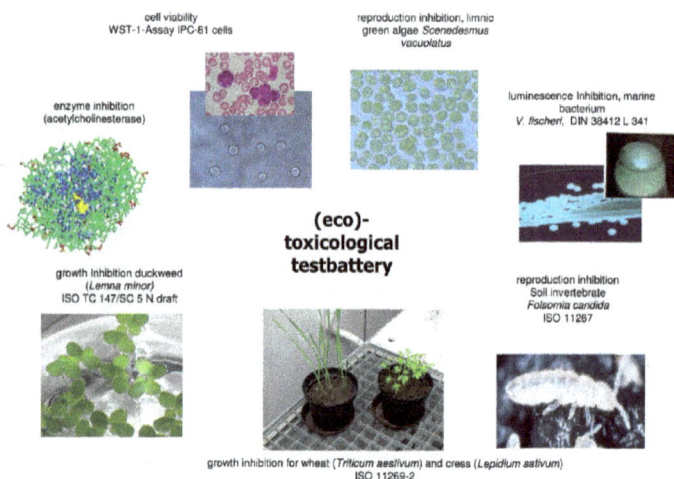

Figure 4. Toxicological test battery [80]

Another important finding of this assay was that increasing the length of alkyl side chains increase the toxicity. This can be explained as long alkyl chain increases lipophilic nature of the ionic liquids, which can then easily incorporate within the biological membrane of nerve cell synapses. [83] Similar trends between the toxicity and length of alkyl chain on luminescence inhibition of *Vibrio fischeri* and *promyelocytic leukemia rat cell line* IPC-81 were reported by Ranke and co-workers. [84] Leukemia rat cell line IPC-81 was also used to observe the cytotoxic effect of commercially available anions. [85] No significant anion effect was found under the test system.

Bernot and co-workers demonstrated that acute toxicity of certain 1-butyl-3-methyl imidazolium ionic liquids on *Daphnia Magna* were mainly due to the cationic part. [86] *Daphnia Magna* has been extensively used for ecotoxicological evaluation of chemicals in invertebrates. Ionic liquids were found to influence the reproduction of *Daphnia Magna*. 1-Butyl-3-methylimidazolium bromide was found to be most toxic in the test system (LC_{50}: 8.03 mg/L). This study demonstrated that the toxicity of ionic liquids was influenced by the cation component, which was confirmed by high LC_{50} values for sodium salts of similar anions. Yu and co-workers reported the toxicity study of 1-alkyl-3-methylimidazolium bromide ionic liquids towards the antioxidant defence system of *Daphnia Magna*. [87] Increasing the length of alkyl side chain was found again to increase toxicity. Toxicity of ionic liquids in this case was due to oxidative stress in *Daphnia Magna*, which was evaluated by measuring the activity of antioxidant defence enzymes, levels of the antioxidant glutathione and malondialdehyde i.e. peroxidation by-product of lipid. [C_{12}MIM][Br] showed very high toxicity with an LC_{50} of 0.05 mg/L under 48h incubation time. Samorì and co-workers reported the toxicity effect of oxygenated alkyl side chain imidazolium ionic liquids in *Daphnia Magna* and *Vibrio Fischeri*. [88] A direct comparison between the toxicity of 1-butyl-3-methylimidazolium tetrafluoroborate ([BMIM][BF_4]) and 1-methoxyethyl-3-methylimidazolium tetrafluoroborate and dicyanamide ([MOEMIM][BF_4] and [MOEMIM][N(CN)$_2$]) proved that incorporation of oxygen functionality helps to lower the toxicity of the ionic liquids (Figure 5). The 50% effective concentration (EC_{50}) for [BMIM][BF_4] towards the inhibition of *Daphnia Magna* and *V. Fischeri* was lower (5.18 and 300 mg/L, respectively) than for [MOEMIM][BF_4] (209-222 and 3196 mg/L, respectively) and [MOEMIM][N(CN)$_2$] (209 and 2406 mg/L, respectively).

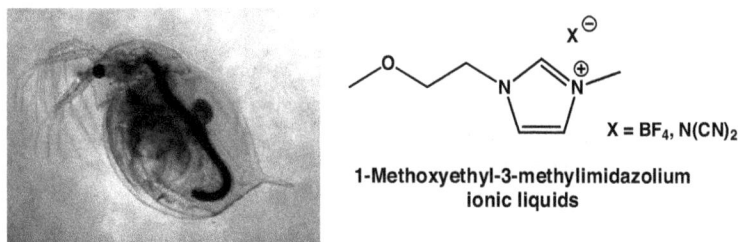

X = BF$_4$, N(CN)$_2$

1-Methoxyethyl-3-methylimidazolium ionic liquids

Figure 5. Schematic of 1-methoxyethyl-3-methylimidazolium ILs towards Daphnia Magna

Gathergood and co-workers further demonstrated that imidazolium based ionic liquids with an oxygen functionality i.e. ester and ether side chains, have reduced antimicrobial activity to a great extent. [89] Four Gram negative bacteria (*Pseudomonas aeruginosa, Escherichia coli, Klebsiella sp., Salmonella sp.*) and three Gram positive bacteria (*Staphylococcus aureus, Enterococcus sp., Bacillus subtilis*) were screened in the assay. A range of long ether and poly ether ester side chain imidazolium ionic liquids showed a huge reduction in the toxicity in this test system, compared with similar number of atoms in long alkyl side chains.

In order to check the toxicity effect of ionic liquids in humans, a cytotoxicity assay with human cell lines was designed. HeLa is one of the most extensively used cell lines in medicinal research. HeLa is a human tumor cell line, which is a prototype of epithelium. Due to the first contact of an organism with toxic materials, HeLa cell line has great importance. Stepnowski and co-workers reported the cytotoxic effect of imidazolium ionic liquids in Hela cell line. [90] The EC_{50} values of a range of 1-alkyl-3-methylimidazolium ionic liquids were evaluated on human epithelium HeLa cells. Ionic liquids with a decyl side chain with tetra-fluoroborate as the anion component demonstrated high toxicity (EC_{50} = 0.07 mM). This was supportive of the results with other test systems. The cytotoxicity of ionic liquids were compared with the known 50% effective concentrations (EC_{50} values) of traditional organic solvents such as dichloromethane (71.43 mM), phenol (42.68 mM), xylene (52.43 mM) and ethanol (1501.43 mM). These studies revealed that the tested ionic liquids had significant toxicity against human cell line HeLa, compared with organic solvents. Lu and co-workers utilised this assay for testing the cytotoxicity of a large range of ionic liquids containing imidazolium, pyridinium, choline, triethylammonium and phosphonium cations with halide, NTf_2^-, and BF_4^- anions. [91] In general, choline and alkyl-triethylammonium ionic liquids were found to be less toxic than their imidazolium and pyridinium salt counterparts.

In an effort to evaluate the eco(toxicity) of ionic liquids, Yun and co-workers reported an assay of freshwater microalgae *Selenastrum capricornutum*. [92] The bromide salts of commonly used 1-butyl-3-methylimidazolium, 1-butyl-3-methylpyridinium, 1-butyl-1-methylpyrrolidinium, tetrabutylammonium, and tetrabutylphosphonium ILs were tested against the *S. capricornutum* and compared with traditional water miscible organic solvents such as dimethylformamide, 2-propanol and methanol. Increase in the toxicity of imidazolium and pyridinium cations were observed with an increase in incubation time, whereas the opposite trend was found in the case of tetrabutylammonium, and tetrabutylphosphonium ILs. The growth inhibition of *S. capricornutum* was higher in ionic liquids than organic solvents. A similar test system was applied to investigate the toxicological effect of anions. [93] Toxicity of various anions incorporated with 1-butyl-3-methylimidazolium cation were compared with their respective sodium and potassium salts. The anions were found to inhibit the growth of freshwater algae *S. capricornutum*. The clear trend in algae toxicity was observed as hexafluoroantimonate (SbF_6^-) > hexafluorophosphate (PF_6^-) > tetrafluoroborate (BF_4^-) > triflate ($CF_3SO_3^-$) > octyl sulphate ($OctOSO_3^-$) > halide (Br^-, Cl^-). Toxicity studies (in fish, aquatic plants/invertebrates) on anionic surfactants have shown that toxicity is dependent on a number of factors such as alkyl chain length, solubility and stability in water. [94] As the length of alkyl chain increases, toxicity increases until certain limits. Further increase in chain length can decrease the hydrophilic nature of these materials, reducing bioavaibility of compound which results in a general decrease in the toxicity. [95]

3.2. Biodegradation of ionic liquids

Ionic liquids are well known for being stable to heating and in a variety of reaction conditions. Although this is an important property in their applications, it can raise issues regarding degradation and bioaccumulation when released in nature. Accumulated data on the anti-microbial toxicity of novel ionic liquids can be used as a preliminary guideline before performing the biodegradation tests. The biological test system has its limitations, such as when reported toxicity data is only available for certain individual organisms, whereas biodegradation assays usually have a large sample group of organisms. Also, breakdown products/intermediates of ionic liquids can be toxic, which can be resistant to further degradation, which leads to the issue of bioaccummulation. Hence it is important to perform biodegradation studies of ionic liquids. [96] Boethling and co-workers in their review article *"Designing Small Molecules for Biodegradability"*, gave useful and general guidelines for the design and synthesis of environmental friendly chemicals. [97] According to their observations, compounds containing unsubstituted alkyl chains, benzene rings, oxygen functionalities such as esters, aldehydes, and carboxylic acids (potential sites for enzymatic hydrolysis) greatly increase biodegradation. Whereas compounds containing halogens, branched chains, heterocycles, functional groups such as nitro, nitroso and arylamines motifs, adversely affect the biodegradation. There are several biodegradation study methods approved by the Organisation for Economic Cooperation and Development (OECD) (See Table 1).

Test No.	Name	Analytical method
OECD 301 A	DOC Die-Away	Dissolved organic carbon
OECD 301 B	CO_2 evolution	CO_2 evolution
OECD 301 C	MITI (Ministry of International Trade and Industry, Japan)	Oxygen consumption
OECD 301 D	Closed bottle	Dissolved oxygen
OECD 301 E	Modified OECD screening	Dissolved organic carbon
OECD 301 F	Manometric respirometry	Oxygen consumption
ISO 14593	CO_2 headspace test	CO_2 evolution
OECD 309	OECD 309	^{14}C labelling
ASTM 5988	ASTM 5988	CO_2 production / Biochemical oxygen demand

Table 1. Biodegradation methods in use

Data collected from all of the tests mentioned in Table 1 can be catagorised according to OECD guidelines as - (a) Ultimate biodegradation: Denotes complete degradation/utilisation of a test compound to produce carbon dioxide (CO_2), water, biomass and inorganic substances. Such biodegradation can achieved due to the mineralisation by microorganisms. This is one of the significant characteristics, before classed as a 'biocompatible' compound. (b) Readily biodegradable: These are positive results showing rapid ultimate degradation of the test compound under aerobic conditions in stringent screening tests. Both mineralisation and elimination/alteration (abiotic process such as hydrolysis, oxidation and photolysis) of the test substance can be observed. (c) Primary biodegradation: An elimination or alteration of the test sample by microorganisms, in order to lose its specific properties. [98]

Gathergood and Scammells reported the synthesis of ester and amide functionalised side chain imidazolium ionic liquids [99] according to the guidelines outlined by Dr. Boeth- ling. [100] All of the novel alkyl ester and amide side chain methylimidazolium ionic liq- uids were subjected to biodegradation studies. A biodegradation study of bromide salts of these ionic liquids along with commonly used [BMIM][BF$_4$] and [BMIM][PF$_6$] ILs was car- ried out under the 'Closed Bottle Test'   (OECD 301D) [101] although none of the tested ionic liquids passed the minimum 60% biodegradation threshold in order to be classified as 'readily biodegradable'. However, [BMIM][BF$_4$] and [BMIM][PF$_6$] did not show biode- gradation in the test system. Ester functionalised side chain ionic liquids demonstrated improved biodegradation. Increasing the length of the ester side chain increased the bio- degradation, for example a methyl ester derivative showed 17% biodegradation, whereas biodegradation of an octyl ester derivative was 32% after 28 days. Another important ob- servation was that amide side chain ionic liquids showed very negligible biodegradation. The effect of different anion on the rate of biodegradation was tested under the same test system. A range of 1-butyl-3-methylimidazolium and ester functionalised 3-methyl-1-(pro- poxycarbonylmethyl)-imidazolium ionic liquids were prepared with different anions such as Br$^-$, BF$_4^-$, PF$_6^-$, NTf$_2^-$, N(CN)$_2^-$ and OctOSO$_3^-$. BMIM$^+$ based ionic liquids showed poor biodegradation, except with an octyl sulphate as anion. The biodegradation of 3-methyl-1- (propoxycarbonylmethyl)-imidazolium ionic liquid was increased from 19% with bromide anion to 49% with octyl sulphate anion after 28 days. Gathergood and co-workers further demonstrated that incorporation of ether and polyether linkages along with ester function- ality in the side chain can increase biodegradation to a great extent. [89] The biodegrada- tion of a large range of ether and polyether ester side chain methylimidazolium ionic liquids were studied using the CO$_2$ Headspace test. Octyl sulphate salts of 1-methylimida- zolium ionic liquids with propoxyethoxy and butoxyethoxy esters were found to be readi- ly biodegradable (> 60% biodegradation in 28 days).

Docherty and co-workers reported biodegradation studies of imidazolium and pyridinium based ionic liquids by OECD dissolved organic carbon Die-Away test. The test was car- ried out with the activated sludge microorganisms from wastewater treatment plant. [102] Denaturing gradient gel electrophoresis (DGGE) was also used to investigate the microbi- al community profile. The results showed that the tested pyridinium based ionic liquids has better biodegradability than the corresponding imidazolium salts. Octyl-3-methyl-pyri- dinium bromide was found to be readily biodegradable with complete degradation within 15-25 days of incubation, whereas hexyl-3-methyl-pyridinium bromide was degraded within 40-50 days. Such complete biodegradation was further supported by the work of Stolte and co-workers in their investigation of the primary biodegradation of a variety of ionic liquids by the modified OECD 301 D test. [103] A range of 1-alkyl-3-methylimidazo-lium, 1-alkylpyridinium and 4-(dimethylamino)pyridinium halide salts were screened un-der the test system. The biodegradation products were identified by HPLC-MS analysis. 1-Octyl-3-methylimidazolium chloride gave a result of 100% primary biodegradation within 31 days. The stepwise degradation by two possible metabolic pathways was predicted based on HPLC-MS analysis. This predicted breakdown pathway was due to enzymatic oxidation of the terminal carbon.

Scammells and co-workers also reported biodegradation of ester functionalised pyridinium ionic liquids using the CO_2 Headspace test (ISO 14593). [104] Pyridinium ionic liquids with alkyl and ester side chains, along with ester derivatives of nicotinic acid with alkyl side chain were tested under aerobic conditions. The halide salts of pyridinium ionic liquids with alkyl side chain (C_4, C_{10} and C_{16}) showed poor biodegradation, whereas it was improved up to a max. of 45% after 28 days by the use of octyl sulfate anion. Switching from alkyl chain to ester side chain in pyridinium ILs increased the biodegradation dramatically. Such substitu-tion made them readily biodegradable, independent of the chosen anionic component. Bro-mide, hexafluorophosphate and octyl sulfate showed a high level of biodegradability (85% to 90%), whereas NTf_2^- gave 64% biodegradation after 28 days. Ionic liquids derived from the nicotinic acid ester derivative i.e. ester group at 3-position of the pyridine ring were found to be readily biodegradable even with methyl or butyl side chain. Amide side chain derivatives showed low biodegradability, even with using octyl sulfate as the anion (30% bi-odegradation after 28 days). Scammells and co-workers further studied the effect of the in-corporation of ester, ether and hydroxyl side chain in phosphonium based ionic liquids. [105] All phosphonium ionic liquids showed poor biodegradation independent of ester, ether and hydroxyl side chains and anions in CO_2 Headspace test. Only heptyl ester side chain ionic liquid with octyl sulfate anion showed highest 30% biodegradation after 28 days.

Most of the ionic liquids prepared are not 'readily biodegradable'; however, several struc-tural modifications have shown a positive improvement in the biodegradability of ionic liq-uids. [96,106] Striving towards compounds with 'Ultimate biodegradation' is preferred over 'readily biodegradable' examples and is a major research area. Hence it is important to study the biodegradation pathways of ionic liquids along with kinetics and metabolite stud-ies to assist the 'benign by design' approach.

3.3. Guidelines for designing 'Green' ionic liquid catalysts/solvents

From the literature, the following observations were made and are summarised in Figure 6:

- Linear alkyl chains in general can increase biodegradation compared to branched hydro-carbon chains.

- Oxygen containing functionalities, such as ester and hydroxyl groups in the side chain of imidazolium cation, not only reduces microbial toxicity but also increases rate of biode-gradation. This is, however, not effective in phosphonium based ionic liquids.

- Ether substitution reduces bactericidal toxicity.

• Ester substitutions at 1 and 3 position of pyridinium cation can improve biodegradation.

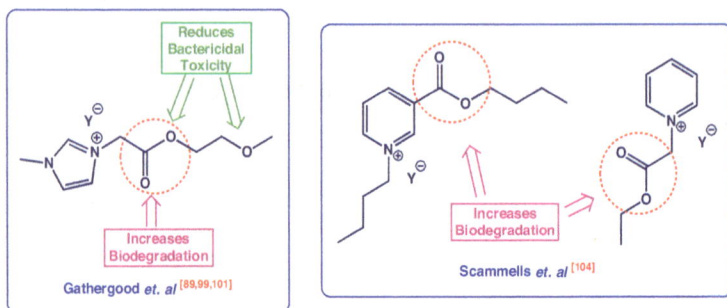

Figure 6. Guidelines for designing 'Green' ionic liquids (left, Gathergood and Scammells; right Scammells)

4. Green chemistry metrics

To achieve 'Green' synthesis of any chemical, the '*12 Principles of Green Chemistry*' given by Anastas and Warner serves as the most useful set of guidelines and gives us the important message that 'Prevention is better than cure'. [107] These principles suggest not only to consider toxicity and biodegradation, but also to measure the 'greenness' of the chemical process. To evaluate the 'greenness' of any process, a number of factors associated with the chemical process has to studied. Green Chemistry metrics can help to measure the efficiency and 'greenness' of the chemical process. There are several well established methods to determine the sustainability of a chemical process under areas such as the economical, technical and social effects of such processes. Economical methods mainly consist of profit related analysis, whereas technical methods analyses quality, productivity and related issues. Social methods concern the society and environmental aspect of the chemical process. [108] Porteous has shown that it is easy to correlate Green Chemistry metrics with the "12 Principles of Green chemistry". [109] There are several metrics available to measure efficiency, use of energy and resources, toxicity, biodegradation, safety and life cycle impact of the chemical process, which are closely related to the 12 Principles.

The Environmental (E) factor is one of the most widely used and efficient methods to measure the amount of waste generated in the process. [110] This is also known as Sheldon's E-factor.

$$\text{E factor} = \frac{\text{Total mass of waste (kg)}}{\text{Mass of product (kg)}}$$

E-factor calculations is one of the important methods to calculate the waste associated with the process, which includes all unwanted side products, reagents, solvents and energy. Water used in the process is usually excluded from the calculations. The higher the E-factor value, the higher the waste generated which has an adverse effect on environment. Economically this adds on to the profit and the cost of disposal. Many modifications on the measurement of E-factor were adopted in industry. GlaxoSmithKline (GSK) has introduced a concept of 'Mass Intensity' based on Sheldon's E-factor. [111]

$$\text{Mass Intensity} = \frac{\text{Mass all materials used (excluding water)}}{\text{Mass of product}}$$

Mass Intensity measures the amount of reagents, solvents and workup reagents used in the process. Hence it takes an account of product yields and stoichiometry of the reagents. In order to measure the synthetic efficiency of the process, the "Atom Economy" concept was found to be useful. Atom economy in chemical reactions is one of the 12 Principles of Green Chemistry, which gives indications as to the overall efficiency of a chemical reaction. [112]

$$\text{Atom Economy} = \frac{\text{Molecular weight of desired product}}{\text{Molecular weight of all products / reactants}} \times 100\%$$

Although this is an important concept, which suggests to minimise the waste. It does not consider the actual mass, yield, solvents and other reagents used in the process. Reaction mass efficiency (RME) has overcome these drawbacks. [111] Reaction mass efficiency considers actual mass of reactants and product. This is one of the commonly used metrics to evaluate the efficiency of the chemical process.

$$\text{Reaction Mass Efficiency} = \frac{\text{Mass of product}}{\text{Mass of all reactants}} \times 100\%$$

Apart from these, real time analysis is also important to analyse the chemical process. Real time analysis is again one of the "12 Principles of Green Chemistry", which enables the chemist to identify the formation of waste along the process. A number of analytical techniques such as HPLC, GC, NMR, FT-IR, and sensors etc. were already found to be useful in real time analysis. It's also important to determine the robustness of the process, which will allow preparing chemicals on a large scale. The measurement of toxicity and biodegradation are also important metrics to evaluate 'greenness' of the chemical products.

5. Case study

In order to design and synthesise 'Green' ionic liquid catalysts, guidelines prepared from toxicity and biodegradation studies can be helpful. Novel ionic liquids have been extensively prepared and used in organic synthetic applications, but only few research groups have published complementary toxicity and biodegradation data to support its environmental impact. [69,104,113]

Connon and co-workers reported an application of pyridinium salts as an effective catalyst in acetalisation reactions. [114-116] Ester group substitution at either the 3 or both 3 and 5 positions of the pyridinium ring displayed excellent catalytic activity with very low catalyst loading in the acetalisation of benzaldehyde with methanol. The remarkable feature of the catalyst was that it is not acidic in nature, but can act as a Brønsted acid in the presence of protic media. The highest catalytic activity pyridinium salts, 3,5-bis(ethoxycarbonyl)-1-(phenylmethyl) bromide, showed excellent catalytic activity in the protection of a variety of aldehydes with methanol. The catalyst was also found to be useful in diol and dithiol protections (Scheme 14). Catalytic activity of the catalyst was predicted based on anticipated nucleophilic attack of the alcohol to the pyridinium to generate the Brønsted acidic active species.

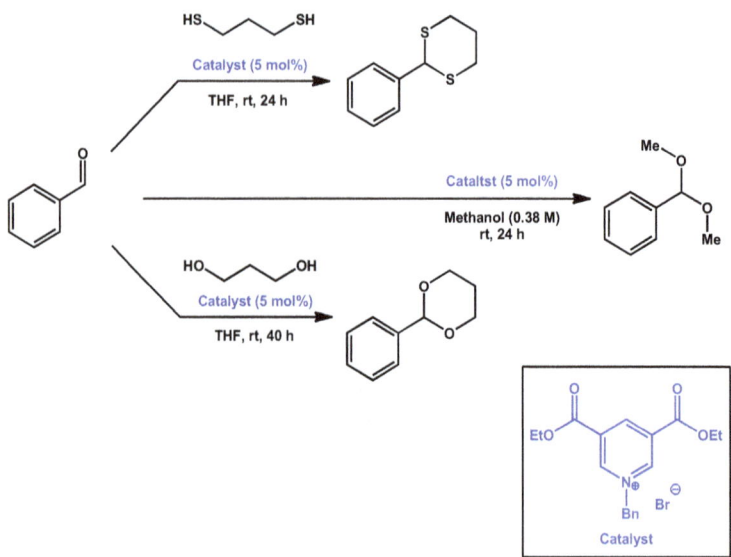

Scheme 14. Acetalisation of benzaldehyde with methanol catalysed by pyridinium IL

On the basis of previous findings, Gathergood, Connon and co-workers reported the design, synthesis and application of imidazolium ionic liquid catalysts in acetalisation reactions. [113] A range of ester and amide side chain imidazolium ionic liquids was prepared. The catalytic activity of all of these imidazolium salts was evaluated in the acetalisation of benzaldehyde in methanol (Scheme 15). The absence of catalyst resulted in no formation of the corresponding acetal after 24 hours. All of the imidazolium bromide salts showed poor catalytic activity independent of ester or amide side chain (9 to 13% conversions). When switched to the NTf$_2$ anion, reaction conversions in amide side chain ionic liquids were increased marginally. Whereas ester side chain ionic liquids with NTf$_2$ anion gave 51% conversion. A tetrafluoroborate salt of a methyl ester side chain imidazolium ionic liquid gave high (85%) conversion towards required product. Hexafluorophosphate and octyl sulfate anions performed poorly in this reaction. The anion exchange from bromide to tetrafluoroborate had greatly influenced the acetalisation reaction of benzaldehyde with methanol, which gave 85% conversion.

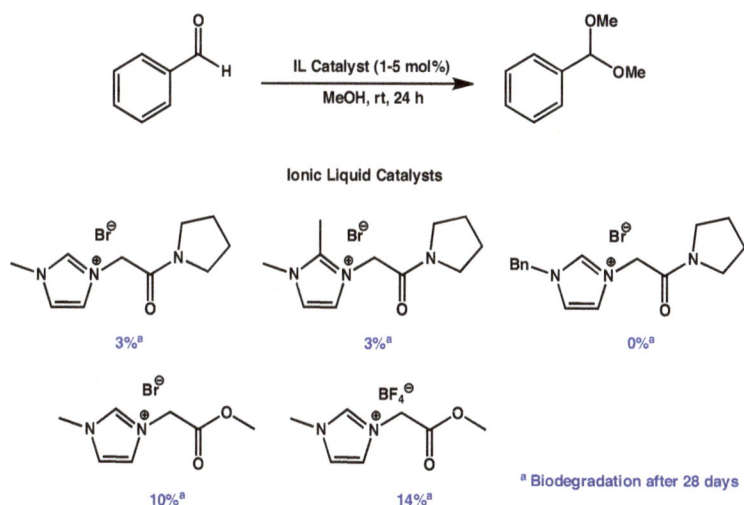

Scheme 15. Acetalisation of benzaldehyde using imidazolium IL as a catalyst. Biodegradation data for IL/Catalyst (ISO 14593) included.

The ionic liquid did not appear to have an acidic nature, but in the presence of a protic medium was proposed to generate a Brønsted acid species to catalyse the acetalisation reaction. (Figure 7) The most active catalyst with tetrafluoroborate anion was further exploited in the acetalisation reactions of a variety of aldehydes with methanol at room temperature. These reactions showed good to excellent conversions with 5-10% catalyst loading. When saturated aldehyde such as 3-phenylpropanal reacted with deuterated methanol in presence of 1 mol% catalyst, the reaction gave quantitative conversion in only 1 minute. Diol and dithiol protection of benzaldehyde showed very good results. The BF$_4$ catalyst promoted protection of benzaldehyde with 1,2-ethanedithiol gave 92% conversion, whereas 1,3-propanedithiol and 1,3-propanediol gave 65% and 86% conversion respectively. Recyclability of the most active BF$_4$ anion catalyst was performed on the 1,3-dithiolane protection of benzaldehyde. The catalyst was recycled and reused 15 times without significant loss of activity.

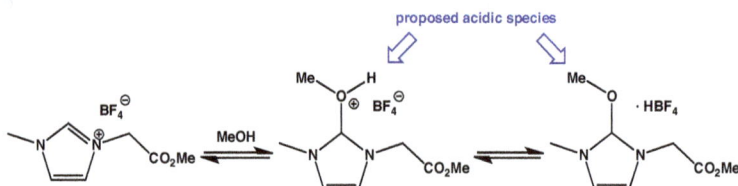

Figure 7. Proposed mode of action of catalytic aprotic imidazolium ions

Biodegradation studies of pyridinium-based ILs have shown that esters substitution at either the 1 or 3-position have a beneficial effect on degradation of the heterocyclic core, independent of the anion. [104] Also biodegradation studies in the literature have shown that only the side chain of the imidazolium ionic liquids undergo degradation, whereas imidazole core was found to persist in most of the OECD tests. [103]

Biodegradation studies of ester and amide side chain ionic liquids along with substituted imidazolium salts was also carried out by using the "CO_2 Headspace" test (ISO 14593) (Scheme 14, 15). [113, 117] All imidazolium ionic liquids prepared failed to pass the minimum 60% biodegradation threshold value in order to be classified as 'readily biodegradable'. Ester functionalised ionic liquids displayed higher biodegradation levels than amide functionalised ionic liquids in 28 days. The first generation imidazolium ionic liquid catalysts showed 10% to 14% biodegradation for methyl ester side chain ionic liquids, where as maximum 3% biodegradation was observed in amide side chain ionic liquids after 28 days.

The toxicity of all ionic liquids was tested in an environmental and medicinally significant microbial assay including 12 fungal and 8 bacterial strains. [113,118] *In vitro* antifungal activities of the compounds were evaluated on a panel of four ATCC strains (*Candida albicans* ATCC 44859, *Candida albicans* ATCC 90028, *Candida parapsilosis* ATCC 22019, *Candida krusei* ATCC 6258) and eight clinical isolates of yeasts (*Candida krusei* E28, *Candida tropicalis* 156, *Candida glabrata* 20/I, *Candida lusitaniae* 2446/I, *Trichosporon asahii* 1188) and filamentous fungi (*Aspergillus fumigatus* 231, *Absidia corymbifera* 272, *Trichophyton mentagrophytes* 445). Whereas *In vitro* antibacterial activities of the compounds were evaluated on a panel of three ATCC strains (*Staphylococcus aureus* ATCC 6538, *Escherichia coli* ATCC 8739, *Pseudomonas aeruginosa* ATCC 9027) and five clinical isolates (*Staphylococcus aureus* MRSA HK5996/08, *Staphylococcus epidermidis* HK6966/08, *Enterococcus* sp. HK14365/08, *Klebsiella pneumoniae* HK11750/08, *Klebsiella pneumoniae* ESBL HK14368/08). All ester and amide side chain ionic liquids shown in Scheme 15 were non-toxic up to the 2000 μM concentration. This is less toxic than values reported for antimicrobial QAC (IC_{95} >10 μM). [80,81]

In order to evaluate the 'greenness' of the synthesis of ionic liquid catalysts, Gathergood and Connon have also applied some important Green Chemistry metrics (Section 4), such as -

- Sheldon E-factor
- GSK Reaction Mass Efficiency,
- Andraos Reaction Mass Efficiency
- Atom economy
- 1 / stoichiom. factor (excess reagents)

These metrics assisted improvements in the synthetic process, by reducing amount of solvents in the work-up and purification procedure, reduce number of steps to make required compounds. For example, as the synthesis of tetrafluoroborate ionic liquids involves preparation of halide salts followed by anion exchange metathesis. In an effort to reduce the number of steps and amount of reagents and solvents in the synthesis, alkyl imidazoles were directly reacted to Meerwein's salt i.e. trimethyloxonium tetrafluoroborate to give tetrafluoroborate ionic liquids in excellent yield. (For full analysis see ref. 117) Hence Green Chemistry metrics assessment is important to achieve green synthesis.

6. Conclusion

In this chapter we have demonstrated that ionic liquids have great potential and versatility in organic synthesis, with the dualistic ability to act as a solvent and as a catalyst. Ionic liquids were found to be possible replacements over traditional volatile organic solvents. Such ionic liquid solvents were found to be useful in transition metal catalysed reactions. They not only enabled catalyst immobilisation, but also increased recyclability of expensive transition metal catalysts. Ease of product separation and their stability against a variety of reagents has proved their important characteristics. We have seen that modification in the cationic part of the ionic liquids, according to the requirements of the aforementioned reactions, enabled them to act as organocatalysts or ligands for transition metal catalysts. These ionic liquid catalysts have shown comparative catalytic activity against known organocatalysts but such materials had a distinct advantage in that the ionic liquids could be recycled, with no discernible loss of activity. In order to increase recyclability, ionic liquid catalysts/ ligands were grafted onto a solid/ polymer support. These modifications helped to separate ionic liquid catalysts from the reaction mixture.

We have also illustrated the efforts attempted by the scientific community to evaluate the 'greenness' of ionic liquids, by using toxicity and biodegradation methods. The majority of ionic liquids are non-natural molecules, hence it's important to check their biocompatibility. Toxicity studies can serve as a "first post" primary evaluation of biodegradation. A variety of test systems including fungi, bacteria, algae, enzymes, rat cell line, human cell line and fish etc. were implemented to check the toxicity of ionic liquids. Most of these test systems have shown that toxicity of the ionic liquids comes from the cationic component. Important observations from such test systems also included that (a) long alkyl chains increase the toxicity with increase in the length of hydrocarbons and (b) incorporation of oxygen functionalities (ether, ester, and hydroxyl etc.) reduces the toxicity.

A number of Organisation for Economic Cooperation and Development (OECD) tests were found to be useful in the estimation of biodegradation. These tests mainly involve calculation of CO_2 evolution and oxygen consumed. These tests showed that most of the 1,3-dialkyl imidazolium ionic liquids are non-biodegradable, in most of the test systems. Although the alkyl side chain can undergo degradation, the imidazole core can still persist during biodegradation studies. Introduction of oxygen functionality such as ether / ester, either in the side chain of the imidazolium cation or at the C1 or C3 position of the pyridinium cation, were found to increase the rate of biodegradation.

In the case study, we have demonstrated the schematic approach towards the design and synthesis of ionic liquid catalysts for acetalisation reactions and evaluation of the biocompatibility of such catalysts, by using toxicity and biodegradation methods. According to the guidelines laid by the toxicity and biodegradation testing, ionic liquid catalysts were designed and prepared. Such ester and amide side chain imidazolium catalysts were found to be useful in acetalisation and thioacetalisation reactions. These catalysts have shown low toxicity against a variety of fungal and bacterial strains, but poor biodegradability in the "CO_2 Headspace Test". Further modifications in the cationic part are underway with the goal of increasing biodegradation and catalytic activity. Although biodegradation was not improved by such modifications, the catalytic activity was modified and increased to a great extent. Green Chemistry metrics had given useful information about the 'greenness' of the synthetic route. Which allowed to modify the process by reducing the number of steps, amount of solvents, etc. This case study was a clear example of the importance, in the design of organocatalysts, of what the potential environmental impact of such compounds might be, and why it is important to understand that a truly "green" catalyst needs to possess a balance of both activity and eco-friendliness.

Acknowledgements

The authors wish to thank Enterprise Ireland (EI), the Irish Research Council for Science, Engineering and Technology (IRCSET) and the Environmental Protection Agency (EPA) in Ireland for funding green chemistry research in Nicholas Gathergood's group. The case study was funded by EPA STRIVE project 2008-ET-MS-6-S2 (Rohitkumar Gore & Nicholas Gathergood (DCU) and Lauren Myles & Stephen Connon (TCD)). We also thank Alan Coughlan for assistance with proof-reading.

Author details

Rohitkumar G. Gore and Nicholas Gathergood*

*Address all correspondence to: Nick.Gathergood@dcu.ie

School of Chemical Sciences and National Institute for Cellular Biotechnology, Dublin City University, Glasnevin, Dublin, Ireland

References

[1] Rogers R. D. and Seddon K. R. (eds.), Ionic Liquids: Industrial Applications for Green Chemistry, ACS Symposium Series 818, American Chemical Society, USA, 2002

[2] Plechkova N. V. and Seddon K. R. Applications of ionic liquids in the chemical industries. Chemical Society Reviews, 2008; 37, 123-150

[3] Wasserscheid P. and Stark A. (eds.), Handbook of Green Chemistry, Volume 6: Ionic Liquids, Wiley-VCH, 2010

[4] Wilkes J. S. A short history of ionic liquids - from molten salts to neoteric solvents. Green Chemistry, 2002; 4, 73-80

[5] Seddon K. R. Ionic Liquids for Clean Technology. Journal of Chemical Technology & Biotechnology, 1997; 68, 351-356

[6] Marsh K. N., Boxall J. A., Lichtenthaler R. Room temperature ionic liquids and their mixtures - a review, Fluid Phase Equilibria, 2004; 219, 93-98

[7] McFarlane J., Ridenour W. B., Luo H., Hunt R. D., Depaoli D. W., Ren R. X. Room Temperature Ionic Liquids for Separating Organics from Produced Water. Separation Science and Technology, 2005; 40, 1245-1265

[8] Sheldon R. A., Green solvents for sustainable organic synthesis: state of the art. Green Chemistry, 2005; 7, 267-278

[9] Freemantle M. Designer solvents - Ionic liquids may boost clean technology development. Chemical English News 76, (30th March), 1998, 32-37

[10] Hallett J. P., Welton T. Room-temperature ionic liquids: Solvents for synthesis and catalysis. Chemical Reviews (Washington, DC, United States), 2011; 111(5), 3508-3576

[11] Hubbard C. D., Illner P., Eldik R. Understanding chemical reaction mechanisms in ionic liquids: successes and challenges. Chemical Society Reviews, 2011; 40, 272-290

[12] Wasserscheid P., Joni J. Green organic synthesis in ionic liquids, in Wasserscheid P., Stark A. (eds.), Handbook of Green Chemistry, Volume 6: Ionic Liquids, Wiley-VCH, Weinheim, Germany 2010; 41-63

[13] Chowdhury S. M., Ram S., Scott J. L. Reactivity of ionic liquids. Tetrahedron, 2007; 63(11), 2363-2389

[14] Stark A. and Seddon K. R. Ionic Liquids, in Kirk-Othmer Encyclopaedia of Chemical Technology, 5th Edit., Ed. A. Seidel, Vol. 26 (John Wiley & Sons, Inc., Hoboken, New Jersey) 2007; 836-920

[15] Opallo M., Lesniewski A. A review on electrodes modified with ionic liquids. Journal of Electroanalytical Chemistry, 2011; 656, 2-16

[16] Shiddiky M. J. A., Torriero A. A. J. Application of ionic liquids in electrochemical sensing systems. Biosensors & Bioelectronics, 2011; 26(5), 1775-1787

[17] Liu H., Liu Y., Li J., Ionic liquids in surface electrochemistry. Physical Chemistry Chemical Physics, 2010; 12(8), 1685-1697

[18] Pitner W. R., Kirsch P., Kawata K., Shinohara H., Applications of Ionic Liquids in Electrolyte Systems, in Wasserscheid P., Stark A. (eds.), Handbook of Green Chemistry, Volume 6: Ionic Liquids, Wiley-VCH, Weinheim, Germany 2010; 191-201

[19] Buzzeo M. C., Evans R. G., Compton R. G. Non-Haloaluminate Room-Temperature Ionic Liquids in Electrochemistry-A Review. ChemPhysChem, 2004; 5, 1106-1120

[20] Poole C. F., Poole S. K. Ionic liquid stationary phases for gas chromatography. Journal of Separation Science, 2011; 34(8), 888-900

[21] Ho T. D., Canestraro A. J., Anderson J. L. Ionic liquids in solid-phase microextraction: A review. Analytica Chimica Acta, 2011; 695, 18-43

[22] Sun P., Armstrong D. W. Ionic liquids in analytical chemistry. Analytica Chimica Acta, 2010; 661(1), 1-16

[23] Pandey S. Analytical applications of room-temperature ionic liquids: A review of recent efforts. Analytica Chimica Acta, 2006; 556, 38-45

[24] Koel M. Ionic liquids in chemical analysis. Critical Reviews in Analalytical Chemistry, 2005; 35, 177-192

[25] Dominguez de Maria P., Maugeri Z. Ionic Liquids in Biotransformations: From proof-of-concept to emerging deep-eutectic-solvents. Current Opinion in Chemical Biology, 2011; 15(2), 220-225

[26] Klingshirn M. A., Rogers R. D., Shaughnessy K. H. Palladium-catalyzed hydroesterification of styrene derivatives in the presence of ionic liquids. Journal of Organometallic Chemistry, 2005; 690, 3620-3626

[27] Mizushima E., Hayashi T., Tanaka M. Palladium-catalysed carbonylation of aryl halides in ionic liquid media: high catalyst stability and significant rate-enhancement in alkoxycarbonylation. Green Chemistry, 2001; 3, 76-79

[28] Earle M. J., McCormac P. B., Seddon K. R. Diels-Alder reactions in ionic liquids. A safe recyclable alternative to lithium perchlorate–diethyl ether mixtures. Green Chemistry, 1999; 1, 23-25

[29] Vijayaraghavan R., MacFarlane D. R. Charge Transfer Polymerization in Ionic Liquids. Australian Journal of Chemistry, 2004; 57, 129-133

[30] Rosa J. N., Afonso C. A. M., Santos A. G. Ionic liquids as a recyclable reaction medium for the Baylis-Hillman reaction. Tetrahedron, 2001; 57, 4189-4193

[31] Yadav J. S., Reddy B. V. S., Baishya G., Reddy K. V., Narsaiah A. V. Conjugate addition of indoles to α,β-unsaturated ketones using Cu(OTf)$_2$ immobilized in ionic liquids. Tetrahedron, 2005; 61, 9541-9544

[32] Johansson M., Linden A. A., Baeckvall J.-E. Osmium-catalyzed dihydroxylation of alkenes by H$_2$O$_2$ in room temperature ionic liquid co-catalyzed by VO(acac)$_2$ or MeReO$_3$. Journal of Organometallic Chemistry, 2005; 690, 3614-3619

[33] Serbanovic A., Branco L. C., Nunes da Ponte M., Afonso C. A. M. Osmium catalyzed asymmetric dihydroxylation of methyl trans-cinnamate in ionic liquids, followed by supercritical CO$_2$ product recovery. Journal of Organometallic Chemistry, 2005; 690, 3600-3608

[34] Picquet M., Stutzmann S., Tkatchenko I., Tommasi I., Zimmermann J., Wasserscheid P. Selective palladium-catalysed dimerisation of methyl acrylate in ionic liquids: towards a continuous process. Green Chemistry, 2003; 5, 153-162

[35] Forsyth S. A., Gunaratne H. Q. N., Hardacre C., McKeown A., Rooney D. W., Seddon K. R. Utilisation of ionic liquid solvents for the synthesis of Lily-of-the-Valley fragrance {β-Lilial®; 3-(4-t-butylphenyl)-2-methylpropanal}. Journal of Molecular Catalysis A: Chemical, 2005; 231, 61-66

[36] Reetz M. T., Wiesenhoefer W., Francio G., Leitner W. Biocatalysis in ionic liquids: batchwise and continuous flow processes using supercritical carbon dioxide as the mobile phase. Chemical Communications, 2002; 992-993

[37] Heck R. F. Palladium-catalyzed reactions of organic halides with olefins. Accounts of Chemical Research, 1979; 12, 146-151

[38] Mizoroki T., Mori K., Ozaki A. Arylation of Olefin with Aryl Iodide Catalyzed by Palladium. Bulletin of the Chemical Society of Japan, 1971; 44, 581

[39] Kaufmann D. E., Nouroozian M., Henze H. Molten Salts as an Efficient Medium for Palladium Catalyzed C-C Coupling Reactions. Synlett, 1996; 11, 1091-1092

[40] Bellina F. and Chiappe C. The Heck Reaction in Ionic Liquids: Progress and Challenges. Molecules, 2010; 15, 2211-2245

[41] Wang L., Li H., Li P. Task-specific ionic liquid as base, ligand and reaction medium for the palladium-catalyzed Heck reaction. Tetrahedron, 2009; 65, 364-368

[42] Xu L., Chen W., Xiao J. Heck Reaction in Ionic Liquids and the in Situ Identification of N-Heterocyclic Carbene Complexes of Palladium. Journal of Organometallic Chemistry, 2000; 19, 1123-1127

[43] Carmichael A. J., Earle M. J., Holbrey J. D., McCormac P. B., Seddon K. R. The Heck Reaction in Ionic Liquids: A Multiphasic Catalyst System. Organic Letters, 1999; 1(7), 997-1000

[44] Deshmukh R. R., Rajagopal R., Srinivasan K. V. Ultrasound promoted C-C bond for-
      mation: Heck reaction at ambient conditions in room temperature ionic liquids.
      Chemical Communications, 2001; 17, 1544-1545

[45] Crudden C. M., Sateesh M., Lewis R. Mercaptopropyl-Modified Mesoporous Silica:
      A Remarkable Support for the Preparation of a Reusable, Heterogeneous Palladium
      Catalyst for Coupling Reactions. Journal of American Chemical Society, 2005; 127,
      10045-10050

[46] Ma X., Zhou Y., Zhang J., Zhu A., Jiang T., Han B. Solvent-free Heck reaction cata-
      lyzed by a recyclable Pd catalyst supported on SBA-15 via an ionic liquid. Green
      Chemistry, 2008; 10, 59-66

[47] Liu G., Hou M., Song J., Jiang T., Fan H., Zhang Z., Han B. Immobilization of Pd
      nanoparticles with functional ionic liquid grafted onto cross-linked polymer for sol-
      vent-free Heck reaction. Green Chemistry, 2010; 12, 65-69

[48] Sonogashira K., Tohda Y., Hagihara N. A convenient synthesis of acetylenes: catalyt-
      ic substitutions of acetylenic hydrogen with bromoalkenes, iodoarenes and bromo-
      pyridines. Tetrahedron Letters, 1975; 16, 4467-4470

[49] Glaser C. Beiträge zur Kenntniss des Acetenylbenzols, Chemische Berichte, 1869; 2,
      422-424

[50] Hay A. S., Oxidative Coupling of Acetylenes and Systems Containing a Silicon-Oxy-
      gen-Vanadium Linkage. Journal of Organic Chemistry, 1962; 27, 3320-3323

[51] Rossi R., Carpita A., Begelli C. A palladium-promoted route to 3-alkyl-4-(1-alkynyl)-
      hexa-1,5-dyn-3-enes and/or 1,3-diynes. Tetrahedron Letters, 1985; 26, 523-526

[52] Liu Q., Burton D. J. A facile synthesis of diynes. Tetrahedron Letters, 1997; 38,
      4371-4374

[53] For a review of alkyne coupling, see: Siemsen P., Livingston R. C., Diederich F. Ace-
      tylenic Coupling: A Powerful Tool in Molecular Construction. Angewandte Chemie
      International Edition (English), 2000, 39, 2632-2657

[54] Fukuyama T., Shinmen M., Nishitani S., Sato M., Ryu I. A Copper-Free Sonogashira
      Coupling Reaction in Ionic Liquids and Its Application to a Microflow System for Ef-
      ficient Catalyst Recycling, Organic Letters, 2002; 4(10), 1691-1694

[55] Zhang J., Dakovic M., Popovic Z., Wu H., Liu Y. A functionalized ionic liquid con-
      taining phosphine-ligated palladium complex for the Sonogashira reactions under
      aerobic and CuI-free conditions. Catalysis Communications, 2012; 17, 160-163

[56] Miyaura N., Yamada K., Suzuki A. A new stereospecific cross-coupling by the palla-
      dium-catalyzed reaction of 1-alkenylboranes with 1-alkenyl or 1-alkynyl halides. Tet-
      rahedron Letters, 1979; 20(36), 3437-3440

[57] Miyaura N. and Suzuki A. Stereoselective Synthesis of Arylated (E) -Alkenes by the Reaction of Alk-1-enylboranes with Aryl Halides in the Presence of Palladium Catalyst. Journal of Chemical Society, Chemical Communications, 1979; 866-867

[58] Miyura N. and Suzuki A., Palladium-Catalyzed Cross-Coupling Reactions of Organoboron Compounds. Chemical Reviews, 1995; 95, 2457-2493

[59] Wei J., Jiao J., Feng J., Lv J., Zhang X., Shi X., Chen Z. PdEDTA Held in an Ionic Liquid Brush as a Highly Efficient and Reusable Catalyst for Suzuki Reactions in Water. Journal of Organic Chemistry, 2009; 74, 6283-6286

[60] Lombardo M., Chiarucci M., Trombini C. A recyclable triethylammonium ion-tagged diphenylphosphine palladium complex for the Suzuki–Miyaura reaction in ionic liquids. Green Chemistry, 2009; 11, 574-579

[61] Papagni A., Trombini C., Lombardo M., Bergantin S., Chams A., Chiarucci M., Miozzo L., Parravicini M. Cross-Coupling of 5,11-Dibromotetracene Catalyzed by a Triethylammonium Ion Tagged Diphenylphosphine Palladium Complex in Ionic Liquids. Organometallics, 2011; 30, 4325-4329

[62] Stille J. K. Angewandte Chemie, 1986; 98, 504-519

[63] Stille J. K. The Palladium-Catalyzed Cross-Coupling Reactions of Organotin Reagents with Organic Electrophiles [New Synthetic Methods (58)]. Angewandte Chemie International Edition (English), 1986; 25, 508-524

[64] Handy S. T. and Zhang X. Organic Synthesis in Ionic Liquids: The Stille Coupling. Organic Letters, 2001; 3(2), 233-236

[65] Louaisil N., Pham P. D., Boeda F., Faye D., Castanet A.-S., Legoupy S. Ionic Liquid Supported Organotin Reagents: Green Tools for Stille Cross-Coupling Reactions with Brominated Substrates. European Journal of Organic Chemistry, 2011; 1, 143-149

[66] Diels O., Alder K. Synthesen in der hydroaromatischen Reihe. Justus Liebigs Annalen der Chemie, 1928; 460, 98-122

[67] Pégot B., Vo-Thanh G. Ionic Liquid Promoted Aza-Diels-Alder Reaction of Danishefsky's Diene with Imines. Synlett, 2005; 9, 1409-1412

[68] Zhou Z., Li Z., Hao X., Dong X., Li X., Dai L., Liu Y., Zhang J., Huang H., Li X., Wang J. Recyclable copper catalysts based on imidazolium-tagged C2-symmetric bis(oxazoline) and their application in D-A reactions in ionic liquids. Green Chemistry, 2011; 13, 2963-2971

[69] Evans D. A., Miller S. J., Lectka T., von Matt P. Chiral Bis(oxazoline)copper(II) Complexes as Lewis Acid Catalysts for the Enantioselective Diels-Alder Reaction. Journal of American Chemical Society, 1999; 121, 7559-7573

[70] Greene T. W. Protective groups in Organic Synthesis, Wiley-Interscience, New York, 1981, p. 178

[71] Bornstein J., Bedell S. F., Drummond P. E., Kosoloki C. F. Alkylations of Diphenyla-
      cetonitrile with Certain Halides by Potassium Amide in Liquid Ammonia. Dehydro-
      cyanations of Polyphenyl Nitriles to Form Olefins. Journal of American Chemical
      Society, 1956; 78, 83-86

[72] McKinzie C. A., Stocker J. H. Preparation of ketals. A Reaction mechanism. Journal of
      Organic Chemistry, 1955; 20, 1695-1701

[73] Cole A. C., Jensen J. L., Ntai I., Tran T., Weaver K. J., Forbes D. C., Davis Jr. J. H. Nov-
      el Brønsted Acidic Ionic Liquids and Their Use as Dual Solvent-Catalysts. Journal of
      American Chemical Society, 2002; 124, 5962-5963

[74] Fang D., Gong K., Shi Q., Liu Z. A green procedure for the protection of carbonyls
      catalyzed by novel task-specific room-temperature ionic liquid. Catalysis Communi-
      cations, 2007; 8, 1463-1466

[75] Du Y. and Tian F. Brønsted Acidic Ionic Liquids as Efficient and Recyclable Catalysts
      for Protection of Carbonyls to Acetals and Ketals Under Mild Conditions. Synthetic
      Communications, 2005; 35, 2703-2708

[76] Anthony J. L., Maginn E. J., Brennecke J. F. Solution thermodynamics of imidazoli-
      um-based ionic liquids and water. Journal of Physical Chemistry B, 2001, 105,
      10942-10949

[77] Wong D. S. H., Chen J. P., Chang J. M., Chou C. H. Experimental study on the trans-
      port properties of fluorinated ethers Fluid. Journal of Phase Equilibria and Diffusion,
      2002; 194-197, 1089-1095

[78] Wood N. and Stephens G. Accelerating the discovery of biocompatible ionic liquids.
      Physical Chemistry Chemical Physics, 2010; 12, 1670-1674

[79] Alfonsi K., Colberg J., Dunn P. J., Fevig T., Jennings S., Johnson T. A., Kleine H. P.,
      Knight C., Nagy M. A., Perry D. A., Stefaniak M., Green chemistry tools to influence
      a medicinal chemistry and research chemistry based organisation. Green Chemistry,
      2008; 10, 31-36

[80] Matzke M., Stolte S., Thiele K., Juffernholz T., Arning J., Ranke J., Welz-Biermann U.,
      Jastorff B. The influence of anion species on the toxicity of 1-alkyl-3-methylimidazoli-
      um ionic liquids observed in an (eco)toxicological test battery. Green Chemistry,
      2007; 9, 1198-1207

[81] Pham T. P. T., Cho C.-W., Yun Y.-S. Environmental fate and toxicity of ionic liquids:
      A review. Water Research, 2010; 44, 352-372

[82] Stock F., Hoffmann J., Ranke J., Störmann R., Ondruschka B., Jastorff B. Effects of ion-
      ic liquids on the acetylcholinesterase - a structure–activity relationship consideration.
      Green Chemistry, 2004; 6, 286-290

[83] Couling D. J., Bernot R. J., Docherty K. M., Dixon J. K., Maginn E. J. Assessing the factors responsible for ionic liquid toxicity to aquatic organisms via quantitative structure–property relationship modeling. Green Chemistry, 2006; 8, 82-90

[84] Ranke J., Mölter K., Stock F., Bottin-Weber U., Poczobutt J., Hoffmann J., Ondruschka B., Filser J., Jastorff B. Biological effects of imidazolium ionic liquids with varying chain lengths in acute Vibrio fischeri and WST-1 cell viability assays. Ecotoxicology and Environmental Safety, 2004; 58, 396-404

[85] Stolte S., Arning J., Bottin-Weber U., Matzke M., Stock F., Thiele K., Uerdingen M., Welz-Biermann U., Jastorff B., Ranke J. Anion effects on the cytotoxicity of ionic liquids. Green Chemistry, 2006; 8, 621-629

[86] Bernot R. J., Brueseke M. A., Evans-White M. A., Lamberti G. A. Acute and chronic toxicity of imidazolium-based ionic liquids on Daphnia magna. Environmental Toxicology and Chemistry, 2005; 21, 87-92

[87] Yu M., Wang S.-H., Luo Y.-R., Han Y.-W., Li X.-Y., Zhang B.-J., Wang J.-J. Effects of the 1-alkyl-3-methylimidazolium bromide ionic liquids on the antioxidant defense system of Daphnia magna. Ecotoxicology and Environmental Safety, 2009; 72, 1798-1804

[88] Samorì C., Pasteris A., Galletti P., Tagliavini E. Acute toxicity of oxygenated and non-oxygenated imidazolium-based ionic liquids to Daphnia magna and Vibrio fischeri, Environmental Toxicology and Chemistry, 2007; 26, 2379-2382

[89] Morrissey S., Pegot B., Coleman D., Garcia M. T., Ferguson D., Quilty B., Gathergood N. Biodegradable, non-bactericidal oxygen-functionalised imidazolium esters: A step towards 'greener' ionic liquids. Green Chemistry, 2009; 11, 475-483

[90] Stepnowski P., Składanowski A. C., Ludwiczak A., Łaczyńska E. Evaluating the cytotoxicity of ionic liquids using human cell line HeLa. Human & Experimental Toxicology, 2004; 23, 513-517

[91] Wang X., Ohlin C. A., Lu Q., Fei Z., Hu J., Dyson P. J. Cytotoxicity of ionic liquids and precursor compounds towards human cell line HeLa. Green Chemistry, 2007; 9, 1191-1197

[92] Cho C.-W., Jeon Y.-C., Pham T. P. T., Vijayaraghavan K., Yun Y.-S. The ecotoxicity of ionic liquids and traditional organic solvents on microalga Selenastrum capricornutum. Ecotoxicology and Environmental Safety, 2008; 71, 166-171

[93] Cho C.-W., Jeon Y.-C., Pham T. P. T., Yun Y.-S. Influence of anions on the toxic effects of ionic liquids to a phytoplankton Selenastrum capricornutum. Green Chemistry, 2008; 10, 67-72

[94] Könnecker G., Regelmann J., Belanger S., Gamon K., Sedlak R. Environmental properties and aquatic hazard assessment of anionic surfactants: Physico-chemical, environmental fate and ecotoxicity properties. Ecotoxicology and Environmental Safety. 2011; 74, 1445-1460

[95] Dyer S. D., Lauth J. R., Morrall S. W., Herzog R. R., Cherry D. S. Development of a Chronic Toxicity Structure–Activity Relationship for Alkyl Sulfates. Environmental Toxicology and Water Quality. 1997; 12, 295-303

[96] Coleman D. and Gathergood N. Biodegradation studies of ionic liquids. Chemical Society Reviews, 2010; 39, 600-637

[97] Boethling R. S., Sommer E., DiFiore D. Designing Small Molecules for Biodegradability. Chemical Reviews, 2007; 107, 2207-2227

[98] Introduction to the OECD guidelines for testing of chemicals, Section 3, 2003

[99] Gathergood N. and Scammells P. J. Design and Preparation of Room-Temperature Ionic Liquids Containing Biodegradable Side Chains. Australian Journal of Chemistry, 2002; 55, 557-560

[100] Boethling R. S. Designing Biodegradable Chemicals. ACS Symposium Series, 1996; 640, 156-171

[101] Garcia M. T., Gathergood N., Scammells P. J. Biodegradable ionic liquids: Part I. Concept, preliminary targets and evaluation. Green Chemistry, 2004; 6, 166-175

[102] Docherty K. M., Dixon J. K., Kulpa Jr. C. F. Biodegradability of imidazolium and pyridinium ionic liquids by an activated sludge microbial community. Biodegradation, 2007; 18, 481-493

[103] Stolte S., Abdulkarim S., Arning J., Blomeyer-Nienstedt A., Bottin-Weber U., Matzke M., Ranke J., Jastorff B., Thoeming J. Primary biodegradation of ionic liquid cations, identification of degradation products of 1-methyl-3-octylimidazolium chloride and electrochemical wastewater treatment of poorly biodegradable compounds. Green Chemistry, 2008; 10, 214-224

[104] Harjani J. R., Singer R. D., Garcia M. T., Scammells P. J. Biodegradable pyridinium ionic liquids: design, synthesis and evaluation. Green Chemistry, 2009; 11, 83-90

[105] Atefi F., Garcia M. T., Singer R. D., Scammells P. J. Phosphonium ionic liquids: design, synthesis and evaluation of biodegradability. Green Chemistry, 2009; 11, 1595-1604

[106] Stolte S., Steudte S., Igartua A., Stepnowski P. The Biodegradation of Ionic Liquids - the View from a Chemical Structure Perspective. Current Organic Chemistry, 2011; 15, 1946-1973

[107] Anastas P. T., Warner J. C. Green Chemistry: Theory and Practice, Oxford University Press, Oxford, UK, 1998

[108] Heaton C. A. An Introduction to Industrial Chemistry, 3rd edn, Blackie Academic and Professional (Springer), London, UK, 1995

[109] Porteous A. Dictionary of Environmental Science and Technology, John Wiley & Sons Ltd, Chichester, UK, 1992

[110] Sheldon R. A. Chemistry & Industry (London), 1992; 903-906

[111] Constable D. J. C., Curzons A. D., Cunningham V. L. Metrics to 'green' chemistry - which are the best? Green Chemistry, 2002; 4, 521-527

[112] Trost B. M. The atom economy-a search for synthetic efficiency. Science, 1991; 254, 1471-1477

[113] Myles L., Gore R., Gathergood N., Connon S. J. Highly recyclable, imidazolium de- rived ionic liquids of low antimicrobial and antifungal toxicity: A new strategy for acid catalysis. Green Chemistry, 2010; 12, 1157-1162

[114] Procuranti B. and Connon S. J. A reductase-mimicking thiourea organocatalyst incor- porating a covalently bound NADH analogue: efficient 1,2-diketone reduction with in situ prosthetic group generation and recycling. Chemical Communications, 2007; 1421-1423

[115] Procuranti B. and Connon S. J. Unexpected catalysis: aprotic pyridinium ions as ac- tive and recyclable Brønsted acid catalysts in protic media. Organic Letters, 2008; 10, 4935-4938

[116] Procuranti B., Myles L., Gathergood N., Connon S. J. Pyridinium Ion Catalysis of Carbonyl Protection Reactions. Synthesis, 2009; 23, 4082-4086

[117] Rohitkumar Gore, Ph.D. thesis, Dublin City University

Chapter 7

Pharmaceutical Salts: Solids to Liquids by Using Ionic Liquid Design

Clarissa P. Frizzo, Izabelle M. Gindri, Aniele Z. Tier,
Lilian Buriol, Dayse N. Moreira and
Marcos A. P. Martins

Additional information is available at the end of the chapter

1. Introduction

Ionic liquids (ILs) have attracted increasing interest lately in several areas such as chemistry, physics, engineering, material science, molecular biochemistry, energy and fuels, among others. Scientific literature has been daily invaded by papers that show a variety of new ionic liquids and new applications. Furthermore, the range of ILs used has been broadened, and there has been a significant increase in the scope of both physical and chemical IL properties [1, 2]. ILs are defined as liquid organic salts composed entirely of ions, and a melting point criterion has been proposed to distinguish between molten salts and ionic liquids (mp< 100 °C) [3].

When ILs based on 1-alkyl-3-methylimidazolium salts were first reported in 1982 by Wilkes et al. as tetrachloroaluminates, they were called ILs of first generation [4]. Replacement of this moisture-sensitive anion by the tetrafluoroborate ion and other anions led, in 1992, to air- and water-stable ILs, called then second generation, [5] which have found increasing applications such as reaction media for various kinds of organic reactions. At the onset of the new millennium, the concept of task-specific ILs, called third generation was introduced by Davis [6] (Figure 1). These compounds are defined as ILs in which the anion, cation, or both covalently incorporate a functional group (designed to endow them with particular properties, such as physical, chemical or in terms of reactivity) as a part of the ion structure [6,7]. Simultaneously, Rogers et al. [8] proposed that the ILs can be grouped into three generations in according to their properties and applications (Figure 1). The use of these compounds as solvents characterized them as ILs of first generation due to a unique and accessible physical property set characterized by low or no volatility, thermal stability, or large liquid ranges. Second generation of ILs has potential application such as energetic materials, lubricants and scavenger materials. In these cases, ILs provide a platform where the properties of both cation and anion can be independently modified, permitting the design of new functional materials, while retaining the desired features of an IL. Third generation of ILs has been described as the one where the biological activity is a primary IL property. Thus, ILs are seen as active pharmacological ingredient (API).

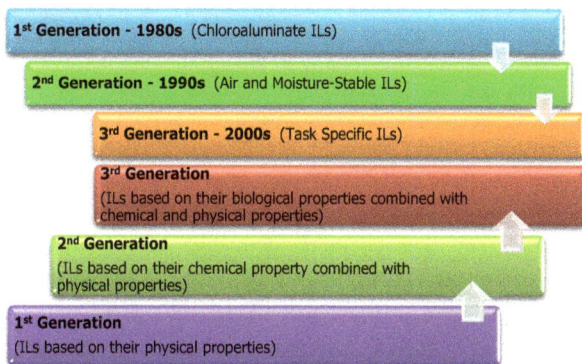

Figure 1. Historical evolution of ILs: chronological and useful development.

Potential pharmaceutical applications of ILs have been showed initially by studies of their toxicity and antimicrobial activity [9,10]. Currently, applications expanded to the use of pharmacologically active ions to develop novel ILs (3rd generation), the use in the formation of microemulsion droplets to transport and release of drugs, and as stabilizing agents of actives, additives and polymers in pharmaceutics [11]. One of the most import pharmaceutical applications is the use of pharmacologically active ions to develop novel liquid salts, since more than 50% of the drugs in the market today are sold as organic salts [12]. Hough and Rogers consider in their review [13] that the conversion of a drug into a salt is a crucial step in the drug development and can have a huge impact on its properties, including solubility, dissolution rate, hygroscopicity, stability, impurity profile and particle characteristics. Thus, the authors believe that an IL approach seems more than appropriate in the design of APIs, where a delicate balance exists between the exact chemical functionality needed for the desired effect in the absence of adverse side effects and the physical properties required for manufacturing, stability, solubility, transport and bioavailability [14,15]. Historically, the pharmaceutical industry depends mostly on crystalline APIs. However, many formulations fail during testing because of issues as, for example, delivery mechanisms such as dissolution, transport, and bioavailability or poor control over polymorphism which can dramatically change properties such as solubility [16,17,18]. In the context of APIs, the counter ions could be selected to synergistically enhance the desired effects or to neutralize unwanted side effects of the active entity. They could also be chosen to pharmacologically act independently [12,19] or to improve the pharmacokinetics properties [20] (Figure 2). Over the past few years, there have been three reviews published in which ILs from APIs occupied a central theme [12,13,20]. In these reviews, the approach of ILs from APIs is discussed from different points of view: i) historical approach of ILs (from solvents to ILs from APIs), ii) focus on the use of ILs from APIs to solve the toxicity of ILs and polymorphism and iii) a good review where all these topics are shortly discussed. However, there is no concern with the issue of the synthesis and physical and chemical characterization of new salts like ILs. Thus, considering the lack of complete and deep survey about all questions (advantages and disadvantages) in the literature and in continuation of our research on ILs [21], we propose this chapter to show the application of IL approach to obtain liquid pharmaceutical salt. This denotes that the material to be covered here includes only papers where pharmaceutical activities (pharmacokinetic and pharmacological) are present at the cation or anion and there is focus on the obtainment of ILs from APIs (Table 1). Here, we consider biologically active the ILs whose components interact with any biological system, and pharmaceutically active the ILs that present any pharmacokinetics and/or pharmacological activities (Table 3). Thus, it was necessary to mention that papers describing ILs with only one of the biologically active components or with no pharmaceutically active were excluded. Thus, the ions alkyimidazolium, phosphoniun, derivatives of non-nutritive sugars, and N-trifluoromethanesulfonate were not included in the scope of this chapter. Another scope limitation of this chapter is about the use of mechanical or thermal methods to the liquefaction of a salt from APIs. This means that pharmacologically active salts such as the procainamide and verapamil hydrochloride that pass from a crystalline state to amorphous state through changes of conditions such as temperature and pressure were excluded [22,23].

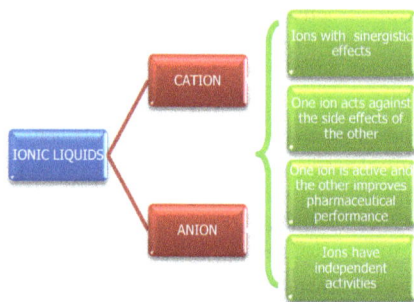

Figure 2. Cation and Anion combination in ILs from APIs and their activities.

Hence, in this chapter we will present the main problems of pharmaceutical industry in relation to solid salt APIs and how the proprieties as well as the limitations of the ILs affect the salification (i.e., salt formation) of APIs. The synthesis, the characterization of physical and chemical properties as well as the pharmaceutical performance of new ILs will be discussed. For this propose, we emphasize that the ILs selected to this chapter (Table 1) present in their structure at least one pharmacologically active entity, and the other (cation or anion) was introduced with the objective to increase the activity, reduce side effects or improve pharmaceutical performance by changing physical or chemical properties (Figure 2).

Compound	Name	Ref.
1	3-hydroxy-1-octyloxymethylpyridinium acesulfamate ([1-(OctOMe)-3-OH-Py][Ace])	[47]
2	3-hydroxy-1-octyloxymethylpyridinium Saccharinate ([1-(OctOMe)-3-OH-Py][Sac])	[47]
3	Benzalkonium Acesulfamate [BA][Ace]	[47]
4	Benzalkonium Saccharinate [BA][Sac]	[47]
5	Benzalkonium Salicylate [BA][Sal]	[37]
6	Benzethonium Acetylsalicylate [BE][Asp]	[37]
7	Benzethonium Aaccharinate[Ben][Sac]	[32]
8	Benzethonium Salicylate [BE][Sal]	[37]
9	Cetylpyridinium Acetylsalicylate [CetPy][Asp]	[37]
10	Cetylpyridinium Salicylate [CetPy][Sal]	[37]
11	Cetylpyridinium Ampicillin [C_{16}pyr][Amp]	[35]
12	Choline Ampicillin [Col][Amp]	[35]
13	Choline Phenytoin[Col][Phe]	[32]
14, 15, 16	Choline-derivative Acesulfamate [Col][Ace]	[46]
17	Didecyldimethylammonium Acesulfamate ([DDA][Ace])	[47]
18	Didecyldimethylammonium Ibuprofenate [DDA][Ibu]	[8]
19	Didecyl-dimethyl-ammonium Saccharinate ([DDA][Sac]	[47]
20	Hexadecylpyridinium Acesulfamate ([Hex][Ace])	[47]
21	Hexadecylpyridinium Aaccharinate ([Hex][Sac])	[47]
22	Hexetidinium Salicylate [Hext][Sal]	[37]
23	Lidocainium Acetylsalicylate [LID][Asp]	[37]
24	Lidocainium Docusate [Lid][Doc]	[8]
25	Lidocainium Salicylate [LID][Sal]	[37]
26	Mepenzolate Acesulfamate[Mep][Ace]	[32]
27	Mepenzolate Saccharinate[Mep][Sac]	[32]
28	Procainium Salicylate [Proc][Sal]	[37]
29	Procainium Amidesalicylate [PA][Sal]	[37]
30	Propantheline Acesulfamate[Pro][Ace]	[32]
31	Propantheline Cyclamate[Pro][Cyc]	[32]
32	Propantheline p-toluenesulfonate[Pro][pTO]	[32]
33	Propantheline Saccharinate[Pro][Sac]	[32]
34	Pyridostigmine Saccharinate[Pyr][Sac]	[32]
35	Ranitidine Docusate [Ran]Doc]	[8]

Compound	Name	Ref.
36	Tetrabutylphosphonium Salicylate [P(BU)₄][Sal]	[37]
37	Tetraethylammonium Ampicillin[TEA][Amp]	[35]
38	Tramadolium Acetyl-salicylate [Tram][Asp]	[37]
39	Tramadolium Salicylate [Tram][Sal]	[37]
40	Trihexyltetradecylphosphonium Ampicillin [P₆,₆,₆,₁₄][Amp]	[35]

Table 1. ILs from active pharmaceutical ingredients found in this chapter.

2. Fundamentals

In this section we will deal with some important points of ILs approach salification of APIs and the strategies used in the search and characterization of new ILs from APIs. The synthetic procedure as well as physical and chemical properties (main thermal properties) of ILs will also be discussed.

2.1. Salification of APIs

The physical form of a drug substance is of great importance since it directly affects the manner in which the material is formulated and presented to the consumer, as well as influence more fundamental characteristics such as solubility and dissolution rate, which, in turn, impact on bioavailability [8, 17, 18].

The drugs converted into salts were found to be more stable and water soluble in comparison to free bases or acid which qualifies them as the preferred forms to use as therapeutic agents [19, 24-26]. Such salts may offer advantages over the corresponding free drug in terms of physical properties such as melting point (thermal stability), crystallinity, hygroscopicity, dissolution rate, or solubility (bioavailability). From a pharmaceutical viewpoint the melting enthalpy, melting temperature and solubility are of particular importance, both because of their routine measurement and their influence on processing and bioavailability [19, 26]. Taking these advantages into account, the pharmaceutical industry relies predominantly on solid, primarily crystalline forms for the delivery of APIs, mainly for reasons of purity, thermal stability, manufacturability, and ease of handling [8, 17, 18]. However, solid forms of APIs often suffer from polymorphic conversion, low solubility, and a variety of factors which affect bioavailability associated with the final solid form [17, 18, 27]. Many phase II trials of new APIs end in failure due to their efficacy, often related to bioavailability and thus solubility [16-18]. These factors motivate the screening for novel solid forms, including salts, polymorphs, pseudopolymorphs (or solvates), and co-crystals (Figure 3).

Liquid drug formulations from salification are rarely found and are usually based on eutectic mixtures [24], however, salt drug generation can result in a liquid salt, known as ionic liquid. The main advantage of ILs in most of the cases is that their salt properties are retained in a wide liquid range. This fundamental property of ILs is because ions are generally organic with low symmetry and diffuse charge. In addition, ILs have properties such as negligible vapor pressure resulting in reduced inhalatory exposure, absence of flammability, and their high variability concerning organic chemical structure in order to optimize technological features like solvation properties. This tunable solubility with several organic compounds, viscosity, conductivity, as well as thermal and electrochemical stability is ideal in terms of technical applicability [3]. The control of the properties of an IL is based on the manipulation of the interactions between the ions. The suppression of these interactions reduces lattice energies and the extreme suppression of these interactions leads to glass formation upon cooling, polymorphism, multiple phase transitions, and ion dissociation [28-31]. An understanding of the physical and chemical properties of ILs allows the proper selection of a specific IL for a given application. Thus, for example, by choosing ionic components capable of solubilizing specific solutes, one can control the critical solubility to crystallization processes [29].

Stoimenovski et al. [12] cited in their work that studies have been suggest that ILs do not dissolve as independent ions but keep a nanostructured organization in aqueous media. This fact, constitute other important advantage of the obtainment of an API as IL. Drugs that are highly ionic have difficulty crossing the membrane in order to reach their site of action. Ion-pair formation enhances the transport of various ionic drugs through the skin and across the absorbing membrane [12]. Therefore, highly ion-associated pharmaceutically active ILs would be highly beneficial forms of the original pharmaceutical active salts, as they could cross the membrane more rapidly [12]. Therefore, targeted alterations of a final drug form based on the various property sets obtainable through an IL approach may help to enhance efficacy, while retaining, improving or even introducing a second activity. The potential of this approach as a drug phase is to date poorly exploited [13, 13, 20].

Clearly, a salt that does not exhibit a crystalline phase will not present polymorphism, but there are further advantages to be realized and exploited in the delivery of the API [32]. In particular, a non-crystalline salt in a liquid or glassy phase will probably exhibit the enhanced solubility exhibited by amorphous phases [17, 18, 32]. This concept bears further discussion as it is potentially one of the most important advantages of the formulation of APIs as ionic liquid phase [32].

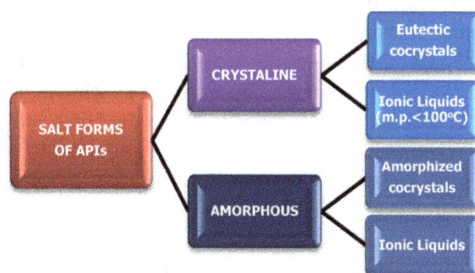

Figure 3. Salt forms of Active Pharmaceutical Ingredients (APIs).

2.2. Synthesis of ILs from APIs

The selection of pairs of ions to form ILs is carried out with candidate ions that have low symmetry and charge diffuse, traits that also characterize several typical APIs. Even the nitrogen-containing heterocycles, commonly used in ILs today, are frequently found in APIs or API precursors [8,33]. The process generally is a simple way to modify the properties of a drug with ionizable functional groups to overcome undesirable features of the parent drug [12].

Care must be taken when choosing appropriate IL-forming ion pairs. Many of the important APIs are not permanent ions, but rather are protonated or deprotonated to form the commonly used salts; thus suitable pKa differences need to be considered [8,34]. MacFarlane and Seddon [8,1] have recently proposed that protic ILs can only be considered ILs if the pKa difference is such that more than 99% of the salt exists in ionized form. For an API this distinction may not be needed, since having a balance between ionized and neutral forms may have advantages. There may be a significant advantage of drugs with low degree of ionisation over the fully ionised ones due to their ability to cross membranes more efficiently. An example of a partially ionized pharmaceutically active IL is 1-methylhexylammonium salicylate [12]. Salicylic acid, an analgesic with a pKa value of 2.98, was reacted with 1-methylhexylamine, a nasal decongestant with a pKa value of 10.5, to produce a liquid at room temperature with a glass transition at −40°C and a ΔpKa of 7.52.

Most of the syntheses found in the papers selected to this chapter consist of metathesis reactions. The cation and anion in their available salt forms were separately dissolved in a solvent (e.g., water, methanol, ethanol, acetone) allowed to stir with heating to ca. 90 °C (if necessary) or at room temperature. Some alternative methods to methatesis reaction to specific ILs were also described. Ferraz et al. [35] used a method to change the anion using ion exchange resin described by Ohno et al. [36]. Ferraz et al. [35] employed Amberlite resin (in the OH form) in order to exchange halides (bromide or chloride) to the hydroxide form and then this basic solution was neutralized by the addition of an adequate acid solution. The acid–base reaction yielded the desired IL. The organic cations were selected from salts which were first transformed into hydroxides by the use of an ionic exchange column (Amberlite IRA-400 OH) in methanol. Next, the β-lactam antibiotic previously dissolved in a moderately basic ammonia solution was used to neutralize the selected cations. In this case, pure ILs were obtained after eliminating the excess ammonia and/or β-lactam antibiotic by evaporation and crystallization, respectively. Bica et al. [37, 38] also showed an alternative synthesis in solvent-free conditions. The compounds **25** and **28** were also prepared by melting a stoichiometric mixture of base and salicylic acid at ~100 °C to obtain a liquid. Similarly, **22** was directly synthesized by the reaction of hexetidine with salicylic acid. This solvent-free preparation is clearly advantageous compared to conventional metathesis, since solvents and stoichiometric NaCl waste are prevented. Furthermore, ILs are obtained in high purity without halide, metal, or solvent impurities, as necessary for pharmaceutical applications. The isolation of the product occurred considering that usually the inorganic salt precipitates. Thus, in most of cases, the product was extracted by filtration of inorganic salt. The solvent was removed with a rotary evaporator. The resulting product was placed on a high vacuum line to remove any residual solvent. In some cases, when inorganic salt is partially soluble, ILs had to undergo a process of extraction typically with chloroform or dichloromethane. Following that step, the organic phase was then washed with water to remove any inorganic salt (e.g., NaCl, which was monitored by a silver nitrate test), and solvent was removed with a rotary evaporator. The resulting product was placed on a high vacuum line to remove any residual solvent. In some cases, an extra purification was described, mainly to remove excess of halides.

2.3. ILs characterization

When searching for an IL from APIs, one has to care about its physical state and properties because it may not be an IL, but a crystalline solid (Figure 3). A significant number of drugs currently on the market are formulated as amorphous materials, and the most common means of preparing the amorphous phase is by quenching: rapid cooling from the melt; rapid precipitation from solution (for example on addition of an antisolvent); spray drying; flash evaporation; lyophilisation, in other words, methods that allow the disordered glassy phase to be "frozen in" before nucleation and growth that would lead to the appearance of crystals [32, 17, 18]. Dean et al. [32] have showed a schematic comparison of the accessible phases and the relative free energy of each phase. The authors highlight that in cases where a stable crystalline form exists at ambient temperature, ΔGf° for the crystalline phase is lower than that for the amorphous material formed by quenching. In this manner, the quenching results in a form that is not the most thermodynamically stable phase at that temperature. The authors also emphasize that in this situation, any event that initiates nucleation may lead to growth of a crystalline phase at the expense of the amorphous glass. Such nucleation initiators may include heating (yielding a plastic phase that then crystallizes) or localized shock, such as that applied during size reduction (grinding) or formulation. In some cases, even the desired increased solubility leads immediately to the crystallization of a stable crystalline phase. In relation to the search by ILs, the authors emphasized that in the preparation of a salt that has an amorphous phase as its most thermodynamically stable form (in the temperature range of interest), a less preferred, but an effective form would be one with fusion temperature below the temperature of interest [32].

In the light of the extensive literature on crystalline drug forms, currently supported by crystal engineering [39,40] by using supramolecular synthons [41] concept and considering the abovementioned data, their benefits are clear to develop strategies for increasing solubility of drug compounds [42] and to chose co-crystal formers [43] for poorly soluble drugs [44]. Therefore, combining an understanding of the effect of robust supramolecular synthons on the process of crystallization, with concepts involving IL, Dean et al. [32] proposed the development of an "anti-crystal engineering" approach to the synthesis of ILs from APIs. This would comprise identifying and intentionally avoiding the pairing of cations and anions that might yield common supramolecular synthons, with the goal of decreasing the likelihood of crystallization of salt. This "anti-crystal engineering" approach is postulated as a means of narrowing the search for ILs from API to cation and anion pairs that have a higher likelihood of not crystallizing. Dean et al. [32] illustrated this approach by preparing and analyzing a series of API salts; some of which crystallized readily, while others were characterized as ILs and remained in an amorphous glass or liquid form in spite of vigorous attempts to bring about crystallization. To achieve the combination, the authors studied the possibility of cations and anions form supramolecular synthons mainly from interactions of hydrogen that usually considered imparting a tendency toward crystallization that should be avoided in the quest for ILs from APIs. As a result, they observed that all cation/anion combinations bearing both hydrogen bond donor and acceptor groups yield crystalline salts (**13**, **26**, **27** and **31**) [32]. For example, **13** (Table 1 and 2) which might be predicted to form the greatest number of strong, directional hydrogen bonds between different ions yields the salt with the highest melting point. Of greater interest, however, are those salts that are crystalline solids and even do not (at first examination of individual ions) exhibit the capability to form hydrogen bonded synthons (**33** and **34**), but have some energy stabilization resulting in crystallization. On the other hand, some salts were observed as salt without a crystalline phase (or with a sub-ambient melting point) indicating that they are in the most thermodynamically stable phase as a liquid or glass at ambient temperature. Thus, these salts (**7**, **30** and **32**) are considered ILs by the authors [32]. In another work in search by ILs from APIs, Bica et al. [38] found that generating oligomeric ions from tautomer's protons has a tremendous influence on physical properties that allow the expansion of the liquid ranges of some salts. They proposed that for this type of liquid salt formulation, the term ionic liquids might be controversial. They suppose (say) this based on initial experiments and further investigation concerning ionicity and simple eutectic behavior are currently ongoing in their laboratories. The authors define oligomeric ions as those that enable liquefaction of solid ILs (or other salts) by simply changing the stoichiometry or complexity of the ions. They highlight that the strategy does not need to employ the parent of the anion or cation in use and that this can be particularly useful for pharmaceutically active salts or ILs. According to the authors, another advantage of the oligomeric ions strategy is that the design of pharmaceutical IL may benefit since it is permitted to modify the physical properties of a given salt form once obtained [38].

Therefore, considering the "anti-crystal strategy" proposed by Dean et al. [32] and the "oligomeric ions" proposed by Bica et al. [38] in the search for ILs from API, the possibility of formation of solid crystals (eutectics), amorphous, or liquid phase is a fundamental question (Figure 3). Thus, one cannot imagine the search for new ILs from API without performing a complete calorimetric characterization of what an IL may be. Currently, unfortunately, in most of the cases, the ions are selected by facility from synthesis or purification routes rather than rational choice or screening.

The calorimetric data reported in papers collected to this chapter showed that most of the salts synthesized can be classified as ILs, while the remaining were crystalline salts. Those who were characterized as IL can be shared in three general types of behavior. The first group of ILs exhibits just melting points below 100 °C, allowing their classification as ILs (**1**, **2**, **4**, **9**, **10**, **14**, **16** and **21**). The second type of behavior is characterized by formation of an amorphous glass. These ILs have no melting, but only glass-transition (**6-8**, **17**, **18**, **23-25**, **28-30**, **32-35** and **38**). The third group of ILs is characterized by compounds that have melting points and glass transition (**2**, **5**, **11**, **12**, **15**, **19**, **20**, **34**, **36** and **37**)). Compounds **13**, **22**, **26**, **27**, **31**, **33** and **39** were found to have high melting not fitting the definition of ILs (Table 2). Other physical and chemical properties also should be evaluated to characterize an organic salt as an IL. Considering the scope of this chapter, some of the properties were found in a few papers (Table 2). These properties were density, solubility and thermal stability (Table 2). Viscosity is an important physical property to characterize an IL. The viscosity of ionic liquids is essentially determined by their tendency to form hydrogen bonding and by the strength of their van der Waals interactions. The structure of the cation strongly influences the viscosity of the IL. However, this property was not reported in any of the selected papers. Density was reported only for two ILs (14, 15). These ILs were denser than water thus the density of comparable ILs decreases as the bulkiness of the organic cation increases. It is evident that the density of such compounds increases with increasing molecular weight of the anion which confirms the results shown by Fredlake et al [45]. Density reported to ILs from APIs in the selected papers to this chapter is characteristic of ILs[3]. At room temperature, the ILs from API reported in this chapter were grouped into miscible in water and other polar organic solvents (hydrophilic) and partially miscible or immiscible in water and hexane (hydrophobic). Water miscible ILs were choline derivatives, ampilicin, saccharinate and water imiscible were choline, ammoniun and pyridiniun derivatives. IL choline derivatives were present in both groups on dependence of anion. Solubility data of other IL derivatives of other cations and anions were not reported by the authors. The thermal stability of ILs is limited by the strength of their heteroatom-carbon and their heteroatom-hydrogen bonds, respectively [3]. The ILs 9, 14, 15 and 23 showed a lower thermal stability (115.14-126.5 °C) while 1-6, 8, 10-12, 17-22, 24, 25, 28, 29 and 35-40 have shown more stable (154.22-307.94 °C).

Compound	Thermal Data (°C) and Density at 25°C (g.mL-1)	Ref.
1	m.p = 79-81[d], TGA = 267	[47]
2	m.p = 95-98[d], TGA = 301	[47]
3	m.p = 90, Tg = -36[a], TGA = 187/249/394[c]	[47]
4	m.p = 74, TGA = 204	[47]
5	m.p = 96.02, Tg = 51.01, TGA = 171.95[b]	[37]
6	Tg =2.84, TGA = 154.22[b]	[37]
7	Tg = -4	[32]
8	Tg = -13.72, TGA = 167.76[b]	[37]
9	m.p = 61.31, TGA = 115.14[b]	[37]
10	m.p = 73.97, TGA = 205.61[b]	[37]
11	m.p = 86.0, Tg = -19.64, TGA = 269.39	[35]
12	m.p = 58.0, Tg = -20.12, TGA = 221.29	[35]
13	m.p = 215 – 217	[32]

Compound	Thermal Data (°C) and Density at 25°C (g.mL-1)	Ref.
14	m.p. = 31-36 and82-84 TGA = 126.5, density = 1.103 - 1.277	[46]
15	m.p. = 85-86,Tg = -49, TGA = 122, density = 1.041-1.270	[46]
16	m.p.= 87-88	[46]
17	Tg = -53, TGA = 232/426^c	[47]
18	Tg = -73, TGA = 168^b	[8]
19	m.p = 16, Tg = -33, TGA = 214	[47]
20	m.p = 57, Tg = -11, TGA = 267/494^c	[47]
21	m.p = 66, TGA = 253/412^c	[47]
22	m.p = 106.81, TGA = 182.14^b	[37]
23	Tg = -13.97, TGA = 120.71^b	[37]
24	Tg = -29, TGA = 222 ^{b}	[8]
25	Tg = 19.78, TGA = 158.46^b	[37]
26	m.p = 135 – 137, Tg = 34	[32]
27	m.p = 187 – 189, Tg = 53	[32]
28	Tg = 13.87, TGA = 187.33^b	[37]
29	Tg = 19.87, TGA = 159.21^b	[37]
30	Tg = -20	[32]
31	m.p = 133 – 137, Tg = 20	[32]
32	Tg = 7	[32]
33	m.p = 133 – 135, Tg = 18	[32]
34	m.p = 94 – 96, Tg = 4	[32]
35	Tg = -12, TGA = 249	[8]
36	m.p = 57.32, Tg = -56.47, TGA = 307.94^b	[37]
37	m.p = 79.0, Tg = -18.64, TGA = 214.75	[35]
38	Tg = 13.78, TGA = 169.64^b	[37]
39	m.p = 176.17, TGA = 177.16^b	[37]
40	TGA = 297.65	[35]

^{a}Solid-solid transition.^{b}T$_{onset\ 5\%}$.^{c}Multiple decomposition steps. ^{d}Visual melting point range via hot-plate apparatus.

Table 2. Physical and Chemical Properties of some ILs from APIs.

3. Pharmaceutical activity assessment

In view of the objective of our chapter, it is important to evaluate the pharmaceutical profile of ILs from API. Pharmaceutical profile includes changes on the pharmacokinetics and/or pharmacological behavior of the salts when they turn into liquids by changing their cation or/anion.

No specific pharmacokinetic property was evaluated. However important observation reporting by Hough et al [8] and related in a review by Stoimenovski et al [12] is that strongly hydrophilic ionic actives often possess insufficient ability to penetrate biological membranes. Combining such an active ion with another of a more lipophilic character may offer a solution to this problem. For example, lidocaine docusate [Lid][Doc], an IL form of the local surface anaesthetic lidocaine, combines the relatively hydrophobic lidocaine cation with a hydropho-bic anion, docusate (an emollient), to produce a hydrophobic IL, which exhibits reduced or controlled water solubility (Figure 2) and thus should exhibit extended residence time on the skin [8].

From the selected papers to this chapter, only three of them bring evaluation of pharmaceutical properties of new API (IL). Pharmacological activities were evaluated in four papers. Phar-macological activity evaluated were antinociception [8], suppression of PC12 neuritic out-growth [8], antibactericidal [46, 47] and antifungal [46, 47] activities and skin irritation [47]. Antinociception activity [8] was assessed using a modification of the tail-withdrawal proce-dure. Two antinociceptive models were used: warm water tail-withdrawal from 49 °C water in intact mice, and warm water tail-withdrawal from a 47 °C bath, following tail injury. In this test it was observed that [Lid][Doc] produced a longer duration of antinociceptive effect than lidocaine hydrochloride [Lid][HCl]. The authors suggest that the high hydrophobicity for [Lid][Doc] in relation to [Lid][HCl] account for the increased duration of [Lid][Doc] over [Lid][HCl] observed *in vivo* models. Also, this may constitute a slow-release mechanism unique to any hydrophobic IL [8].

In the same paper [8], the evaluation of the suppression of PC12 neuritic outgrowth by [Lid][Doc] and [Lid][HCl] was evaluated. The suppression of PC12 neuritic outgrowth is related to the local anesthetic effects. These anesthetics suppress nerve growth factor (NGF) mediated neuronal differentiation in rat pheochromocytoma (PC12) cells. This was used as a bioassay for detecting that killed PC12 cells treated with [Lid][HCl] was higher than [Lid][Doc]. Authors suggest that the PC12-NGF data indicate potential differences between [Lid][Doc] and [Lid][HCl] at the cellular level and showed a mechanism of action entirely different for [Lid][Doc] than that for [Lid][HCl]. Docusate may enhance membrane permeability which may suggest at least one mechanism associated with the apparent increase in [Lid][Docl] efficacy *in vivo*. However, while an increase in permeability may enhance transdermal transport and account for the longer duration and greater efficacy of [Lid][Doc]*in vivo*, the longer duration of [Lid][Doc] on the mouse tail-withdrawal indicates an alternative mechanism.

Antimicrobial, antifungical and antibactericidal activities were evaluated to ILs **14** and **15** [46], **3, 4, 17** and **19** [47]. Results were expressed in terms of mean minimum inhibitory concentration (MIC) and minimum bactericidal concentration (MBC). The efficacies of **14** and **15** were low (but still high enough to be effective) in comparison with that of the widely used benzalkonium chloride. No MIC and MBC values could be established for the [Ace] salts **16**, because of their hydrophobic characters. Salts **3, 4, 17** and **19** were also evaluated; benzalkonium and didecyl-methylamonnium chloride which inherently exhibit anti-microbial, anti-bacterial and anti-fungal activities were used as standard for comparison. ILs activities are similar to those of commercially available, although the ILs were not found to be limited to a specific class of bacteria or fungi. Skin irritation of salts **17** and **19** [47] was also determined. Each IL was tested on 3 male New Zealand albino rabbits, where the fur was previously removed from the back of the rabbit. Half a milliliter of the ILs (100%, pure) was distributed on two 6 cm³ sites of the same animal. The application site was then covered with a porous gauze dressing and secured in place with tape. After a 4h exposure, the dressing was removed and the application site was gently washed with water. Observations were then conducted at 1, 24, 48, and 72 h, where the test sites were evaluated for erythema and edema using a prescribed scale. The skin irritation of these ILs is defined as category 4 (the highest) by standard organization for economic co-operation and development (OECD) grading.

Finally, the acute oral toxicities of salts **17** and **19** were determined. The toxicity was tested according to the method of acute toxic class. Wistar rats male and female were used for each IL tested. Results indicated the acute toxicity range for both ILs was between 300–2000 mg/kg in male and female rats. Thus, these ILs would be classified as category 4 (harmful) toxins according to standard organization for economic co-operation and development (OECD) grading. Table 3 depicted the cation and anion covered in this review. Structures and pharmacological activities of each of them are also showed.

Structure	Name	Activity
	Ranitidine [Ran]	Histamine H2- receptor antagonist
	Lidocainum [Lid]	local anesthetic,

Structure	Name	Activity
	Didecylmethylammonium [DDA]	antibacterial
R^1=H, R^2=from Et to $C_{14}H_{29}$ and $C_{12}H_{23}$(4) R^1=Ac, R^2=from Et to $C_{14}H_{29}$ and $C_{12}H_{23}$(5) R^1 =C_9H_{19},R^2 = from Et to $C_{12}H_{25}$ and $C_{12}H_{23}$(6)[a]	Cholinederivatives[Col]	acetylcholine precursor
	Benzalkonium [BA]	antibacterial
	Hexadecylpyridinium [Hex]	antibacterial
	3-Hydroxy-1-octyloxymethylpyridinium [1-(OctOMe)-3-OH-Py]	antimicrobial
	Pyridostigmine [Pyr]	reversible acethylcholinesteterase inhibitor
	Benzethonium[Ben]	antibacterial

Structure	Name	Activity
	Mepenzolate[Mep]	skeletalmusclerelaxant
	Propantheline[Pro]	antimuscarinic
	Hexetidinium[Hext]	antibacterial
	Procainiunamide[PA]	antiarrhythmic
	Tramadolium[Tram]	analgesic
	Procainiun [Proc]	localanesthesic
	Docusate [Doc]	emolient

Structure	Name	Activity
	Ibuprofenate [Ibu]	anti-inflammatory
	Acesulfamate [Ace]	noncaloric sugar
	Saccharinate[Sac]	noncaloric sugar
	Phenytoin[Phe]	antiepileptic
	Cyclamate [Cyc]	noncaloric sugar
	Acetylsalicylate [Asp] salicylate [Sal]	anti-inflammatory, analgesic, anti-pyretic

Structure	Name	Activity
	Ampicillin [Amp]	antibacterial

[a]Number **4**, **5** and **6** are of the ILs formed from these cations (See Table 1).

Table 3. Structure and activity of ions found in this chapter.

4. Conclusion

After having examined the literature in ILs from APIs it is possible to conclude that: *(i)* this is a research area developed by few groups; *(ii)* a complete and elaborated work including synthesis, physical and chemical properties studies and pharmacological activity estimation is necessary to produce significant results in this area (some groups have already performed more elaborated works); *(iii)* evaluation of physical and chemical properties, main thermal behavior is fundamental to develop new liquid APIs from IL approach; *(iv)* modification in the physical state of an API can result in modification or modulation of pharmaceutical properties of drugs. For example, co-formation of two separate solid actives in a solid dosage form significantly differ from a dual functional IL formulation. The ions in an IL dissolve in the body fluids exactly the same way—since one ion cannot dissolve without the other; this is not true to separate solid forms administered at the same time since each may dissolve at quite different rates. In addition, increase in solubility and bioavailability can enable a new formulation and/or a new doseage. Consequently, pharmacokinetic and pharmacological profiles studies necessarly lead to a potential patent protection for each of new forms of drugs [12].

In this chapter, we hope to have given a clear idea of the use of IL approach in the obtainment of liquid or amorphous API. We would like to conclude with an optimistic view for the future expansion of the development of new drug profiles. This positive view comes from the certainty that the results reported here are the beginning of a great advance in this promising field in the near future.

List of Abbreviations

API – Active Pharmacological Ingredient

IL –Ionic Liquid

MBC - Minimum Bactericidal Concentration

MIC - Minimum Inhibitory Concentration

NGF - Nerve Growth Factor

OECD - Organization Economic Co-operation and Development

PC12 - Pheochromocytoma Cells

$\Delta Gf°$– Free Energy of Fusion

Acknowledgements

The authors are greatful to Conselho Nacional de Desenvolvimento Científico e Tecnológico (CNPq/Universal Proc. No. 578426/2008-0; 471519/2009-0), Fundação de Amparo à Pesquisa do Estado do Rio Grande do Sul (FAPERGS/CNPq-PRONEX Edital No. 008/2009, Proc. No. 10/0037-8), and Coordenação de Aperfeiçoamento de Pessoal de Nível Superior (CAPES/ PROEX) for financial support. The fellowships from CNPq (M.A.P.M., D.N.M., A.Z.T.) and CAPES (C.P.F., L.B., I.M.G.) are also acknowledged.

Author details

Clarissa P. Frizzo*, Izabelle M. Gindri, Aniele Z. Tier, Lilian Buriol, Dayse N. Moreira and Marcos A. P. Martins

*Address all correspondence to: clarissa.frizzo@yahoo.com.br

Department of Chemistry, NUQUIMHE, Federal University of Santa Maria, Santa Maria, RS, Brazil

References

[1] MacFarlane D R., Seddon K R.Ionic Liquids- Progress on the Fundamental Is-sues.Australian Journal of Chemistry (2007). http://www.publish.csiro.au/paper/ CH06478.htmaccessed 20 june 2012)., 60(1), 3-5.

[2] Torimoto, T, Tsuda, T, Okazaki, K, & Kuwabata, S. New Frontiers in Materials Sci-ence Opened by Ionic Liquids. Advanced Materials (2010). http://onlineli-brary.wiley.com/doi/10.1002/adma.200902184/abstractaccessed 20 june 2012)., 22(11), 1196-1221.

[3]   Wilkes, J. S, Wasserscheid, P, & Welton, T. Ionic Liquids in Synthesis, Weinheim: Wiley-VCH;   (2007).   http://onlinelibrary.wiley.com/doi/10.1002/9783527621194.ch1/ summary(accessed june 2012).

[4]   Wilkes, J S, Levisky, J A, Wilson, R A, & Hussey, C L. Dialkylimidazoliumchloroaluminate melts: a new class of room-temperature ionic liquids for electrochemistry, spectroscopy and synthesis. Inorganic Chemistry1982;http://pubs.acs.org/doi/abs/ 10.1021/ic00133a078(accessed june (2012)., 21(3), 1263-1264.

[5]   Wilkes, J S, & Zaworotko, M J. Air and water stable 1-ethyl-3-methylimidazolium based ionic liquids. Journal of the Chemical Society, Chemical Communications1992; (13):   965-967.   http://pubs.rsc.org/en/content/articlelanding/1992/c3/ c39920000965(accessed june (2012).

[6]   Davis, J. H Jr. Task-Specific Ionic Liquids. Chemistry Letters 2004;33(9), 1072-1077. http://www.csj.jp/journals/chem-lett/cl-cont/cl2004-9.html(accessed june (2012).

[7]   Visser, A E, Swatloski, R P, Reichert, W M, Mayton, R, Sheff, S, Wierzbicki, A, & Davis, J. H Jr, Rogers R D. Task-specific ionic liquids for the extraction of metal ions from aqueous solutions. Chemical Communications(2001).

[8]   Hough, W L, Smiglak, M, Rodríguez, H, Swatloski, R P, Spear, S K, Daly, D T, Pernak, J, Grisel, J E, Carliss, R D, Soutullo, M D, & Davis, J. H Jr, Rogers R D. The third evolution of ionic liquids: active pharmaceutical ingredients. New Journal of Chemistry (2007). http://pubs.rsc.org/en/content/articlelanding/2007/nj/b706677paccessed 20 june 2012)., 31(8), 1429-1436.

[9]   Carter, E B, Culver, S L, Fox, P A, Goode, R D, Ntai, I, Tickell, M D, Traylor, R K, Hoffman, N W, & Davis, J. H Jr. Sweet succes: ionic liquids derived from non-nutritive sweeteners. Chemical Communications (2004). http://pubs.rsc.org/en/content/ articlelanding/2004/cc/b313068aaccessed 20 june 2012).

[10] Pernak, J, Stefaniak, F, & Weglewski, J. PhosphoniumAcesulfamate Based Ionic Liquids.   European   Journal   of   Organic   Chemistry   (2005).   http:// onlinelibrary.wiley.com/doi/10.1002/ejoc.200400658/abstractaccessed 20 june 2012).

[11] Moniruzzamana, M, Kamiya, N, & Goto, M. Ionic liquid based microemulsion with pharmaceutically accepted components: Formulation and potential applications. Journal of colloid and Interface Science (2010). http://www.sciencedirect.com/science/ article/pii/S0021979710009306(accessed june 2012).

[12] Stoimenovski, J. MacFarlane D R, Bica K, Rogers R D. Crystalline vs. Ionic Liquid Salt Forms of Active Pharmaceutical Ingredients: A Position Paper. Pharmaceutical Research (2010). http://www.springerlink.com/content/4g14306n81x7t8n0/accessed 20 june 2012)., 27(4), 521-526.

[13] Hough, W L, & Rogers, R D. Ionic Liquids Then and Now: From Solvents to Materials to Active Pharmaceutical Ingredients. Bulletin of the Chemical Society of Japan

(2007). https://www.jstage.jst.go.jp/article/bcsj/80/12/80_A7019/_article(accessed june 2012)., 80(12), 2262-2269.

[14] Bennett, B, & Cole, G. Pharmaceutical Production- An Engineering Guide. United Kingdom: Institution of Chemical Engineers; (2003). http://www.knovel.com/web/portal/browse/display?_EXT_KNOVEL_DISPLAY_bookid=1113(accessed june 2012).

[15] Byrn, S. R, Pfeiffer, R. R, & Stephenson, G. Grant WJD., GleasonWB. Solid-State Chemistry of Drugs. West Lafayette: SSCI; (1999).

[16] Schuster, D, Laggner, C, & Langer, T. Why Drugs Fail- A Study on Side Effects in New Chemical Entities. Current Pharmaceutical Design (2005). http://www.bentham-direct.org/pages/b_viewarticle.php?articleID=3137721(accessed june 2012)., 11(27), 3545-3559.

[17] Datta, S, & Grant, D. J W. Crystal structures of drugs: advances in determination, prediction and engineering. Nature Reviews Drug Discovery (2004). http://www.nature.com/nrd/journal/3n1/abs/nrd1280.html(accessed june 2012).

[18] Yu, L, Reutzel, S M, & Stephenson, G A. Physical characterization of polymorphic drugs: an integrated characterization strategy. Pharmaceutical Science & Technology Today (1998). http://www.sciencedirect.com/science/article/pii/S1461534798000315(accessed june 2012)., 1(3), 118-127.

[19] Kumar, V, & Malhotra, S. V. IonicLiquids as PharmaceuticalSalts: A Historical Perspective. In: Malhotra SV. (ed.) Ionic Liquid Applications: Pharmaceuticals, Therapeutics, and Biotechnology. Washington: American Chemical Society; (2010). http://pubs.acs.org/978-0-84122-547-3(accessed june 2012).

[20] Ferraz, R, Branco, L C, Prudêncio, C, Noronha, J P, & Petrovski, Z. IonicLiquids as Active PharmaceuticalIngredients. ChemMedChem (2011). http://www.ncbi.nlm.nih.gov/pubmed/21557480(accessed june 2012)., 6(6), 975-985.

[21] Martins, M, Frizzo, P, Moreira, C P, Zanatta, D N, & Bonacorso, N. H G. Ionic Liquids in Heterocyclic Synthesis. Chemical Reviews.(2008). http://pubs.acs.org/doi/abs/10.1021/cr078399y(accessed june 2012)., 108(6), 2015-2050.

[22] Wojnarowska, Z, Roland, C M, Swiety-pospiech, A, Grzybowska, K, & Paluch, M. Anomalous Electrical Conductivity Behavior at Elevated Pressure in the Protic Ionic Liquid Procainamide Hydrochloride. Physical Review Letters (2012). http://prl.aps.org/abstract/PRL/108i1/e015701accessed 20 june 2012).

[23] Wojnarowska, Z, Paluch, M, Grzybowski, A, Adrjanowicz, K, Grzybowska, K, Kaminski, K, Wlodarczyk, P, & Pionteck, J. Study of molecular dynamics of pharmaceutically important protic ionic liquid-verapamil hydrochloride. The Journal of Chemical Physics (2009). http://jcp.aip.org/resource/1/jcpsa6/131i10/104505s1accessed 20 june 2012).

[24] Stahl, P. H, & Wermuth, C. G. Handbook of Pharmaceutical Salts; Properties, Selection, and Use. Zurich: Verlag Helvetica ChimicaActa; (2008).

[25] Paulekuhn, G S, Dressman, J B, & Saal, C. Trends in Active Pharmaceutical Ingredient Salt Selection based on Analysis of the Orange Book Database. Journal of Medicinal Chemistry (2007). http://pubs.acs.org/doi/abs/10.1021/jm701032y(accessed june 2012)., 50(26), 6665-6672.

[26] Brodin, A, Nyqvist-mayer, A, Broberg, F, Wadsten, T, & Forslund, B. Phase diagram and aqueous solubility of the lidocaine-prilocaine binary system. Journal of Pharmaceutical Sciences (1984). http://onlinelibrary.wiley.com/doi/10.1002/jps.2600730413/abstract(accessed june 2012)., 73(4), 481-484.

[27] Karpinski, P H. Polymorphism of Active Pharmaceutical Ingredients. Chemical Engineering & Technology (2006). http://onlinelibrary.wiley.com/doi/10.1002/ceat.200500397/abstract(accessed june 2012)., 29(2), 233-237.

[28] Lu, J, & Rohani, S. Polymorphism and Crystallization of Active Pharmaceutical Ingredients (APIs) Current Medicinal Chemistry (2009). http://www.benthamdirect.org/pages/content.php?CMC/2009/00000016/00000007/0006C.SGM(accessed june 2012)., 16, 884-905.

[29] Reichert, W M, Holbrey, J D, Vigour, K B, Morgan, T D, Broker, G A, & Rogers, R D. Approaches to Crystallization from Ionic Liquids: Complex Solvents-Complex Results, or, a Strategy for Controlled Formation of New Supramolecular Architectures? ChemInform (2007). http://onlinelibrary.wiley.com/doi/10.1002/chin.200712247/abstract?deniedAccessCustomisedMessage=&userIsAuthenticated=true(accessed june 2012)., 38(12), 4767-4779.

[30] Golding, J, & Forsyth, S. MacFarlane D R, Forsyth M, Deacon G B. Methanesulfonate and p-toluenesulfonate salts of the N-methyl-N-alkylpyrrolidinium and quaternary ammonium cations: novel low cost ionic liquids. Green Chemistry (2002). http://pubs.rsc.org/en/content/articlelanding/2002/gc/b201063a(accessed june 2012)., 4(3), 223-229.

[31] Abdul-sada, A K, Elaiwi, A E, & Greenway, A M. European Journal of Mass Spectrometry (1997). http://www.impublications.com/content/abstract?code=E03_0245(accessed june 2012)., 3(3), 245-247.

[32] Dean, M P, Turanjanin, J, & Yoshizawa-fujita, M. MacFarlane D R, Scott J L. Exploring an Anti-Crystal Engineering Approach to the Preparation of Pharmaceutically Active Ionic Liquids. Crystal Growth & Design (2009). http://pubs.acs.org/doi/abs/10.1021/cg8009496accessed 20 june 2012)., 9(2), 1137-1145.

[33] Higasio, Y S, & Shoji, T. Heterocyclic compounds such as pyrroles, pyridines, pyrrolidins, piperdines, indoles, imidazol and pyrazins. Applied Catalysis A (2001). http://www.sciencedirect.com/science/article/pii/S0926860X01008158(accessed june 2012).

[34] Belieres, J-P, & Angell, C A. Protic Ionic Liquids: Preparation, Characterization, and Proton Free Energy Level Representation. The Journal of Physical Chemistry B (2007). http://pubs.acs.org/doi/abs/10.1021/jp067589u(accessed june 2012)., 111(18), 4926-4937.

[35] Ferraz, R, Branco, L C, Marrucho, I M, Araújo, J, Rebelo, M, Ponte, L P N, Prudêncio, M N, Noronha, C, & Petrovski, J P. Z. Developmentof novel ionicliquidsbasedonampicillin. Medicinal Chemistry Communication (2012). http://pubs.rsc.org/en/content/articlelanding/2012/md/c2md00269haccessed 20 june 2012)., 3(4), 494-497.

[36] Fukumoto, K, Yoshizawa, M, & Ohno, H. Room Temperature Ionic Liquids from 20 natural Amino Acids. Journal of the Americal Chemical Society (2005). http://pubs.acs.org/doi/abs/10.1021/ja043451iaccessed 20 june 2012)., 127(8), 2398-2399.

[37] Bica, K, Rijksen, C, Nieuwenhuyzen, M, & Rogers, R D. In search of pure liquid salt forms of aspirin: ionic liquid approaches with acetylsalicylic acid and salicylic acid. Physical Chemistry Chemical Physics (2010). http://pubs.rsc.org/en/content/articlelanding/2010/cp/b923855gaccessed 20 june 2012)., 12(8), 2011-2017.

[38] Bica, K, & Rogers, R D. Confused ionic liquid ions-a "liquification" and dosage strategy for pharmaceutically active salts. Chemical Communications (2010). http://pubs.rsc.org/en/content/articlelanding/2010/cc/b925147b(accessed june 2012)., 46(8), 1215-1217.

[39] Thalladi, V R, Goud, B S, Hoy, V J, Allen, F H, Howard, J, & Desiraju, K. G R. Supramolecularsynthons in crystal engineering. Structure simplification, synthon robustness and supramolecularretrosynthesis. Chemical Communications (1996). http://pubs.rsc.org/en/content/articlelanding/1996/cc/cc9960000401(accessed june 2012).

[40] Desiraju, G R. Crystal Engineering: A Holistic View. AngewandteChemie International Edition (2007). http://onlinelibrary.wiley.com/doi/10.1002/anie.200700534/abstract(accessed june 2012)., 46(44), 8342-8356.

[41] Desiraju, G R. SupramolecularSynthons in Crystal Engineering- A New Organic Synthesis. AngewandteChemie International Edition in English (1995). http://onlinelibrary.wiley.com/doi/10.1002/anie.199523111/abstract(accessed june 2012)., 34(21), 2311-2327.

[42] Blagden, N, Matas, M, Gavan, P T, & York, P. Crystal engineering of active pharmaceutical ingredients to improve solubility and dissolution rates. Advanced Drug Delivery Reviews (2007). http://www.sciencedirect.com/science/article/pii/S0169409X07000828(accessed june 2012)., 59(7), 617-630.

[43] Trask, A V, & Jones, W. Crystal Engineering of Organic Cocrystals by the Solid-State Grinding Approach. Topics in Current Chemistry (2005). http://www.springerlink.com/content/j54wclkmeug3y254/(accessed june 2012)., 254, 41-70.

[44] Vishweshwar, P, Mcmahon, J A, Bis, J A, & Zaworotko, M J. Pharmaceutical co-crystals. Journal of Pharmaceutical Sciences (2006). http://onlinelibrary.wiley.com/doi/10.1002/jps.20578/abstract(accessed june 2012)., 95(3), 499-516.

[45] Fredlake, C P, Crosthwaite, J M, Hert, D G, Aki, S, & Brennecke, V K. J F. Thermophysical Properties of Imidazolium-Based Ionic Liquids. Journal of Chemical & Engineering Data (2004). http://pubs.acs.org/doi/abs/10.1021/je034261aaccessed 20 june 2012)., 49(4), 954-964.

[46] Pernak, J, Syguda, A, Mirska, I, Pernak, A, Nawrot, J, Pradzynska, A, Griffin, S T, & Rogers, R D. Choline-Derivative-Based Ionic Liquids. Chemistry A European Journal (2007). http://onlinelibrary.wiley.com/doi/10.1002/chem.200700285/abstractaccessed 20 june 2012)., 13(24), 6817-6827.

[47] Hough-troutman, W L, Smiglak, M, Griffin, S, Reichert, W M, Mirska, I, Jodynis-liebert, J, Adamska, T, Nawrot, J, Stasiewicz, M, Rogers, R D, & Pernak, J. Ionic liquids with dual biological function: sweet and anti-microbial, hydrophobic quaternary ammonium-based salts. New Journal of Chemistry (2009). http://pubs.rsc.org/en/content/articlelanding/2009/nj/b813213paccessed 20 june 2012)., 33(1), 26-33.

[48] Viau, L, Tourne-peteilh, C, Devoisselle, J-M, & Vioux, A. Ionogels as drug delivery system: one-step sol-gel synthesis using imidazoliumibuprofenate ionic liquid. Chemical Communications. (2010). http://pubs.rsc.org/en/content/articlelanding/2010/cc/b913879jaccessed 20 june 2012)., 46(2), 228-230.

Chapter 8

New Generations of Ionic Liquids Applied to Enzymatic Biocatalysis

Ana P.M. Tavares, Oscar Rodríguez and
Eugénia A. Macedo

Additional information is available at the end of the chapter

1. Introduction

Ionic liquids are salts in a liquid state, combinations of cations and anions that are liquid at temperatures below 100 °C. Thus, they have been called Room-Temperature Ionic Liquids (RTILs, or just ILs) in order to differentiate them from traditional salts, which melt at much higher temperatures and receive the name of "molten salts". In contrast to conventional organic solvents, ILs usually have extremely low volatility. Indeed, vapor pressures for ILs are scarce in the literature exactly because they are extremely low (< 1 Pa) and have to be obtained at high temperatures (400-500 K) [1]. For this "negligible" vapor pressure, ILs are often said to be "green" solvents when compared to traditional, environmentally harmful volatile organic compounds (VOCs). A big goal in the use of ILs in enzymatic reactions is the replacement of VOCs by ILs. In addition, ILs have other potential advantageous properties such as reasonable thermal stability; ability to dissolve a wide range of organic, inorganic and organometallic compounds; controlled miscibility with organic solvents (which is relevant for applications in biphasic systems) among others. All these properties make them very attractive non-aqueous solvents for biocatalysis. As they have been extensively described, ILs offer new possibilities for the application of solvent engineering to enzymatic reactions. Biocatalysis with ILs as reaction medium was first showed in the beginning of 2000 [2-4]. During the last decade, ILs have fast increased their attention as reaction media for enzymes with some remarkable results [2-4]. The advantage of using ILs in enzymatic biocatalysis, as compared to VOCs, is the enhancement in the solubility of substrates or products without inactivation of the enzymes, high conversion rates and high activity and stability [5]. ILs are also being used as co-solvents in aqueous biocatalytic reactions, since ILs help to dissolve nonpolar substrates, while avoiding enzyme inactivation like water-miscible organic solvents, as DMSO or acetonitrile, often do [6].

Another mentioned characteristic of ILs is the possibility of obtaining the desired physico-chemical properties by selecting combinations of cations and anions ("tunability"), which makes them "designer solvents". For example, ILs can be produced to be water-miscible, partially miscible or totally immiscible, and can also be synthesized with different viscosities. These interesting properties make them a very important reaction media for enzyme stabilization and reaction. The use of organic solvents in bioprocess presents a number of further problems. The main concerns are the toxicity of the organic solvents to both the process operators and the environment (eco-toxicity), and also the volatile and flammable nature of these solvents, which make them a potential explosion hazard [7]. Thus, ILs have emerged as a potential replacement for organic solvents in biocatalytic processes at both laboratory and industrial scale. The negligible vapor pressure means that they emit no volatile compounds, and also introduces the additional possibility of removal of products by distillation without further contamination by the solvent. It also facilitates the recycling of ILs, decreasing operation costs. All these properties make ILs very important for the stabilization and activation of enzymes; therefore, numerous enzymatic reactions have been investigated in different types of ILs as will be shown in the next sections. Several topics about biocatalysis in ILs will be reviewed: their effect on the activity and stability of enzymes, toxicity of ILs, new generation of ILs and methods to stabilize enzymes will be discussed.

2. Enzymatic activity and stability in ionic liquids

The most important criteria for selecting an enzyme-IL system are the activity and stability of the enzyme within the reaction medium. ILs have been reported to be an appropriated medium to increase the stability and activity of enzymes, as opposed to common organic solvents [8-11]. However, depending on the enzyme nature, the IL can be or not suitable for the reaction [12].

2.1. Lipases

Lipases, and *Candida Antarctica* Lipase B (CALB) in particular, are the most studied enzymes in ILs. Most of these reactions in ILs are carried out with, no or low content of water as co-solvent. Therefore, hydrophobic ILs are used and the enzyme activity and stability is dependent on the IL. Several studies in the literature show that enzymes, majority lipases, exhibit greater stability in pure ILs than in traditional organic solvents [13, 14]. A review of Zhao 2005 [15] shows that ILs with larger cations are better for enzyme activity than smaller cations. The reasons for that are the longer hydrophobic alkyl chains in the cation presents less tendency to take away the essential water molecules from the enzyme. In fact, one of the most interest conclusions of Zhao is that hydrophobic ILs maintain lipase activity and stability better than hydrophilic ILs, as the latter will take water molecules away from enzyme structure. According to Diego et al. [16] the enzyme stabilization by water immiscible ILs (such as $[(CF_3SO_2)_2N]^-$ types) can be explained by a more compact enzyme conformation/ confinement formed from the evolution of α-helix to β-sheet secondary structure of the enzyme. On the other hand, hydrophobic ILs may decrease the stability and activity of the enzyme due to: (1) the interaction with the substrates or products, as organic solvents [17]; (2) interaction by electrostatic forces [18] and (3) removing essential water molecules from the enzyme [17]. Lau et al. [19] observed that enzyme activity in ILs was related with the conformation of enzyme; the hydrogen bonding could be the key to understanding the interactions of enzymes and ILs. Another work of Lozano et al. [20] showed that lipase and α-chymotrypsin were strongly stabilized in two ILs ([btma][NTf$_2$] and [emim][NTf$_2$]) due to the maintenance of the native structure of the enzymes, as observed by both fluorescence and circular dichroism spectroscopy.

2.2. Cellulases

ILs are also used in the pretreatment of cellulose hydrolysis by cellulase for production of biofuels and other products. However, cellulases can be inactivated in the presence of ILs, even when present at low concentrations. In order to explore these ILs abilities, it is important to find a compatible cellulose-IL system [10]. The IL must solubilize the lignocellulosic biomass and at the same time, keep the enzyme active. It was shown that pretreatment of cellulose with ILs such as [bmim][Cl], [mmim][Cl], and [HEMA] resulted in faster conversion to glucose and thermostability than hydrolysis with cellulose that was not pretreated [22, 23]. A similar behavior was found for cellulases from different sources with imidazolium-based ILs, which enhanced the enzyme and thermal stability [24]. The stability of cellulases from *Penicillium janthinellum* mutants was evaluated in 10-50% (v/v) of [bmim][Cl] and the enzymes were significantly stable in 10% (v/v) of IL [25]. Another work investigated the stability and activity of commercial cellulases in aqueous solutions of 1-ethyl-3-methylimidazolium acetate [emim][OAc]. Cellulases retained 77% of their original activity in 15% and 20% (w/v) of IL and presented an avicel (a model substrate for cellulose) conversion efficiency of 91% [26].

2.3. Oxidoreductases

Several oxidoreductases, such as laccase, peroxidase, chloroperoxidase, D-amino acid oxidase and alcohol dehydrogenases, have been reported as active enzymes in aqueous solution with ILs [27]. When compared to organic solvents, these enzymes are more active and stable in the presence of ILs [27].

Laccases and peroxidases are the most effective enzymes capable to catalyze the degradation of phenolic compounds. Phenolics such as hydroquinone, catechols, guaiacol, and 2,6-dimethoxyphenol are good substrates for these enzymes in either aqueous and non-aqueous media. Recent reports have been addressing the activity and stability of both enzymes in ILs [6, 28-31]. For example, laccase activity and stability was well maintained in the presence of several imidazolium-based ILs [31] such as [C$_4$mim][Cl], [emim][MDEGSO$_4$], [emim][EtSO$_4$] and [emim][MeSO$_3$] [6], but are inactivated in the presence of [C$_{10}$mim][Cl] [29]. Peroxidase was also described to mantain its activitiy and stability in imidazolium-based ILs for concentrations up to 25 % v/v [30].

Alcohol dehydrogenases are enzymes that catalyze the reduction of ketones. Due to the vast field of aplication of alcohol dehydrogenases, the study of this enzyme in ILs is promising. A recent work presented the effect of 10 different ILs (with either imidazolium or ammonium cations) on the enzyme stability. Improved storage stabilities and improved enzyme activities were found in the most promising, ammonium-based, AMMOENG™ 101 IL [32]. Later, the same group [33] proved the feasibility of continuous production using the previously recommended IL, combined with product separation using a membrane bioreactor (the so-called process integration). Hussain and co-workers [34] showed that the use of 10% (v/v) [bmp][NTf₂] facilitated the conversion of ketone to the chiral alcohol. Dabirmanesh et al. [21], showed the influence of different imidazolium based ILs on the structure and stability of alcohol dehydrogenase and the results exhibited that the ILs could affected the enzyme stability, but not the tertiary structure, suggesting that the enzyme was reversibly inhibited.

There are only few reports investigating the enzyme activity of D-amino acid oxidase in ILs. This enzyme catalyzes the deamination of various d-amino acids into imino acids. The activity and stability of free and immobilized d-amino acid oxidase in five imidazolium ILs were evaluated, and the most promising ILs were [bmim][BF₄] and [mmim][MMPO₄]. Total conversion of substrate in presence of 20% [mmim][MMPO₄] was obtained [35].

3. Factors affecting enzymes in ionic liquids

The section before showed the stabilization and activation of enzymes in ILs. However, it is also very important to understand the factors affecting the enzymes activity and stability in IL media. It has been reported that enzyme reactions in ILs can be affected by several factors such as the water activity, pH, excipients and impurities [36]. Several properties of ILs have also been related to the activity and stability of enzymes. The most important include: polarity, hydrogen-bonding capacity, viscosity, kosmotropicity/chaotropicity and hydrophobicity, among others. It is clear from this set of properties that the type and strength of interactions ILs can establish with enzyme molecules will certainly influence their 3D structure. Such influence may produce or not changes in enzyme activity.

A few works have related the ILs polarity with the activity of enzymes. Lozano and co-workers [37] observed that in less polar IL, lower activities of α-chymotrypsin were obtained. The same behavior was obtained for lipase, the enzyme activity increased with the increase in IL polarity during the acetylation of racemic 1-phenylethanol with vinyl acetate [38] and for the synthesis of methylglucose fatty acid esters [39].

The negative effect of hydrogen-bonding on the enzyme activity in the presence of ILs can be associated with the anions effect and their action as hydrogen-bonding acceptors for the protein (lipase) [40]. Another work suggested a similar reasoning for the effect of anions: the decrease of lipase activity in [bmim][lactate] was caused by secondary structure changes of the protein, due to hydrogen-bonding interactions between lactate anions and peptide chains [19]. However, due to the limited number of ILs and enzymes investigated, deeper studies are required for a better understanding of this interaction.

As the majority of ILs are viscous fluids, the mass transfer limitations should be considered when the reaction is rapid and the IL is relatively viscous. Many enzymatic reactions in pure ILs can be heterogeneous due to the low solubility of the enzymes in ILs. Some studies have reported that the activity of enzymes is dependent on the IL viscosity: Bose et al. [23] attributed the lower activity of cellulase to the high IL ([HEMA]) viscosity. Lozano et al. [37] indicated that the activity of α-chymotrypsin was dependent on the IL viscosity, and thus higher enzyme activities were observed in less viscous ILs. On the other hand, the work of Zhao et al. [41] suggested that IL viscosity was not directly related to the lipase activity, but mass transfer limitations. The high viscosity may reduce the reaction rate, however the IL structure was responsible for lipase stabilization. So the author concludes that IL viscosity could influence the enzymatic reaction rates, however it is not the principal factor for the enzyme stabilization. Basso et al. [42] suggested that in the reactions for amide synthesis by immobilized penicillin G amidase, the high viscosities of the ILs did not affect the initial rates. Concluding, the effect of IL viscosity can affect the reaction rate, but this behavior is not the same for all enzymatic reactions in ILs, specially when reaction rates are measured in equilibrium instead of kinetics [42].

The kosmotropicity/chaotropicity (Hofmeister series) is related with the effect of water structure (and thus, protein salting in/out). There are reports in the literature that try to correlate the ion kosmotropicity with the enzyme behavior in aqueous solutions of ILs [43-48]. The reviews by Zhao et al. [15] and by Yang [49] discuss the probable mechanisms of Hofmeister effects of ILs. Kosmotropic anions (PO_4^{3-}, CO_3^{2-}, SO_4^{2-}, ...) and chaotropic cations (Cs$^+$, Rb$^+$, K$^+$, NH^{4+}, ...) stabilize enzymes, while chaotropic anions (NO_3^-, I$^-$, BF_4^-, PF_6^-,...) and kosmotropic cations ((C_4H_9)N$^+$, (C_3H_7)$_4$N$^+$, (C_2H_5)$_4$N$^+$,...) destabilize it [15].

Attending to the solubility of ILs in water, they can be divided into hydrophobic (water immiscible) and hydrophilic (water miscible). Most often, water miscibility depends on the ILs anions rather than the cations [50]. The hydrophobicity in ILs is generally determined by the log P scale, based on the partition coefficient of ILs between 1-octanol and water [51]. The stablility of enzymes can also be related to the log P. Usually, enzymes are more stable in solvents with a larger log P (>3) [106]. Many works from literature have reported that for lipases, activity increases with the increase in the IL hydrophobicity [13, 51-55]. Nevertheless, this conclusion is in contradiction with the polarity effect mentioned at the beggining of the section (more polar ILs promote enzyme stability). In our opinion, the vast pool of ILs and enzymes, and the large differences in their chemical structure, make it very difficult to extract general trends and conclusions. Just as an example, several authors [37] have proved that hydrophobic ILs (thus, less polar) maintain better immobilized lipase activity in the pure IL (low water content). At the same time, our group has shown that laccase and peroxidase activities are best maintained in more polar (less hydrophobic, more hydrophilic) ILs, when used in aqueous solutions [6; 29]. Both statements are correct, because the reaction conditions are completely different: very low water content in the former study, and water excess in the latter. Thus, the IL affinity for water will be dramatic at low water content, but not when there is plenty of water.

4. Green aspects of ionic liquids

The interest in the development of biocatalytic processes in ILs media is desired to obtain green technologies and unconventional properties to replace organic solvents (namely VOCs). ILs appear free of many problems associated with the use of VOCs due to their non-volatility, non-flammable character and both high thermal and chemical stability. However, the use of certain ILs raises some concerns regarding environmental impact, attending to their potential toxicity and biodegradability. As the use of ILs has been increasing in different fields from biology to electrochemistry, the assessment of their environmental, health and safety impact is highly required. In recent years, environmental aspects related to ILs have been strongly addressed, stating that many ILs commonly used cannot be regarded as 'green solvents'. In general, ILs used in biocatalysis have not been designed for biocompatibility and harmless. There are some recent reports showing that the ecotoxicity of alkylmethylimidazolium cations (the most used in biocatalysis) is undesirable, and ecotoxicity increases with the length of alkyl chains in cation [56-58]. Thus, for future applications it is necessary to improve the green aspects of ILs. These improvements are currently going on. The best examples are the choline-derived cations (which are based on food grade choline chloride) or imidazolium derivatives designed for biodegradability (*e.g.*, adding ether groups in the alkyl side chains) [59], and ILs based on amino acids [60-61]. It is expected that much improved and green ILs will become available soon. Currently, three different generations of ILs can be identified, as described below.

4.1. First generation of ionic liquids

The first IL known was ethylammonium nitrate, reported in 1914 by Walden [62], but attracted little interest. The first generation of ILs with widespread utilization was mainly composed of cations like dialkylimidazolium and alkylpyridinium derivatives, and anions like chloroaluminate and other metal halides which have been described as toxic and non-biodegradable [57]. The most common anions are chloroaluminate or other metal halide anions that react with water and thus are not suitable for biotransformations. This generation of ILs was also oxygen-sensitive [63] and can only be handled under inert-gas atmosphere due to the hygroscopic nature of $AlCl_3$ [64]. In the 1980s, Wilkes et al. started the extensive research on first generation ILs [65]. However, due to these limitations, the progress in their use was limited. For this reason, research was directed towards the synthesis of air- and water-stable ILs, the second generation of ILs.

4.2. Second generation of ionic liquids

After one decade the second generation of ILs [66] appeared. The water- and oxygen-reactive anions were replaced by halides (Cl⁻, Br⁻, I⁻) or anions such as BF_4^-, PF_6^- and $C_6H_5CO_2$, which are stable to water and air. Cations such as dialkylimidazolium or alkylpyridinium were maintained, and ammonium and phosphonium were added. These ILs present interesting properties such as lower melting points, different solubilities in classic organic solvents, viscosities, etc. Due to these properties, the second generation attracted a great interest in various fields, and research in ILs experienced an important boost from the 1990's. The first reports of biocatalysis with ILs were published in the beginning of 2000's [2, 4, 38, 67]. One of the disadvantages of these ILs is the high cost. According to Gorke et al. [66], the high costs are related to starting materials (namely fluorinated components) and purification of final product required in the preparation. The most important disadvantage of the second generation is the toxicity, which in general is similar to those of chlorinated and aromatic solvents [56]. However, this second generation of ILs attracted the attention of the wide scientific community and has been providing interesting and novel applications in differents areas. This generation of ILs is the most studied and a great number of applications in biocatalysis have been published. The activity, stability, kinetic and thermal stability of different enzymes such as oxidases, lipases or cellulases has been studied, and sinthesis of various products has been carried out.

4.3. Third generation of ionic liquids

The third generation of ILs (advanced ILs) is based on more hydrophobic and stable anions such as $[(CF_3SO_2)_2N^-]$, sugars, amino or organic acids, alkylsulfates, or alkylphosphates and cations such as choline. The cations and/or anions used are biodegradable, readily available, and present lower toxicities. Besides, a new class of solvent systems, called deep eutectic solvents (DES), is more hydrophilic than the second generation, and in general is water-miscible [66]. DES are mixtures of salts (in general they are not liquids at room temperature) such as choline chloride, and uncharged hydrogen bond donors such as amines, amides, alcohols, carboxylic acids, urea, or glycerol [28]. A typical example is the choline chloride/urea mixture, which produces a DES with a melting point of 12°C at concentrations around 50% [66]

The advantages of the third generation are: lower costs (similar to organic solvents), simple to prepare, biodegradable, do not require purification, the purity of the starting materials determines the final purity and uses anions and cations with low toxicity. As this generation is recent, few works have been published [68, 69]. The transesterification of ethyl valerate with 1-butanol, showed good activity in DES, and in choline chloride: glycine the activity was similar to activity in toluene for all lipases [69]. The third generation will reach the commercial level soon [70].

5. Methods for stabilization of enzymes in ionic liquids

Stabilization of enzymes in ILs is one of the keys for the development of more efficient biocatalytical processes for industrial, environmental, or biomedical applications. As discussed in previous sections, stabilization of enzymes in ILs is one of the keys for the development of more efficient biocatalytic processes for industrial, environmental, or biomedical applications. The use of enzymes in ILs presents different advantages when compared to conventional organic solvents. On the other hand, in some cases the application of enzymes can be limited by the low solubility, activity or stability in ILs. The improvement of enzyme functionality is crucial for large-scale applications in order to be economically viable. The methods to stabilize and activate enzymes in ILs can be divided into two different strategies: the modification of enzymes and/or the modification of the solvent (ILs). The modification of enzymes includes lyophilization (to change the morphology of the solid enzyme), chemical modification (for the chemical addition of functionalities into the enzyme biomolecule) and immobilization in a suitable support. The second strategy includes the modification of the IL reaction media, such as IL coating, additives or use of microemulsions with ILs. These methodologies have been used with promissory results [5].

5.1. Modification of solvent media

In order to avoid the enzyme insolubility, some works have reported the introduction of functional groups in IL structure such as hydroxyl, ether, and amide (which present high affinity for enzymes) [19]. For enzymes that are active in pure solvents, such as lipases, the most hydrophilic ILs can remove enzyme-bound water molecules that are essential to maintain protein structure and active function. In such case, these ILs (hydrophilic) are not adequate.

Another strategy is the addition of water in IL (co-solvent), but the enzyme may present low catalytic activity due to a changed conformation in ILs [71]. Several researchers have reported enzymatic reactions, especially for oxidative enzymes, in hydrophilic ILs with a high concentration of water (in the range 5 – 50%) and promissory results have been found [6, 29-31].

Water-in-IL microemulsions, or reverse micelles, have been used as a very efficient technique for solubilizing enzymes in hydrophobic ILs. The advantage of this approach is that the enzyme is protected of the contact with the solvent by a layer of water and surfactant molecules. As an example, the use of water-in-IL microemulsions was reported by Moniruzzaman et al. [50] as a new medium for dissolving various enzymes and proteins. Additionally, several authors have reported the use of different microemulsions systems with good results for enzyme stability [72-74].

5.2. Modification of the enzyme

The most common methodology for enzyme modification is immobilization. It is well known that immobilization of enzymes presents excellent advantages for biocatalysis, namely in the recovery of the enzyme for reutilization, product separation and recovery from the reaction media, application in continuous systems, and for enzyme stabilization. Indeed, enzyme immobilization increases thermal and operational stability of the biocatalysts compared to the free enzyme.

The use of immobilized enzymes in IL media has been reported by many research groups, using different methods of immobilization and supports. The most frequently used enzyme immobilization techniques are: physical adsorption, covalent attachment, entrapment in polymeric matrixes and cross-linking of enzyme molecules. For lipases, it was found that reaction rates in ILs were comparable or higher than in organic solvents and also immobilized lipase was more active than its free form [75-79]. The same behavior was found for proteinase [80], papain [81] and for heme-containing proteins [82].

The chemical modification of enzymes with poly(ethylene glycol) (PEG) is a well-known method (the so-called PEGylation) for enzyme stabilization in denaturing environments. PEG presents both hydrophilic and hydrophobic properties, so the modified enzymes can increase their solubility in some ILs [83]. Turner et al. [84] also reported higher activity of PEGylated cellulase than free cellulase in IL solutions.

Another method for activating and stabilizing enzymes in non-aqueous media is co-lyophilization of the enzyme. Maruyama et al. [85] lyophilized lipase with poly(ethylene glycol) (PEG) to prepare PEG–lipase complexes, finding that the activity of lipase in ILs increased more than 14-fold. Wang and Mei [86] also lyophilized lipase with cyclodextrins, and the activity of lipase in ILs ([bmim][PF$_6$] and [bmim][BF$_4$]) was improved.

6. Applications of ionic liquids in biocatalysis

The use of ILs as solvents or co-solvents for reaction media of enzymes is well recognized in biocatalysis. Examples available in the open literature include: polymerizations, biosensors, production of biofuels, synthesis of sugar- and ester-derivatives, among many others.

A large number of examples of the use of ILs for the enzymatic production of esters by lipase have been published [87]. The common esters synthetized in ILs are aliphatic and aromatic esters, for applications in polymers, biodiesel, and in the perfume, flavour and pharmaceutical industries. The synthesis of a wide range of aliphatic organic esters was carried out by transesterification from vinyl esters and alcohols and catalyzed by lipase in different 1,3-dialkylimidazolium ILs [88, 89]. Aromatic esters have also been synthetized with lipase in two ILs, [bmim][PF$_6$] and [bmim][BF$_4$] [90]. The esterification of 2-substituted-propanoic acids with 1-butanol was catalyzed by lipase in ILs [bmim][PF$_6$] and [omim][PF$_6$] [91]. Yuan et al. [92] studied the enantioselective esterification of menthol with propionic anhydride using lipase in [bmim][PF$_6$] and [bmim][BF$_4$]. The resolution of (R,S)-ibuprofen by esterification with lipases in the same ILs is another interesting example [93]. The aliphatic polyester synthesis by lipase, also in [bmim][PF$_6$], was reported by Nara et al. [94]. Later, the enzymatic preparation of polyesters by ring-opening polymerization and by polycondensation with lipase in [bmim][Tf$_2$N], [bmim][PF$_6$] and [bmim][BF$_4$] was also investigated [95]. According to these authors, the use of ILs could be an advantage in the polymerization of highly polar monomers with low solubility in organic solvents.

The production of biofuels, such as biodiesel (fatty acid methyl esters) has been also investigated in ILs through the transesterification of a triglyceride with methanol. Biodiesel is a renewable and environmentally-friendly fuel. Several ILs have been utilized for biodiesel production. Most often, the synthesis of biodiesel by enzymatic reactions in ILs is based on a short-chain 1,3-dialkylimidazolium cation, such as [bmim][PF$_6$] or [bmim][NTf$_2$], and the reaction is carried out in a biphasic system with lipase and using an adequate substrate (e.g., soybean oil) [96]. For homogeneous one-phase systems, imidazolium ILs with long alkyl chains such as [C$_{16}$mim][NTf$_2$] and [C$_{18}$mim][NTf$_2$] have been used [97,98]. These long chain, lipophilic ILs create a nonaqueous system suitable for oil transesterification. Ha et al. [99] studied the biodiesel production using immobilized lipase in 23 ILs. Among the ILs tested, it was found that highest biodiesel production yield was obtained in [emim][TfO]. But it is important to highlight that several works have been published for biodiesel production by lipases [100-102].

In recent years, a significant number of publications have showed the direct electron-transfer reaction between redox proteins or enzymes and IL-based composite electrodes. Biosensors are small devices which convert the biological recognition event into an electrical signal, so it can be used for selective analysis [103]. Several composite electrodes based on ILs have been prepared. Many of them can be found in a recent review by Shiddiky and Torriero[104], such as: hemoglobin biosensor; myoglobin and cytochrome c biosensors; catalase biosensors; glucose oxidase biosensors; horseradish peroxidase biosensors.

Sugar-based compounds are widely used in pharmaceuticals, cosmetics, detergents and food. A recent review by Galonde et al. [105] shows the synthesis of glycosylated compounds in ILs.

7. Conclusion

Ionic liquids have demonstrated to be suitable solvents for enzymatic reactions. They can be beneficial regarding to activity, (enantio)selectivity and stability of enzymes. The use of enzymes in ILs opens new possibilities for non-aqueous enzymology with high efficiency in several areas. Here, it was shown that a large variety of enzymes tolerate ILs or aqueous-IL mixtures as reaction medium. Moreover, the development of green and biodegradable ILs is reinforcing enzymatic applications of ILs, as stated in this work. Indeed, it is expected to become a standard in biotransformations, thus contributing to a greener chemical and biochemical industries.

Nomenclature

tris-(2-hydroxyethyl)-methylammonium methylsulfate [HEMA]

Cations:

[mmim] = 1,3-dimethylimidazolium

[emim] = 1-ethyl-3-methylimidazolium

[bmim] or [C_4mim] = 1-butyl-3-methylimidazolium

[bmp] = butylmethylpyrrolidinium

[btma] = butyl-trimethylammonium

[omim] = 1-octyl-3-methylimidazolium

[C_{10}mim] = 1-decyl-3-methylimidazolium

[C_{16}mim] = 1-hexadecyl-3-methylimidazolium

[C_{18}mim] = 1-octadecyl-3-methyl- imidazolium

Anions

[OAc] = acetate

[MDEGSO$_4$] = 1-ethyl-3-methylimidazolium 2-(2-methoxyethoxy)

[MeSO$_3$] = methanesulfonate

[TfO] = trifluoromethanesulfonate

[BF$_4$] = tetrafluoroborate

[MMPO$_4$] = dimethylphosphate

[Cl] = chloride

[EtSO$_4$] = ethyl sulfate

[(CF$_3$SO$_2$)$_2$N$^-$] = bis(trifluoromethylsulfonyl)amide

[MeSO$_4$] = methyl sulfate

[PF$_6$] = hexafluorophosphate

[NTf$_2$] = bis(trifluoromethylsulfonyl)imide

Acknowledgements

This work was supported by project PEst-C/EQB/LA0020/2011, financed by FEDER through COMPETE - Programa Operacional Factores de Competitividade and by Fundação para a

Ciência e a Tecnologia (FCT, Portugal). A.P.M. Tavares and O. Rodríguez acknowledge the financial support (Programme Ciência 2008 and Programme Ciência 2007, respectively) from FCT.

Author details

Ana P.M. Tavares*, Oscar Rodríguez and Eugénia A. Macedo

LSRE - Laboratory of Separation and Reaction Engineering - Associate Laboratory LSRE/ LCM, Faculdade de Engenharia, Universidade do Porto, Porto, Portugal

References

[1] Esperança JMSS, Canongia JNL, Tariq M, Santos LMNBF, Magee JW, Rebelo LPN. Volatility of Aprotic Ionic Liquids A Review. Journal of Chemical Engineering Data 2010;55 3–12.

[2] Cull SG, Holbrey JD, Vargas-Mora V, Seddon KR, and Lye GJ. Room-temperature ionic liquids as replacements for organic solvents in multiphase bioprocess operations. Biotechnology and Bioengineering 2000;69(2) 227-233.

[3] Erbeldinger M, Mesiano AJ, Russell AJ. Enzymatic catalysis of formation of Z-aspartame in ionic liquid - an alternative to enzymatic catalysis in organic solvents. Biotechnology Progress 2000;16 1129-1131.

[4] Lau RM, van Rantwijk F, Seddon KR, Sheldon RA. Lipase-catalyzed reactions in ionic liquids. Organic Letters 2000;2 4189-4191.

[5] Moniruzzaman M, Kamiya N, Nakashima K, Goto M. Water-inionic liquid microemulsions as a new medium for enzymatic reactions. Green Chemistry 2008;10 497–500.

[6] Tavares APM, Rodriguez O, Macedo EA. Ionic liquids as alternative co-solvents for laccase: Study of enzyme activity and stability. Biotechnology and Bioengineering 2008;101 201-207.

[7] Schmid A, Kollmer A, Mathys RG, Withot B. Development toward large-scale bacterial bioprocesses in the presence of bulk amounts of organic solvents. Extremophiles 1998;2 249–256.

[8] Datta S, Holmes B, Park JI, Chen Z, Dibble DC, Hadi M, Blanch HW, Simmons BA, Sapra R. Ionic liquid tolerant hyperthermophilic cellulases for biomass pretreatment and hydrolysis. Green Chemistry 2010;12(2) 338–345.

[9] Lozano P, Diego TD, Carrie D, Vaultier M, Iborra JL. Continuous green biocatalytic processes using ionic liquids and supercritical carbon dioxide. Chemical Communication 2002;7: 692–693.

[10] Wang Y, Radosevich M, Hayes D, Labbe N. Compatible ionic liquidcellulases system for hydrolysis of lignocellulosic biomass. Biotechnology and Bioengineering 2011;108(5) 1042–1048.

[11] Persson M, Bornscheuer UT. Increased stability of an esterase from Bacillus stearothermophilus in ionic liquids as compared to organic solvents. Journal of Molecular Catalysis B: Enzymatic 2003;22 21–27.

[12] Mantarosie L, Coman S, Parvulescu VI. Comparative behavior of various lipases in benign water and ionic liquids solvents. Journal of Molecular Catalysis A: Chemical 2008;279 223–229.

[13] Kaar JL, Jesionowski AM, Berberich JA, Moulton R, Russell AJ. Impact of ionic liquid physical properties on lipase activity and stability. Journal of the American Chemical Society 2003;125 4125–4131.

[14] Shan H, Li Z, Li M, Ren G, Fang Y. Improved activity and stability of pseudomonas capaci lipase in a novel biocompatible ionic liquid, 1-isobutyl-3-methylimidazolium hexafluorophosphate. Journal of Chemical Technology and Biotechnology 2008;83 886–891.

[15] Zhao H. Effect of ions and other compatible solutes on enzyme activity, and its implication for biocatalysis using ionic liquids. Journal of Molecular Catalysis B: Enzymatic 2005;37 16–25.

[16] De Diego T, Lozano P, Gmouh S, Vaultier M, Iborra JL. Understanding structure-stability relationships of Candida antartica Lipase B in ionic liquids. Biomacromolecules 2005;6 1457-1464.

[17] Yang Z, Russell AJ in: Koskinen AMP, Kilbanov AM (Ed) Enzymatic Reactions in Organic Media. Blackie Academic & Professional, New York; 1996. p43.

[18] Park C, Raines RT. Quantitative Analysis of the Effect of Salt Concentration on Enzymatic Catalysis. Journal of the American Chemical Society. 2001;123 11472-11479.

[19] Lau RM, Sorgedrager MJ, Carrea G, van Rantwijk F, Secundo F, Sheldon RA. Dissolution of Candida antarctica lipase B in ionic liquids: effects on structure and activity. Green Chemistry 2004;6 483–487.

[20] Lozano P, De Diego T, Gmouh S, Vaultier M, Iborra JL. Dynamic structure_/function relationships in enzyme stabilization by ionic liquids. Biocatalysis and Biotransformation 2005;23 169-176.

[21] Dabirmanesh B, Khajeh K, Ranjbar B, Ghazi F, Heydari A. Inhibition mediated stabilization effect of imidazolium based ionic liquids on alcohol dehydrogenase. Journal of Molecular Liquids 2012;170 66–71.

[22] Bose S, Armstrong DW, Petrich JW. Enzyme-catalyzed hydrolysis of cellulose in ion-
      ic liquids: A green approach toward the production of biofuels. The Journal of Physi-
      cal Chemistry B 2010;114 8221–8227.

[23] Bose S, Barnes CA, Petrich JW. Enhanced stability and activity of cellulase in an ionic
      liquid and the effect of pretreatment on cellulose hydrolysis. Biotechnology and Bio-
      engineering 2012;109 434-443.

[24] Ilmberger N, Meske D, Juergensen J, Schulte M, Barthen P, Rabausch U, Angelov A,
      Mientus M, Liebl W, Schmitz RA, Streit WR. Metagenomic cellulases highly tolerant
      towards the presence of ionic liquids—linking thermostability and halotolerance.
      Applied Microbiology and Biotechnology 2012;95 135–146.

[25] Adsul MG, Terwadkar AP, Varma, AJ, Gokhale DV. Cellulases from penicillium jan-
      thinellum mutants: Solid-state production and their stability in ionic liquids. BioRe-
      sources 2009;4 1670-1681.

[26] Wang Y, Radosevich M, Hayes D, Nicole L. Compatible Ionic liquid-cellulases sys-
      tem for hydrolysis of lignocellulosic biomass. Biotechnology and Bioengineering
      2011;108 1042-1048.

[27] Tavares APM, Rodriguez O, Raquel Cristóvão, Macedo EA. Ionic Liquids: Alterna-
      tive Reactive Media for Oxidative Enzymes. In: Kokorin A (Ed) Ionic Liquids: Appli-
      cations and Perspectives. Rijeka: InTech; 2011 p449-516.

[28] Domínguez de María P. Ionic Liquids in Biotransformations and Organocatalysis:
      Solvents and Beyond. John Wiley & Sons; 2012.

[29] Rodríguez O, Cristóvão RO, Tavares APM, Macedo EA. Effect of the alkyl chain
      length on enzymatic kinetic with imidazolium ionic liquids. Applied Biochemistry
      and Biotechnology 2011;164 524-533.

[30] Carneiro AP, Rodríguez O, Mota FL, Tavares APM, Macedo EA. Kinetic and stability
      study of the peroxidase inhibition in ionic liquids. Industrial & Engineering Chemis-
      try Research 2009; 10810-10815.

[31] Domínguez A, Rodríguez O, Tavares APM, Macedo EA, Longo MA, Ma. Sanromán
      A. Studies of laccase from Trametes versicolor in aqueous solutions of several meth-
      ylimidazolium ionic liquids. Bioresource Technology 2011;102 7494–7499.

[32] Kohlmann C, Robertz N, Leuchs S, Dogan Z, Lütz S, Bitzer K, Naamnieh S, Greiner
      L. Ionic liquid facilitates biocatalytic conversion of hardly water soluble ketones.
      Journal of Molecular Catalysis B: Enzymatic 2011;68 147–153.

[33] Kohlmann C, Leuchs S, GreinerL. Walter Leitner Continuous biocatalytic synthesis of
      (R)-2-octanol with integrated product separation. Green Chemistry 2011;13
      1430-1436.

[34] Hussain W, Pollard DJ, Truppo M, Lye GJ. Enzymatic ketone reductions with co-fac-tor recycling: Improved reactions with ionic liquid co-solvents. Journal of Molecular Catalysis B: Enzymatic 2008;55 19–29.

[35] Lutz-Wahl S, Trost EM, Wagner B, Manns A, Fischer L. Performance of d-amino acid oxidase in presence of ionic liquids. Journal of Biotechnology 2006;124 163–171.

[36] Yang Z, Pan W. Ionic liquids: green solvents for nonaqueous biocatalysis. Enzyme Microbiology and Technology 2005;37 19–28.

[37] Lozano P, de Diego T, Guegan J-P, Vaultier M, Iborra JL. Stabilization of α-chymo-trypsin by ionic liquids in transesterification reactions. Biotechnology and Bioengin-eering 2001;75 563–569.

[38] Park S, Kazlauskas RJ. Improved preparation and use of room-temperature ionic liq-uids in lipasecatalyzed enantio- and regioselective acylations. Journal of Organic Chemistry 2001;66 8395-8401.

[39] Mutschler J, Rausis T, Bourgeois J-M, Bastian C, Zufferey D, Mohrenz. Ionic liquid-coated immobilized lipase for the synthesis of methylglucose fatty acid esters. Green Chemistry 2009; 1793–1800.

[40] Ventura SPM, Santos LDF, Saraiva JA, Coutinho JAP. Concentration effect of hydro-philic ionic liquids on the enzymatic activity of Candida antarctica lipase B. World Journal of Microbiology and Biotechnology 2012;28 2303–2310.

[41] Zhao H, Baker GA, Song Z, Olubajo O, Zanders L, Campbell SM. Effect of ionic liq-uid properties on lipase stabilization under microwave irradiation. Journal of Molec-ular Catalysis B: Enzymatic 2009;57: 149–157.

[42] Basso A, Cantone S, Linda P, Ebert C. Stability and activity of immobilised penicillin G amidase in ionic liquids at controlled. Green Chemistry 2005;7 671-676.

[43] Zhao H, Olubajo O, Song Z, SimsAL , Person TE, Lawal RA, Holley LA. Effect of kos-motropicity of ionic liquids on the enzyme stability in aqueous solutions. Bioorganic Chemistry 2006;34 15–25.

[44] Fujita K, MacFarlane DR, Forsyth M, Yoshizawa-Fujita M, Murata K, Nakamura N. Solubility and stability of cytochrome c in hydrated ionic liquids: effect of oxo acid residues and kosmotropicity. Biomacromolecules 2007;8 2080–2086.

[45] Constantinescu D, Weingartner H, Herrmann C, Protein denaturation by ionic liq-uids and the Hofmeister series: a case study of aqueous solutions of ribonuclease A. Angewandte Chemie International Edition 2007;6 8887–8889.

[46] Kaftzik N, Wasserscheid P, Kragl U. Use of ionic liquids to increase the yield and en-zyme stability in the β-galactosidase catalyzed synthesis of N-acetyllactosamine. Or-ganic Process Research and Development 2002;6 553–557.

[47] Lang M, Kamrat T, Nidetzky B. Influence of ionic liquid cosolvent on transgalactosy-lation reactions catalyzed by thermostable β-glycosylhydrolase CelB from Pyrococ-cus furiosus. Biotechnology and Bioengineering 2006;95 1093–1100.

[48] Zhao H, Song Z. Nuclear magnetic relaxation of water in ionic-liquid solutions: De-termining the kosmotropicity of ionic liquids and its relationship with the enzyme enantioselectivity. Journal of Chemical Technology and Biotechnology 2007;82 304-312.

[49] Yang Z. Hofmeister effects: an explanation for the impact of ionic liquids on biocatal-ysis. Journal of Biotechnology 2009;144 12–22.

[50] Moniruzzaman M, Nakashima K, Kamiya N, Goto M. Recent advances of enzymatic reactions in ionic liquids. Biochemical Engineering Journal 2010;48 295-314.

[51] Zhao H, Baker GA, Song Z, Olubajo O, Zanders L, Campbell SM. Effect of ionic liq-uid properties on lipase stabilization under microwave irradiation. Journal of Molec-ular Catalysis B: Enzymatic 2009;57 149–157.

[52] de los Rios AP, Hernandez-Fernandez FJ, Martinez FA, Rubio M, Villora G. The ef-fect of ionic liquid media on activity, selectivity and stability of Candida antarctica lipase B in transesterification reactions. Biocatalysis Biotransformation 2007;25 151–156.

[53] Hernandez-Fernandez FJ, de los Rios AP, Tomas-Alonso F, Gomez Dand Villora G. Stability of hydrolase enzymes in ionic liquids. The Canadian Journal of Chemical Engineering 2009;87 910–914.

[54] Nara SJ, Harjani JR and Salunkhe MM. Lipase-catalysed transesterification in ionic liquids and organic solvents: a comparative study. Tetrahedron Letter 2002;43 2979–2982.

[55] Shen Z-L, Zhou W-J, Liu Y-T, Ji S-J, Loh T-P. One-potchemo enzymatic syntheses of enantiomerically-enrichedO-acetylcyanohydrins from aldehydes in ionic liquid. Green Chemistry 2008;10 283–286.

[56] Docherty KM, Kulpa CF. Toxicity and antimicrobial activity of imidazolium and pyr-idinium ionic liquids. Green Chemistry 2005;7 185-189.

[57] Wells AS; Coombe VT. On the freshwater ecotoxicity and biodegradation properties of some common ionic liquids. Organic Process Research and Development 2006;10 794-798.

[58] Stolte S, Arning J, Bottin-Weber U, Matzke M, Stock F, Thiele K, Uerdingen M, Welz-Biermann U, Jastorff B, Ranke J. Anion effects on the cytotoxicity of ionic liquids. Green Chemistry 2006;8 621-629.

[59] Gathergood N, Scammels PJ, Garcia MT. Biodegradable ionic liquids. Part III. The first readily biodegradable ionic liquids. Green Chemistry 2006;8 156-160.

[60] Fukumotu K, Yoshizawa M, Ohno H. Room Temperature Ionic Liquids from 20 Natural Amino Acids. Journal of the American Chemical Society 2005;127 2398-2399.

[61] Tao G, He L, Liu W, Xu L, Xiong W, Wang T, Kou Y. Preparation, characterization and application of amino acid-based green ionic liquids. Green Chemistry 2006;8 639-646.

[62] Walden P. Molecular weights and electrical conductivity of several fused salts. Bulletin of the Imperial Academy of Sciences (St. Petersburg) 1914;1800 405-422.

[63] Endres F, Abedinw SZ. Air and water stable ionic liquids in physical chemistry. Physical Chemistry Chemical Physics 2006,8 2101–2116.

[64] Moustafa EM, Abedin SZ, Shkurankov A, Zschippang E, Saad AY, Bund A, Endres F. Electrodeposition of Al in 1-Butyl-1-methylpyrrolidinium Bis(trifluoromethylsulfonyl) amide and 1-Ethyl-3-methylimidazolium Bis(trifluoromethylsulfonyl)amide Ionic Liquids: In Situ STM and EQCM Studies. The Journal of Physical Chemistry B 2007;111 4693-4704.

[65] Wilkes JS, Levisky JA, Wilson RA, Hussey CL. Dialkylimidazolium chloroaluminate melts: a new class of room-temperature ionic liquids for electrochemistry, spectroscopy and synthesis. Inorganic Chemistry 1982;21 1263-1264.

[66] Gorke J, Srienc F, Kazlauskas RJ. Toward advanced ionic liquids. Polar, enzyme-friendly solvents for biocatalysis. Biotechnology Bioprocess Engineering 2010;15 40-53.

[67] Itoh T, Akasaki E, Kudo K, Shirakami S. Lipase-catalyzed enantioselective acylation in the ionic liquid solvent system: reaction of enzyme anchored to the solvent. Chemical Letter 2001;30 262-263.

[68] Lindberg D, de la Fuente Revenga M, Widersten M. Deep eutectic solvents (DESs) are viable cosolvents for enzyme-catalyzed expoxide hydrolysis. Journal of Biotechnology 2010;147 169-171.

[69] Gorke J, Srienc F, Kazlauskas RJ. Hydrolasecatalyzed biotransformations in deep euthetic solvents. Chemical Communication 2008 1235-1237.

[70] Domínguez PM, Maugeri Z. Ionic liquids in biotransformations: from proof-of-concept to emerging deep-eutectic-solvents. Current Opinion in Chemical Biology 2011;15 220–225.

[71] Eckstein M, Sesing M, Kragl U, Adlercreutz P. At low water activity α-chymotrypsin is more active in an ionic liquid than in non-ionic organic solvents. Biotechnology Letters 2002;24 867–872.

[72] Zhou G-P, Zhang Y, Huang X-R, Shi C-H, Liu W-F, Li Y-Z. Catalytic activities of fungal oxidases in hydrophobic ionic liquid 1-butyl-3- methylimidazolium hexafluorophosphate-based microemulsion. Colloids and Surfaces B: Biointerfaces 2008;66 146–149.

[73] Zhang Y, Huang X, Li Y. Negative effect of [bmim][PF6] on the catalytic activity of alcohol dehydrogenase: mechanism and prevention. Journal of Chemical Technology and Biotechnology 2008;83 1230–1235.

[74] Pavlidis IV, Gournis D, Papadopoulos GK, Stamatis H. Lipases in water-in-ionic liquid microemulsions. Structural and activity studies. Journal of Molecular Catalysis B: Enzymatic 2009;60 50–56.

[75] Toral AR, de los Rios AP, Hernandez FJ, Janssen MHA, Schoevaart R, van Rantwijk F. Cross-linked Candida antarctica lipase B is active in denaturing ionic liquids. Enzyme Microbiology and Technology 2007;40 1095–1099.

[76] Shah S, Solanki K, Gupta MN. Enhancement of lipase activity in non-aqueous media upon immobilization on multi-walled carbon nanotubes. Chemistry Central Journal 2007 1-30.

[77] Lee SH, Dang DT, Ha SH, Chang WJ, Koo YM. Lipase-catalyzed synthesis of fatty acid sugar ester using extremely supersaturated sugar solution in ionic liquids. Biotechnology and Bioengineering 2008;99 1-8.

[78] Schofer SH, Kaftzik N, Wasserscheid P, Kragl U. Enzyme catalysis in ionic liquids: lipase catalysed kinetic resolution of 1-phenylethanol with improved enantioselectivity. Chemical Communications 2001 425-426.

[79] Zhao H, Jones CL, Cowins JV. Lipase dissolution and stabilization in ether-functio nalized ionic liquids Green Chemistry 2009;11 1128-1138.

[80] Eker B, Asuri P, Murugesan S, Linhardt RJ, Dordick JS. Enzyme carbon nanotube conjugates in room-temperature ionic liquids. Applied Biochemistry and Biotechnology 2007;143 153–163.

[81] Bian W, Yan B, Shi N, Qiu F, Lou, LL, Qi B, Liu S. Room temperature ionic liquid (RTIL)-decorated mesoporous silica SBA-15 for papain immobilization: RTIL increased the amount and activity of immobilized enzyme. Materials Science and Engineering 2012;32 364-368.

[82] Du P, Liu S, Wu P, Cai C. Preparation and characterization of room temperature ionic liquid/single-walled carbon nanotube nanocomposites and their application to the direct electrochemistry of heme-containing proteins/enzymes. Electrochimica Acta 2007;52 6534–6547.

[83] Nakashima K, Okada J, Maruyama T, Kamiya N,Goto M. Activation of lipase in ionic liquids by modification with comb-shaped poly(ethylene glycol). Science and Technology of Advanced Materials 2006;7 692–698.

[84] Turner MB, Spear SK, Huddleston JG, Holbrey JD, Rogers RD. Ionic liquid salt-induced inactivation and unfolding of cellulase from Trichoderma reesei. Green Chemistry 2003;5 443–447.

[85] Maruyama T, Nagasawa S, Goto M. Poly(ethylene glycol)-lipase complex that is catalytically active for alcoholysis reactions in ionic liquids. Biotechnol Letters 2002;24 1341–1345.

[86] Wang Y, Mei L. Lyophilization of lipase with cyclodextrins for efficient catalysis in ionic liquids. Journal of Bioscience and Bioengineering 2007;103 345-349.

[87] Hernández-Fernández FJ, Rios AP, Lozano-Blanco LJ, Godíinez C. Biocatalytic ester synthesis in ionic liquid media. Journal of Chemical Technology and Biotechnology 2010;85 1423–1435.

[88] Rios AP, Hernandez-Fernandez FJ, Tomas-Alonso F, Gomez D and Villora G. Synthesis of esters in ionic liquids. The effect of vinyl esters and alcohols. Process Biochemistry 2008;43 892–895.

[89] Rios AP, Hernandez-Fernandez FJ, Tomas-Alonso F, Gomez D and Villora G. Synthesis of flavour esters using free Candida antarctica lipase B in ionic liquids. Flavour and Fragrance Journal 2008;23 319–322.

[90] De Diego T, Lozano P, Abada MA, Steffenskya K, Vaultier M and Iborra JL. On the nature of ionic liquids and their effects on lipases that catalyze ester synthesis. Journal of Biotechnology 2009;140 234–241.

[91] Ulbert O, Frater T, Belafi-Bako K, Gubicza L. Enhanced enantioselectivity of Candida rugosa lipase in ionic liquids as compared to organic solvents. Journal of Molecular Catalysis B: Enzymatic 2004;31 39–45.

[92] Yuan Y, Bai S, Sun Y. Comparison of lipase-catalysed enantioselective esterification of (9)-menthol in ionic liquids and organic solvents. Food Chemistry 2006;97 324–330.

[93] Contesini FJ, de Oliveira Carvalho P. Esterification of (RS)- Ibuprofen by native and commercial lipases in a two-phase system containing ionic liquids. Tetrahedron Asymmetry 2006;17 2069–2073.

[94] Nara SJ, Harjani JR, Salunkhe MM, Mane AT, Wadgaonkar PP. Lipase-catalysed polyester synthesis in 1-butyl-3- methylimidazolium hexafluorophosphate ionic liquid. Tetrahedron Letter 2003;44 1371–1373.

[95] Marcilla R, de Geus M, Mecerreyes D, Duxbury CJ, Koning CE, Heise A. Enzymatic polyester synthesis in ionic liquids. European Polymer Journal 2006;42 1215–1221.

[96] Gamba M, Lapis AAM, Dupont J. Supported ionic liquid enzymatic catalysis for the production of biodiesel. Advanced Synthesis and Catalysis 2008;350 160–4.

[97] De Diego T, Manjón A, Lozano P, Vaultier M, Iborra JL, Vaultier M. An efficient activity ionic liquid–enzyme system for biodiesel production. Green Chemistry 2011;13 444–51.

[98] Lozano P, Bernal JM, Piamtongkam R, Fetzer D, Vaultier M. One-phase ionic liquid reaction medium for biocatalytic production of biodiesel. ChemSusChem 2010;3 1359–63.

[99] Ha SH, Lan MN, Lee SH, Hwang SM, Koo Y-M. Lipase-catalyzed biodiesel production from soybean oil in ionic liquids. Enzyme and Microbial Technology 2007;41 480–483.

[100] De Diego T, Arturo M, Pedro L, Iborra JL. A recyclable enzymatic biodiesel production process in ionic liquids. Bioresource Technology 02 6336-6339

[101] Ruzich NI, Bassi AS. Investigation of enzymatic biodiesel production using ionic liquid as a co-solvent. Canadian Journal of Chemical Engineering 2010;88 277-282.

[102] Lai J-Q, Hu Z-L, Wang P-W, Yang Z. Enzymatic production of microalgal biodiesel in ionic liquid [BMIm][PF6]. Fuel 2012;95 329-333.

[103] Wang J. Real-Time Electrochemical Monitoring: Toward Green Analytical Chemistry. Accounts of Chemical Research 2002;35 811–816.

[104] Shiddiky MJA, Torriero AAJ. Application of ionic liquids in electrochemical sensing systems. Biosensors and Bioelectronics 2011;26 1775–1787.

[105] Galonde N, Nott K, Debuigne A, Deleu M, Jerome C, Paquot M, Wathelet J-P. Use of ionic liquids for biocatalytic synthesis of sugar derivatives. Journal of Chemical Technology and Biotechnology 2012;87 451–471.

[106] Laane C, Boeren S, Vos K, Veeger C. Rules for optimization of biocatalysis in organic solvents. Biotechnology and Bioengineering 1987;30 81-87.

Chapter 9

Increase in Thermal Stability of Proteins by Aprotic Ionic Liquids

Hidetaka Noritomi

Additional information is available at the end of the chapter

1. Introduction

Proteins are biomolecules of great importance in the biochemical processes such as the medical, pharmaceutical, and food fields, since they exhibit their outstanding biological activities under mild condition. However, most of proteins dissolved in an aqueous solution are immediately denatured and inactivated at high temperatures due to the disruption of weak interactions including ionic interactions, hydrogen bonds, and hydrophobic interactions, which are prime determinants of protein tertiary structures [5-3]. In particular, protein aggregation easily occurs upon the exposure of the hydrophobic parts of proteins, which are usually located in the inside of native proteins, and this phenomenon becomes the major problem because of the fast irreversible inactivation. Thermal denaturation of proteins is a serious problem not only in the separation and storage of proteins but also in the processes of biotransformation, biosensing, drug production, and food manufacturing. Several strategies have so far been proposed in order to prevent thermal denaturation of proteins [4-14]. They include chemical modification, immobilization, genetic modification, and addition of stabilizing agents. The addition of stabilizing agents is one of the most convenient methods for minimizing thermal denaturation, compared to other methods. It has been reported that inorganic salts, polyols, sugars, amino acids, amino acid derivatives, chaotropic reagents, and water-miscible organic solvents are available for improving protein stability. However, these additives do not sufficiently prevent irreversible protein aggregation or some of them are no longer stable at high temperatures.

Ionic solvent that is liquid at room temperature has attracted increasing attention as a green solvent for the chemical processes because of the lack of vapor pressure, the thermal stability, and the high polarity [15, 16]. Chemical and physical properties of ionic liquids can be changed by the appropriate modification of organic cations and anions, which are constituents of ionic liquids. It has recently been reported that protic ionic liquids such as alkylam-monium salts keep the stability of proteins in an aqueous solution at high temperatures [17, 18], and amyloid fibrils of proteins are dissolved in protic ionic liquids and are refolded by dilution with an aqueous solution [19]. On the other hand, biotransformation in ionic liquids has increasingly been studied [20-2]. Aprotic ionic liquids such as immidazolium salts have mainly been employed as reaction media, since the high activity of enzymes is exhibited as usual. We have found that the activity of protease is highly maintained not only in water-immiscible aprotic ionic liquids but also in water-miscible aprotic ionic liquids [23, 24].

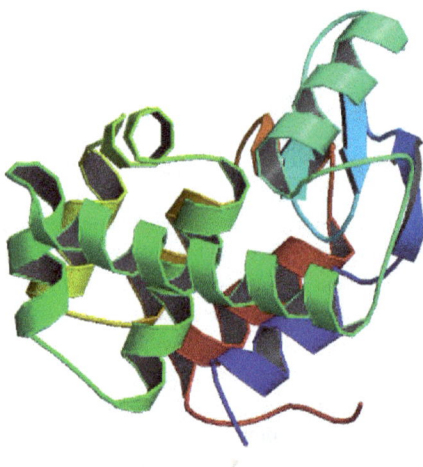

Figure 1. Structure of hen egg white lysozyme.

Ionic liquid	Structure	m.p.	Water miscibility
[emim][Tf]	H₃C—N⁺—N—CH₃ [CF$_3$SO$_3$]	-9	Miscible
[emim][BF$_4$]	H₃C—N⁺—N—CH₃ [BF$_4$]	14.6	Miscible
[emim][Cl]	H₃C—N⁺—N—CH₃ [Cl]	78	Miscible

Figure 2. Structures of ionic liquids used in the present work.

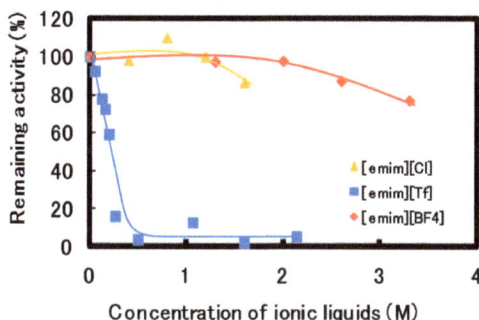

Figure 3. Effect of concentration of ionic liquids on remaining activity of lysozyme during incubation at 25 °C. The aqueous solution of 100 μM lysozyme with requisite concentration of ionic liquids was incubated in a water bath thermostated at 25 °C for 30 min.

However, despite such potential capability of aprotic ionic liquids, there have not been any works on the thermal stability of proteins in aqueous solutions containing water-miscible aprotic ionic liquids.

In the chapter, the effect of water-miscible aprotic ionic liquids consisting of 1-ethyl-3-methylimidazolium cations and several kinds of anions on thermal stability of proteins in aqueous solutions is mainly discussed [25].

2. Dependence of the remaining activity of lysozyme on the concentration of ionic liquids after the incubation at 25 °C

As a model protein, hen egg white lysozyme has been employed as shown in Fig. 1, since it has been well investigated regarding its structure, properties, and functions [26]. Lysozyme is a compact protein of 129 amino acids which folds into a compact globular structure. The molecular weight of lysozyme is 14,300, and the structure of lysozyme includes α-helices, β sheets, random coils, β turns, and disulfide bonds, which are typical structures of proteins. Lysozyme attacks peptidoglycans in the cell walls of Gram-positive bacteria, and catalyzes hydrolysis of 1,4-beta-linkages between N-acetylmuramic acid and N-acetyl-D-glucosamine residues in a peptidoglycan. Accordingly, lysozyme has been used as an anti-inflammatory agent, a preservative, a freshness-keeping agent, an antibacterial agent, a disinfectant, and so on.

Room temperature ionic liquids of alkyl imidazolium cations are widely used, and are commercially available. Figure 2 shows structures and properties of ionic liquids introduced in this chapter. 1-Ethyl-3-methylimidazolium trifluoromethanesulfonate, 1-ethyl-3-methylimidazolium tetrafluoroborate, and 1-ethyl-3-methylimidazolium chloride are abbreviated to [emim][Tf], [emim][BF₄], and [emim][Cl], respectively. Their properties such as melting point alter by switching from one anion to another.

It has been well known that ions and other compatible solutes affect enzyme activity [22]. Figure 3 shows the plot of the remaining activity of lysozyme against the concentration of ionic liquids after the incubation at 25 °C for 30 min. The remaining activity in the presence of [emim][Cl] or [emim][BF$_4$] was independent on the concentration of ionic liquids till 1.2 M [emim][Cl] or 2.0 M [emim][BF$_4$] and gradually dropped, while it in the presence of [emim][Tf] dramatically decreased with an increase in the concentration of [emim][Tf]. These results indicate that [emim][Tf] tends to strongly function as a denaturant, compared with [emim][Cl] and [emim][BF$_4$]. Electrolytes promote or inhibit the stability of proteins according to the kind of electrolytes [27-28].

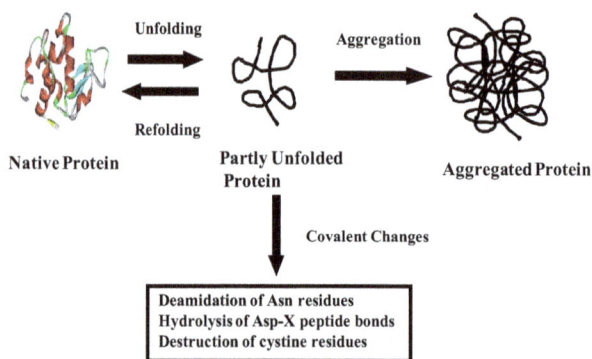

Figure 4. Schematic illustration of thermal denaturation of proteins.

3. Thermal stabilization of lysozyme by aprotic ionic liquids

3.1. Thermal inactivation of lysozyme

When proteins dissolved in an aqueous solution are placed at high temperatures, most of proteins are immediately unfolded due to the disruption of weak interactions including ionic effects, hydrogen bonds, and hydrophobic interactions, which are prime determinants of protein tertiary structures. In addition, the intermolecular aggregation among unfolded proteins and the chemical deterioration reactions in unfolded proteins proceed as shown in Fig. 4 [2, 29, 30]. In particular, protein aggregation easily occurs upon the exposure of the hydrophobic surfaces of a protein, and this phenomenon becomes the major problem because of the fast irreversible inactivation. On the other hand, when a heated solution of denatured proteins without protein aggregation is slowly cooled back to its normal biological temperature, the reverse process, which is renaturation with restoration of protein function, tends to occur. Accordingly, if stabilizing agents can sufficiently prevent irreversible aggregation of unfolded proteins, it is expected that unfolded proteins are refolded by cooling treatment, and the high remaining activity is obtained.

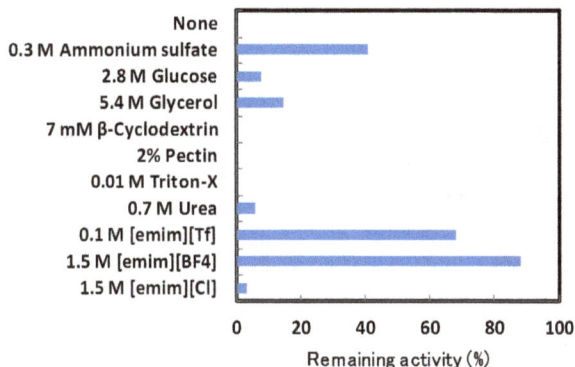

Figure 5. Remaining activities of lysozyme in the presence of various kinds of additives after heat treatment at 90 °C for 30 min.

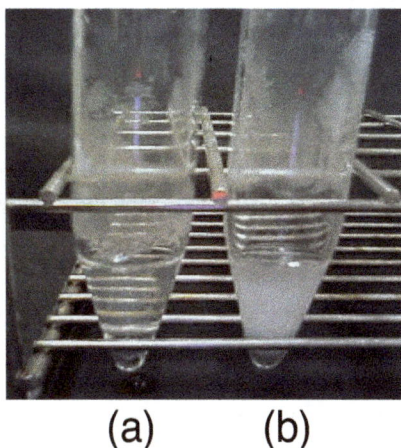

Figure 6. Photographs of lysozyme solutions after heat treatment at 90 °C for 30 min: (a) lysozyme solution with 1.5 M [emim][BF$_4$], (b) lysozyme solution without [emim][BF$_4$].

Figure 5 represents the remaining activities of lysozyme in the presence of various kinds of additives after heat treatment at 90 °C for 30 min as an accelerated test. Lysozyme without additives lost its activity perfectly after heat treatment. Native lysozyme solution immediately became turbid due to the aggregation of thermally-denatured proteins, as soon as heat treatment was carried out, as shown in Fig. 6(b). It has been reported that the precipitation due to protein aggregation at high temperatures is observed above 10 µM lysozyme [31]. As lysozyme concentration in the present work was 100 µM (1.4 mg/mL) which was ten times higher than that, the formation of protein aggregation was dramatically accelerated. Ammonium sulfate, which was an inorganic salt, glucose and glycerol, which were polyols, and urea, which was a chemical denaturant, inhibited the formation of protein aggregation, and exhibited thermal stabilization to some extent. β-Cyclodextrin, which was an inclusion compound, pectin, which was a thickener, and Triton-X, which was a nonionic surfactant, could not maintain the stability of lysozyme at high temperatures, although they were widely used as a stabilizer. On the other hand, [emim][BF$_4$] and [emim][Tf] showed high remaining activities. The lysozyme solution in the presence of ionic liquids was transparent after heat treatment, as seen in Fig. 6(a). When lysozyme solution in the presence of protic ionic liquids such as alkylammonium formates is heated at 90 °C, protein aggregation is prevented, and any cloudy appearance is absent [18]. The hydrophobic core of lysozyme unfolded by heat interacts with the cation of ionic liquids, and cation adsorption results in acquisition of a net positive charge preventing aggregation via electrostatic repulsion [17].

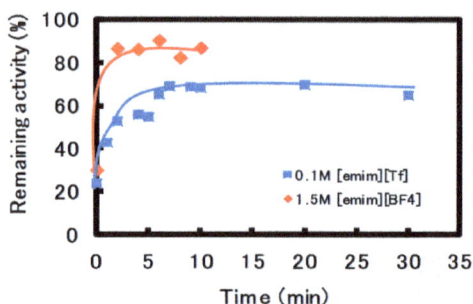

Figure 7. Time dependence of remaining activity of lysozyme with ionic liquids under cooling at 25 °C after heat treatment at 90 °C for 30 min. After heat treatment, the aqueous solution of 100 µM lysozyme with 0.1 M [emim][Tf] or 1.5 M [emim][BF$_4$] was incubated in a water bath thermostated at 25 °C.

3.2. Refolding of lysozyme by ionic liquids

When the formation of protein aggregation is inhibited at high temperatures by ionic liquids, and thermally-denatured proteins are individually dispersed in an aqueous solution, it is probably that denatured proteins are gradually refolded under cooling conditions. Figure 7 shows the time course of remaining activity in the presence of ionic liquids under cooling treatment at 25 °C after the heat treatment at 90 °C for 30 min. The remaining activities of lysozyme in the presence of 1.5 M [emim][BF$_4$] and 0.1 M [emim][Tf] exhibited 30 and 24%, respectively, just after heat treatment. The remaining activity of lysozyme with 1.5 M

[emim][BF$_4$] or 0.1 M [emim][Tf] rapidly increased with incubation time at 25 °C, and reached a plateau around 2 and 7 min, respectively. In sufficiently low concentration of proteins, where protein aggregation is not formed, when the hydrophobic core of proteins is exposed, but the disulfide bonds keep intact, denatured proteins gradually tend to refold to their native structures on cooling after thermal denaturation [32-36]. The refolding of thermally-denatured proteins is enhanced in the presence of protic ionic liquids such as alkylammonium nitrate and alkylammonium formates [32, 21]. Moreover, N'-alkyl and N'-(ω-hydroxyalkyl) N-methylimidazolium chlorides refold denatured proteins such as hen egg white lysozyme and the single-chain antibody fragment ScFvOx [37].

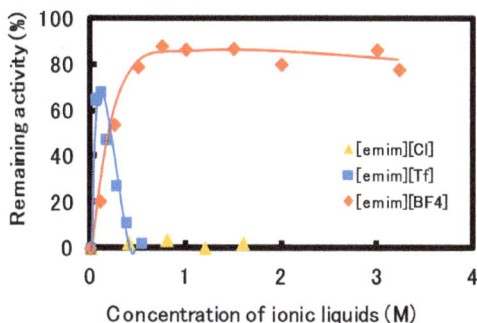

Figure 8. Effect of concentration of ionic liquids on remaining activity of lysozyme after heat treatment at 90 °C for 30 min. The aqueous solution of 100 µM lysozyme with requisite concentration of ionic liquids was incubated in a silicone oil bath thermostated at 90 °C for 30 min.

3.3. Dependence of the remaining activity of lysozyme on the concentration of ionic liquids via heat treatment

The stability of proteins depends upon the kind and concentration of electrolytes [27, 28]. Figure 8 shows the plot of the remaining activity of lysozyme against the concentration of ionic liquids after the heat treatment at 90 °C for 30 min. The remaining activity was strongly dependent on the concentration of [emim][BF$_4$] or [emim][Tf], while the effect of concentration of [emim][Cl] was not observed. The remaining activity in the presence of [emim][BF$_4$] increased with an increase in the concentration of [emim][BF$_4$] and reached a plateau around 0.8 M. The remaining activity in the presence of [emim][Tf] dramatically increased with increasing the concentration of [emim][Tf], the maximal remaining activity was obtained at 0.1 M [emim][Tf], and then decreased steeply. As seen in Fig. 3, the remaining activity decreased at 25 °C with an increase in the concentration of [emim][Tf]. Chemical denaturants, such as urea and guanidine hydrochloride, can promote dissolution of inclusion bodies, which are protein aggregation formed by prokaryotic expression systems [38]. Similarly, [emim][Tf] inhibits the formation of protein aggregation at low [emim][Tf] concentrations, but it mainly denatures proteins at higher [emim][Tf] concentrations. Moreover, it has been reported that after heat treatment the remaining activity of lysozyme increases with an increase in the concentration of ethylammonium formate and 2-methoxyethylammonium formate, while the remaining activity increases at low concentration of propylammonium formate, but at higher concentrations of propylammonium formate the protein spontaneously denatures [18]. Thus, the dependence of concentration of ionic liquids on the remaining activity of proteins changes by switching from one ionic liquid to another.

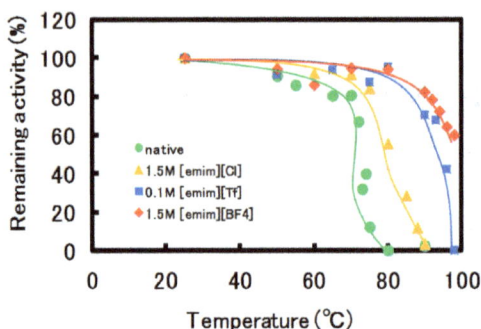

Figure 9. Thermal denaturation curves of lysozyme with or without ionic liquids. The aqueous solution of 100 µM lysozyme with or without ionic liquids was incubated in a silicone oil bath thermostated at requisite temperature for 30 min.

3.4. Dependence of the remaining activity of lysozyme on the temperature of heat treatment

The thermal inactivation of proteins more rapidly proceeds by higher temperatures. Figure 9 shows the relationship between the temperature of heat treatment and the remaining activity of lysozyme in aqueous solutions containing water-miscible ionic liquids after the heat treatment for 30 min. As seen in the figure, the remaining activity of lysozyme without ionic liquids gradually decreased with an increase in temperature below 70 °C, accompanied with the formation of precipitation due to protein aggregation, drastically dropped in the range from 70 to 80 °C, and was then lost at temperatures of 80 °C or higher. The transition temperature was exhibited around 75 °C, similar to the case measured by differential scanning calorimetry [17]. On the other hand, the remaining activity of lysozyme with 1.5 M [emim][Cl] gradually decreased with an increase in temperature below 75 °C, and drastically dropped in the range from 75 to 90 °C. The remaining activity of lysozyme with 1.5 M [emim][BF$_4$] was highly maintained below 80 °C, gradually decreased with temperature, and the remaining activity depicted 60% at 98 °C. Similarly, the remaining activity of lysozyme with 0.1 M [emim][Tf] was highly retained below 80 °C, gradually decreased with temperature below 92 °C, drastically dropped in the range from 92 to 98 °C, and was then lost at 98 °C. These results indicate that the addition of aprotic ionic liquids to an aqueous solution of lysozyme effectively improves the thermal stability of lysozyme at high temperatures.

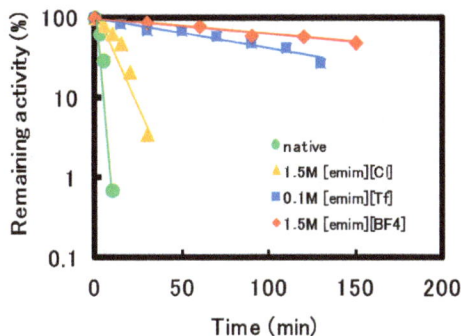

Figure 10. Time dependence of remaining activity with or without ionic liquids after heat treatment at 90 °C. The aqueous solution of 100 μM lysozyme with or without ionic liquids was incubated in a silicone oil bath thermostated at 90 °C.

Ionic liquid	Rate constant (min⁻¹)	Half line (min)
none	0.43	1.6
1.5 M [emim][Cl]	0.065	22
0.1 M [emim][Tf]	0.0081	86
1.5 M [emim][BF₄]	0.0049	141

Table 1. Rate constants and half lives of inactivation of lysozyme at 90 °C.

3.5. Time course of remaining activity of lysozyme via heat treatment with or without ionic liquids

Heating time directly enhances the thermal inactivation of proteins. Figure 10 shows time course of remaining activity of lysozyme with or without ionic liquids through the heat treatment at 90 °C. The remaining activity of lysozyme without ionic liquids dramatically decreased with an increase in time, accompanied with the formation of protein aggregation, and was almost lost at 10 min. It has been reported that the remaining activity in the thermal denaturation process accompanied with the formation of protein aggregation follows first-order kinetics [31]. As seen in the figure, the relationship of the remaining activity of proteins in the absence of ionic liquids with the time of heat treatment could be correlated by first-order kinetics. On the other hand, 1.5 M [emim][BF₄] or 0.1 M [emim][Tf] effectively prevented the thermal inactivation of lysozyme. In the presence of ionic liquids the turbidity of solutions due to protein aggregation was not observed during heat treatment. This indicates that the thermal inactivation mainly results from the covalent change as shown in Fig. 4. The plots of remaining activity versus heat treatment time on thermal inactivation of lysozyme in the presence of ionic liquids followed first-order kinetics on linearity. It has been reported that the thermal inactivation of lysozyme obeyed first-order kinetics when it irreversibly proceeded by the covalent change without the formation of protein aggregation [39]. Table 1 represents rate constants and half lives of inactivation of lysozyme with or without ionic liquids calculated from the fitting curves in Fig. 10. The half lives with 1.5 M [emim][BF$_4$], 0.1 M [emim][Tf], and 1.5 M [emim][Cl] were 88, 54, or 6.9 times longer than that without ionic liquids, respectively.

4. Conclusion

In this chapter the effect of water-miscible aprotic ionic liquids on thermal stability of lysozyme has been described. Aprotic ionic liquids could sufficiently prevent thermally-denatured proteins from aggregating. The activity of lysozyme in the presence of aprotic ionic liquids was kept to some extent, just after heat treatment at high temperatures. Moreover, thermally-denatured lysozyme was effectively refolded by cooling. Consequently, the high remaining activity of lysozyme was obtained. These results indicate that aprotic ionic liquids act not only as an inhibitor of protein aggregation, but also as a protective agent in the native structure of protein and an accelerator in the refolding of thermally-denatured proteins. The remaining activity of lysozyme markedly depended upon the kind of anions and the concentration of ionic liquids. Especially, [emim][Tf] exhibited the thermostabilization effect of proteins at low concentrations, but mainly worked as a denaturant of proteins at high concentrations. The effect of [emim][BF$_4$] and [emim][Tf] upon thermal stabilization at high temperatures was much superior to that of [emim][Cl]. As chemical and physical properties of ionic liquids can be changed by the appropriate modification of organic cations and anions, which are constituents of ionic liquids, it is expected that the ionic liquid, which is more suitable for the thermal stabilization of proteins, is prepared by tailoring the constituents of ionic liquids.

Author details

Hidetaka Noritomi[1]

1 Tokyo Metropolitan University Japan

References

[1] Ballesteros, A., Plou, F. J., Iborra, J. L., & Halling, P.J. (1998). *Stability and Stabilization of Biocatalysts*, Elsevier, Amsterdam.

[2] Volkin, D.B., & Klibanov, A.M. (1989). Minimizing protein inactivation. *Protein Function: Practical Approach*, 1-24, IRL Press, Oxford.

[3] Klibanov, A.M. (1983). Stabilization of enzymes against thermal inactivation. *Adv. Appl. Microbiol.*, 29, 1-28.

[4] Illanes, A. (1999). Stability of biocatalysts. *Electro. J. Biotechnol.*, 2, 1-9.

[5] Noritomi, H., Kai, R., Iwai, D., Tanaka, H., Kamiya, R., Tanaka, M., Muneki, K., & Kato, S. (2011b). Increase in thermal stability of proteins adsorbed on biomass charcoal powder prepared from plant biomass wastes. *J. Biomedical, Sci. Eng.*, 4, 692-698.

[6] Gerlsma, S.Y. (1968). Reversible denaturation of ribonuclease in aqueous solutions is influenced by polyhydric alcohols and some other additives. *J. Biological Chem.*, 243, 957-961.

[7] Kaushik, J.K., & Bhat, R. (1998). Thermal stability of proteins in aqueous polyol solutions: role of the surface tension of water in the stabilizing effect of polyols. *J. Phys. Chem. B*, 102, 7058-7066.

[8] Back, J. F., Oakenfull, D., & Smith, M. B. (1979). Increased thermal stability of proteins in the presence of sugars and polyols. *Biochemistry*, 18, 5191-5196.

[9] Lee, J. C., & Timasheff, S.N. (1981). The stabilization of proteins by sucrose. *J. Biological Chem.*, 256, 7193-7201.

[10] Santoro, M.M., Liu, Y., Khan, S.M.A., Hou, L.X., & Bolen, D.W. (1992). Increased thermal stability of proteins in the presence of naturally occurring osmolytes. *Biochemistry*, 31, 5278-5283.

[11] Yancey, P.H., Clark, M.E., Hand, S. C., Bowlus, R.D., & Somero, G.N. (1982). Living with water stress: evolution of osmolyte systems. *Science,*, 217, 1214-1222.

[12] Arakawa, T., Bhat, R., & Timasheff, S.N. (1990). Why preferential hydration does not always stabilize the native structure of globular proteins. *Biochemistry*, 29, 1924-1931.

[13] Ikegaya, K. (2005). Kinetic analysis about the effects of neutral salts on the thermal stability of yeast alcohol dehydrogenase. *J. Biochemistry,*, 137, 349.

[14] Cioci, F., & Lavecchia, R. (1998). Thermostabilization of proteins by water-miscible additives. *Chem. Biochem. Eng. Q.*, 12, 191-199.

[15] Welton, T. (1999). Room-temperature ionic liquids. Solvents for synthesis and calalysis. *Chem. Rev.*, 99, 2071-2083.

[16] Greaves, T.L., & Drummond, C.J. (2008). Protic ionic liquids: properties and applications. *Chem. Rev.*, 108, 206-237.

[17] Summers, C.A., & Fowers, I. I. R. A. (2000). Protein renaturation by the liquid organic salt ethylammonium nitrate. *Protein Sci.*, 9, 2001-2008.

[18] Mann, J.P., Mc Cluskey, A., & Atkin, R. (2009). Activity and thermal stability of lysozyme in alkylammonium formate ionic liquids- influence of cation modification. *Green Chem.*, 11, 785-792.

[19] Byrne, N., & Angell, C. A. (2009). Formation and dissolution of hen egg white lysozyme amyloid fibrils in protic liquids. *Chem Comm.*, 1046-1048.

[20] Moniruzzaman, M., Nakashima, K., Kamiya, N., & Goto, M. (2010). Recent advances of enzymatic reactions in ionic liquids. *Biochem. Eng. J.*, 48, 295-314.

[21] Yang, Z., & Pan, W. (2005). Ionic liquids: Green solvents for nonaqueous biocatalysis. *Enz. Microbial Technol.*, 37, 19-28.

[22] Zhao, H. (2011a). Effect of ions and other compatible solutes on enzyme activity, and its implication for biocatalysis using ionic liquids. *J. Mol. Catal. B: Enzymatic*, 37, 16-25.

[23] Noritomi, H., Nishida, S., & Kato, S. (2007). Protease-catalyzed esterification of amino acid in water-miscible ionic liquid. *Biotechnol Lett.*, 29, 1509-1512.

[24] Noritomi, H., Suzuki, K., Kikuta, M., & Kato, S. (2009). Catalytic activity of α-chymotrypsin in enzymatic peptide synthesis in ionic liquids. *Biochem. Eng. J.*, 47, 27-30.

[25] Noritomi, H., Minamisawa, K., Kamiya, R., & Kato, S. (2011b). Thermal stability of proteins in the presence of aprotic ionic liquids. *J. Biomedical Sci. Eng.*, 4, 94-99.

[26] Jollès, P. (1996). Lysozymes:. *Model Enzymes in Biochemistry and Biology.*, Birkhäuser Verlag, Basel.

[27] Von Hippel, P.H., & Schleich, T. (1969). The effects of neutral salts on the structure and conformational stability of macromolecules in solution. *Structure and Stability of Biological Macromolecules*, 417-574, Marcel-Dekker, New York.

[28] Nostro, P.L., & Ninham, B.W. (2012). Hofmeister phenomena: an update on ion specificity in biology. *Chem. Rev.*, 112, 2286-2322.

[29] Lumry, R., & Eyring, H. (1954). Conformation changes of proteins. *J. Phys. Chem.*, 58, 110-120.

[30] Zale, S.E., & Klibanov, A.M. (1983). On the role of reversible denaturation (unfolding) in the irreversible thermal inactivation of enzymes. *Biotechnol. Bioeng*, 25, 2221-2230.

[31] Nohara, D., Mizutani, A., & Sakai, T. (1999). Kinetic study on thermal denaturation of hen egg-white lysozyme involving precipitation. *J. Biosci. Bioeng.*, 87, 199-205.

[32]   Ibara-Molero, B., & Sanchez-Ruiz, J.M. (1997). Are there equilibrium intermediate states in the urea-induced unfolding of hen egg-white lysozyme? *Biochemistry,,* 36, 9616-9624.

[33]   Griko, Y.V., Freire, E., Privalov, G., Dael, H.V., & Privalov, P.L. (1995). The unfolding thermodynamics of c-type lysozyme- a calorimetric study of the heat denaturation of equine lysozyme. *J. Mol. Biol.,* 252, 447-459.

[34]   Privalov, P.L., & Khechinashvili, N.N. (1974). A thermodynamic approach to the problem of stabilization of globular protein structure. *J. Mol. Biol.,* 86, 665-684.

[35]   Khechinashvili, N.N., Privalov, P.L., & Tiktopulo, E.I. (1973). Calorimetric investigation of lysozyme thermal denaturation. FEBS Lett., , 30, 57-60.

[36]   Anfinsen, C.B. (1973). Principles that govern the folding of protein chains. *Sci.,* 181, 223-230.

[37]   Lange, C., Patil, G., & Rudolph, R. (2005). Ionic liquids as refolding additivesN'-alkyl and N'-(ω-hydroxyalkyl) N-methylimidazolium chlorides. *Protein Sci.,* 14, 2693-2701.

[38]   Rudolph, R., & Lilie, H. (1996). In vitro folding of inclusion body proteins. *FASEB J.,* 10, 49-56.

[39]   Ahern, T.J., & Klibanov, A.M. (1985). The mechanism of irreversible enzyme inactivation at 100 °C. *Science,* 228, 1280-1284.

Materials and Processing

Chapter 10

Ionic Liquids as Components in Fluorescent Functional Materials

Jun-ichi Kadokawa

Additional information is available at the end of the chapter

1. Introduction

In recent years, photo functional materials have been increasing much attention because of their attractive characteristics such as good specificity, excellent sensitivity, and easy handling [1]. Fluorescent-emitting materials are one of the most practically used photo functional materials in the many application fields such as color sensors and probes in biological science, key elements in color devises and displays, organic light-emitting diodes, and organic field-effect transistors [2]. Furthermore, a variety of polymers bearing covalently linked fluorescent dye moieties, exampling polymethacrylate, polyacrylamide, and conjugated polymer, have been synthesized to provide novel polymeric fluorescent materials [3-5]. To develop new fluorescent functional materials, the author has noted ionic liquids (ILs) as material components. ILs are low-melting-point molten salts, defined as which form liquids at room temperature or even at temperatures lower than a boiling point of water. The property is owing to that the liquid state is thermodynamically favorable due to the large size and conformational flexibility of the ions, in which these behaviors lead to small lattice enthalpies and large entropy changes that favor the liquid state [6]. In the past more than a decade, ILs have attracted much attention due to their specific characteristics such as a negligible vapor pressure, excellent thermal stabilities, and controllable physical and chemical properties [7]. Beyond these traditional properties of ILs, recently, interests and applications on ILs have been extended to the researches related to functional materials as designer substrates with controllable physical and chemical properties or even specific functions [8], so-called 'task-specific ILs' [9,10]. As one of the unique and specific properties of ILs, it has been reported that imidazolium-type ILs exhibit excitation-wavelength-dependent fluorescent behavior due to the presence of energetically different associated species [11-14]. For example, 1-butyl-3-methylimidazolium chloride (BMIMCl) typically exhibits emissions maxima at around 450-600 nm depending on the excitation wavelengths (Figure 1). The imidazolium-type ILs which form such different species have potential as components to contribute to developing new fluorescent functional materials.

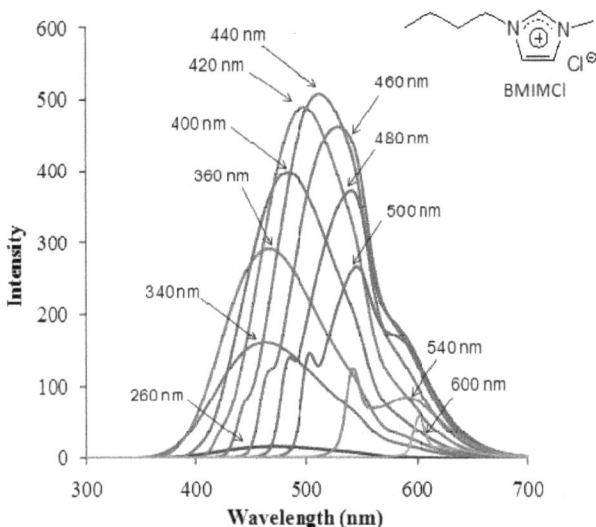

Figure 1. Fluorescence spectra of a liquid BMIMCl by excitation at 260-600 nm.

In this chapter, the author describes the use of the imidazolium-type ILs as components to prepare new fluorescent functional materials. A first topic deals with the appearance of fluorescent resonance-energy-transfer (FRET) in solutions of fluorescent dyes with BMIMCl. As a second topic, on the basis of this unique FRET system, the preparation and FRET functions of polymeric IL (PIL) films carrying fluorescent dye moieties are disclosed. Furthermore, a third topic deals with the preparation of ion gel materials from BMIMCl which exhibit the FRET function and other unique fluorescent properties.

2. Fluorescent Properties in Solutions of Rhodamine 6G with Ionic Liquid

2.1. FRET

Besides exhibiting emission by excitation at a characteristic wavelength of each fluorescent dye, the fluorescent materials in practical applications are often required to exhibit fluorescent emissions by excitation at different wavelength areas. For the purpose to develop such dye materials, the author has noted the FRET technique [15], which has been used in designed fluorescent materials to obtain a large shift of the excitation wavelength from that the dyes natively show. FRET is an interaction between the electronic excited states of two fluorescent substrates, a donor and an acceptor, in which excitation energy is transferred from the former to the latter without emission of a photon (Figure 2) [16]. By means of FRET, new high performance biosensors, fluorescence imaging, and quantification systems of selective interaction have been developed for targets of biological molecules, such as proteins and lipids [17,18].

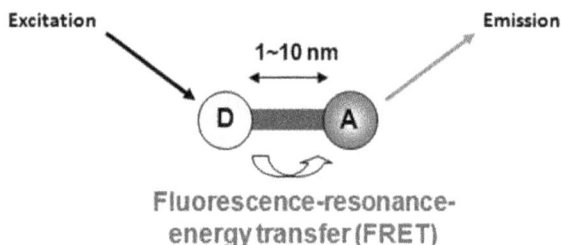

Figure 2. Image of FRET from a donor to an acceptor.

2.2. FRET in Systems of Rhodamine 6G with BMIMCl

To develop the basic technique for the preparation of new functional fluorescent materials, the author found a unique FRET system in a solution of rhodamine 6G with BMIMCl, where the former and latter acted as an acceptor and a donor, respectively [19]. Rhodamine 6G is a representative red fluorescent dye and exhibits emission maxima at ca. 540-610 nm by excitation at around 520 nm [20]. When the fluorescence spectra of the solution of rhodamine 6G with BMIMCl (2.5 mmol/L) were measured by excitation at 260-600 nm, emissions at ca. 608 nm due to the dye were observed in all the spectra, whereas fluorescence peaks due to BMIMCl were not detected (Figure 3). From these results, the occurrence of FRET from BMIMCl to rhodamine 6G in the solution was supposed. Indeed, all the fluorescence spectra of a sole BMIMCl liquid excited at various wavelengths were overlapped with an absorption peak of rhodamine 6G at 545 nm. The occurrence of FRET in the solution of rhodamine 6G with BMIMCl was confirmed further using the Stern-Volmer relation [21].

On the other hand, the fluorescence spectra of a solution of another fluorescent dye, pyrene, with BMIMCl by excitation at 260-600 nm showed the emissions due to BMIMCl. This was owing to no occurrence of FRET in the solution because an absorption of pyrene was not overlapped with the emissions of BMIMCl. Moreover, when the fluorescence spectra of a solution of a dye with no fluorescent emission, that is, Congo red, were measured by excitation at various wavelengths, the emissions due to BMIMCl were not observed. This result was explained by the energy transfer from BMIMCl to Congo red because an absorption of the dye overlapped with the emissions of BMIMCl.

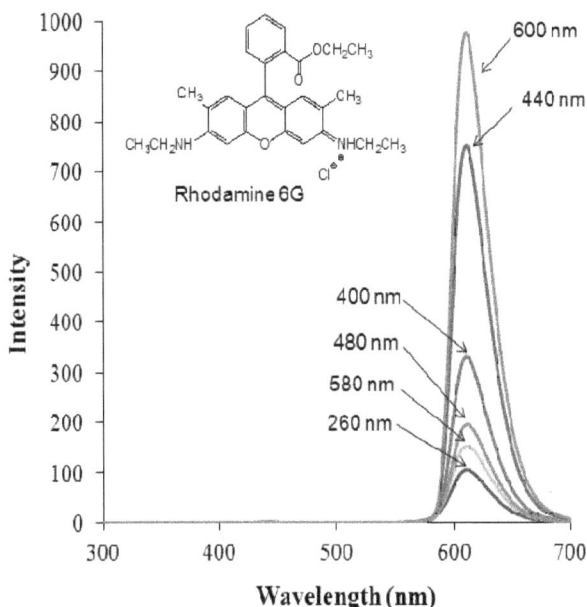

Figure 3. Fluorescence spectra of rhodamine 6G/BMIMCl solution by excitation at 260-600 nm.

3. Tunable Multicolor Emissions of Polymeric Ionic Liquid Films Carrying Fluorescent Dye Moieties

3.1. Polymeric Ionic Liquids

Polymeric ionic liquids (PILs) are defined as the polymers obtained by polymerization of ILs having polymerizable groups (polymerizable ILs) [22,23]. Thus, 'PILs' are termed just the polymeric forms of ILs, but they are not necessary to show liquid forms at room temperature or even at some ambient temperatures. The polymeric ILs, therefore, are often called 'polymerized ILs' too. The major advantages for providing the PILs are to be enhanced stability, and improved processability and feasibility in application as practical materials. Polymerizable ILs as a source of the PILs can be available by incorporating the polymerizable groups at both anionic and cationic sites (Figure 4). In the former case, polymerizable anions are ionically exchanged from some anions of general ILs (Figure 4(a)), giving the polymerizable ILs. In the latter case, vinyl, meth(acryloyl), and vinylbenzyl groups have typically been appeared as the polymerizable group (Figure 4(b)). Because 1-vinylimidazole is a commercially available, the vinylimidazolium-type polymerizable ILs are prepared by quaternization of 1-vinylimidazole with a variety of alkyl halides. The reaction of vinylbenzyl halides or haloalkyl (meth)acrylates with 1-alkylimidazoles gives the corresponding imidazolium-type polymerizable ILs having the vinylbenzyl or (meth)acryloyl polymerizable group (Figure 5). Furthermore, when vinylbenzyl halides or haloalkyl (meth)acrylates are reacted with 1-vinylimidazole, the polymerizable ILs having two polymerizable groups are produced. Because these polymerizable ILs can be converted into insoluble and stable PILs with the cross-linked structure by the radical polymerization (Figure 6), they have a highly potential as the source of the components in the practical materials.

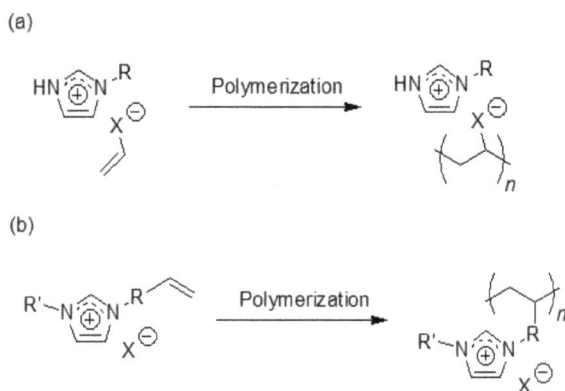

Figure 4. Polymerization of polymerizable ILs having a polymerizable group at anionic site (a) and cationic site (b).

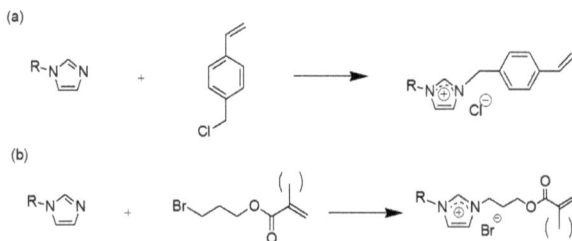

Figure 5. Typical synthetic schemes for polymerizable ILs having vinylbenzyl (a) and (meth)acrylate (b) groups.

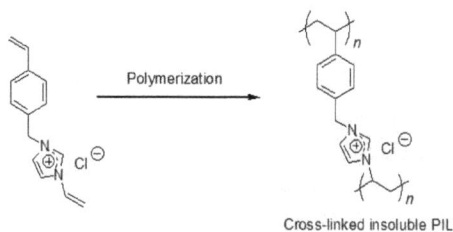

Figure 6. Polymerization of a polymerizable IL having two polymerizable groups to produce a cross-linked insoluble PIL.

3.2. Preparation of Transparent Polymeric Ionic Liquid Films

To incorporate the aforementioned unique FRET function into a film material, the prepara-tion of a transparent PIL film was attempted by radical polymerization of the appropriate polymerizable ILs [24]. For this purpose, the two imidazolium-type polymerizable ILs, 1-methyl-3-(4-vinylbenzyl)imidazolium chloride (1) and 1-(3-methacryloyloxypropyl-3-vinyli-midazolium bromide (2) were employed to obtain a cross-linked PIL (Figure 7). For the preparation of the film form of PIL, a solution of 1 and 2 (10:1), and AIBN as a radical initia-tor (1mol% for 1+2) was sandwiched between two glass plates. Then, the system was heated at 65 °C for 30 min and subsequently at 75 °C for 2 h to occur the copolymerization. The re-sulting material had the film form with transparent property.

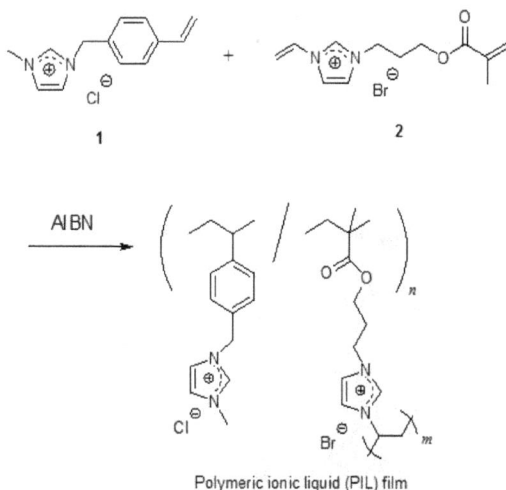

Figure 7. Radical copolymerization of 1 with 2 by AIBN to give PIL film.

The UV-vis spectrum of the film showed small absorptions at 280-550 nm, which were prob-ably related to the fluorescent emissions of the imidazolium-type ILs, besides large absorp-tions at the wavelengths below 280 nm. The fluorescence spectra of the film showed excitation-wavelength-dependent fluorescent emission maxima at around 430-470 nm by ex-citation at 260-400 nm (Figure 8). Indeed, the film exhibited blue emission by UV light irra-diation at 365 nm (Figure 9). The fluorescent behavior of the film was similar as that of the general imidazolium-type IL such as BMIMCl.

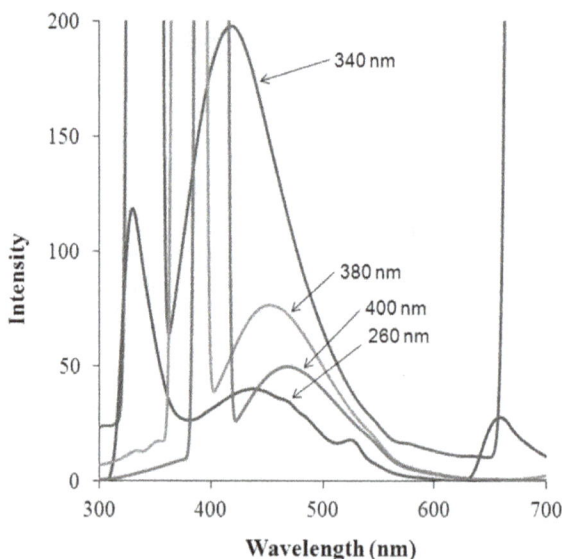

Figure 8. Fluorescence spectra of PIL film by excitation at 260-400 nm.

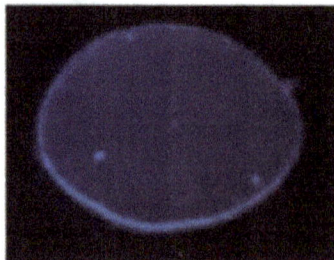

Figure 9. Photograph of PIL film under UV light irradiation at 365 nm.

3.3. Preparation and Multicolor Emissions of Fluorescent Polymeric Ionic Liquid Films

On the basis of the principle of three primary colors, the PIL films which exhibit multicolor emissions depending on combinations of the primary colors have considerably been designed [24,25]. For this purpose, three fluorescent dyes, rhodamine (red emission), 7-(diethylamino)coumarin-3-carboxylic acid (DEAC, green emission), and pyranine (blue emission) were selected, and thus, polymerizable rhodamine, DEAC, and pyranine derivatives (3, 4, and 5) having a methacrylate group were synthesized. Then, radical copolymerization of 1, 2, with 3, 4, or 5 was conducted by a similar procedure as aforementioned for PIL film to produce the PIL films 6, 7, and 8 carrying respective dye moieties (Figure 10).

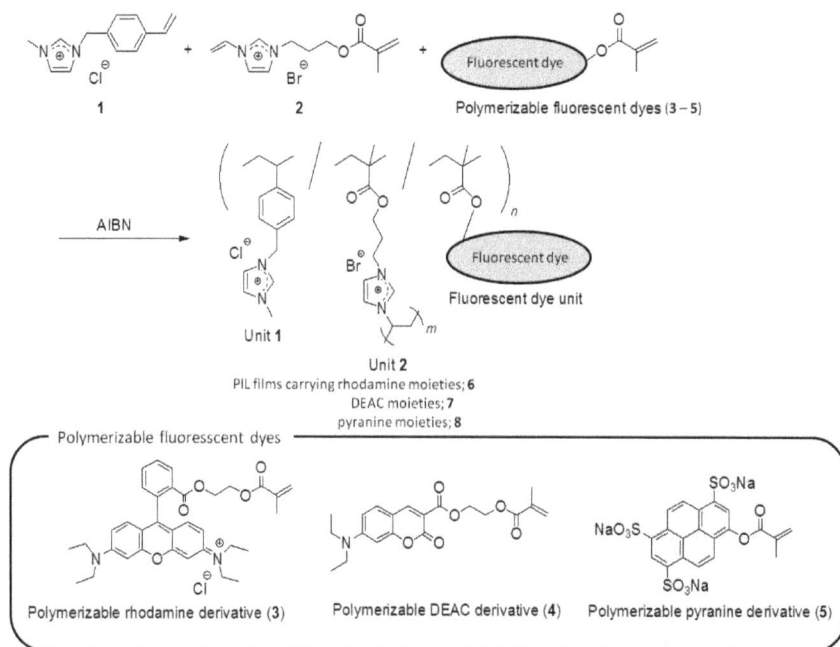

Figure 10. Radical copolymerization of 1, 2, with 3, 4, or 5 to give PIL films carrying primary color fluorescent dye moieties.

When the fluorescence spectra of the film 6 were measured by excitation at 260-400 nm, emissions at ca. 620 nm due to the rhodamine group in addition to scattering peaks of excitation lights were observed in all the spectra (Figure 11(a)). On the other hand, fluorescent emissions at around 430-470 nm due to the units 1 and 2 did not appear. These results suggested the occurrence of FRET from the units 1 and 2 to the rhodamine group in the film. Indeed, all the emissions of the PIL film composed of the units 1 and 2 (without fluorescent dye moieties; hereafter, this film is named the basic PIL film) excited at various wavelengths were partially overlapped with an absorption peak of the film 6 at wavelength areas of around 450-600 nm.

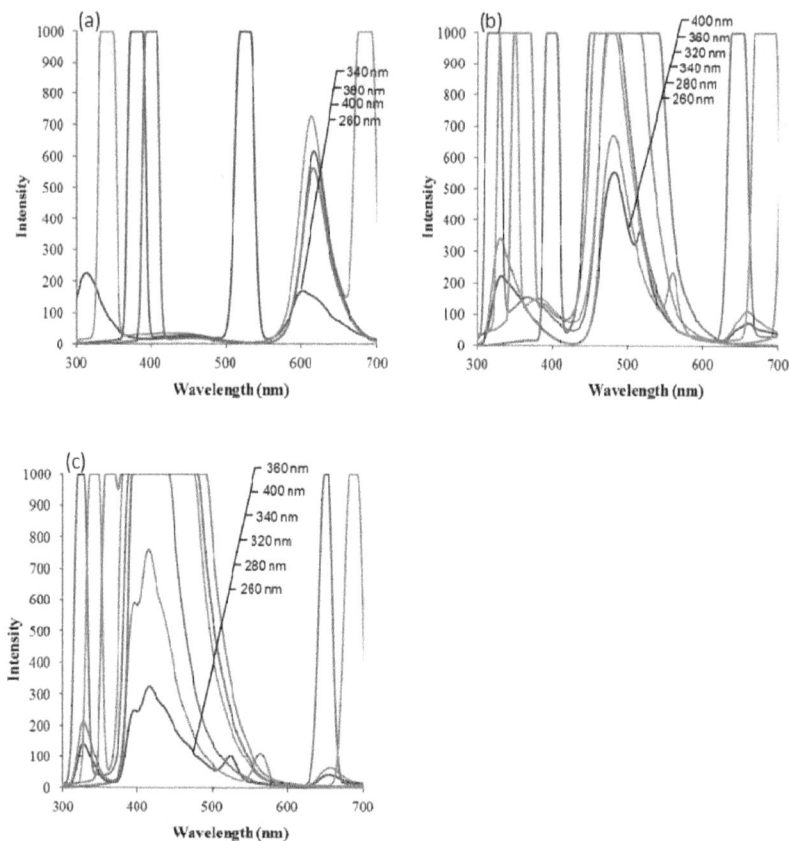

Figure 11. Fluorescence spectra of PIL films 6, 7, and 8 ((a)-(c), respectively) by excitation at 260-400 nm.

When the fluorescence spectra of the film 7 were also measured by excitation at 260-400 nm, emissions at ca. 470 nm due to the DEAC group were observed (Figure 11(b)). Furthermore, all the emissions of the basic PIL film excited at various wavelengths were totally or even partially overlapped with absorptions of the film 7. Taking the UV-vis spectrum of the film 7 into consideration, it was also supposed that the DEAC moieties in 7 emitted by excitation at around the wavelengths areas where the absorptions of 7 appeared. Therefore, the above re-sults suggested that the emissions due to the DEAC group in 7 excited at wide wavelength areas were owing to either direct excitation of the DEAC group or FRET from the units 1 and 2 to the DEAC group in the film.

Similarly, emissions due to pyranine moieties were observed at ca. 420 nm in the fluorescence spectra of the film 8 excited at 260-400 nm (Figure 11(c)). The fluorescent emissions of the basic PIL film excited at shorter wavelength areas, i.e., 260-360 nm were partially overlapped with absorptions of the film 8 at around 300-400 nm. On the other hand, the emissions of the basic PIL film by excitation at longer wavelength area such as 380 and 400 nm appeared at wavelengths longer than ca. 400 nm, which were not mostly overlapped with the absorptions of the film 8. Moreover, the pyranine moieties in 8 emitted by excitation at around the wavelength areas where the absorptions of 8 appeared. Therefore, it was supposed that the emissions of the pyranine group in the film 8 by excitation at shorter wavelength area were owing to either direct excitation of the pyranine group or FRET from the units 1 and 2 to the pyranine group, whereas those excited at longer wavelength areas were probably caused by only direct excitation of the pyranine group in the film.

Actually, the film 6, 7, and 8 showed the red, green, and blue emissions, respectively, by the UV-vis light irradiations at 365 nm (Figure 12).

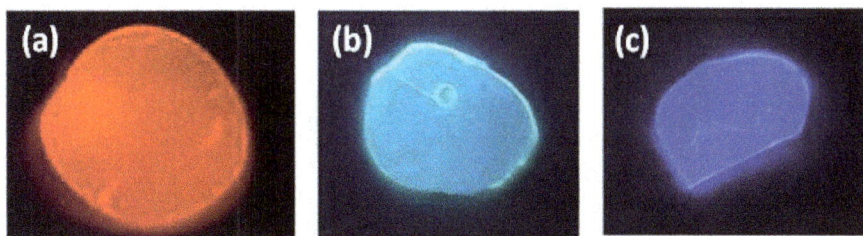

Figure 12. Photographs of PIL films 6, 7, and 8 ((a)-(c), respectively) under UV light irradiations at 365 nm.

By means of possible combinations among the rhodamine, DEAC, and pyranine dyes, which emitted the three primary colors, the PIL films exhibiting tunable color emissions were prepared. Three combinations of polymerizable dyes, that is, 3 and 4, 3 and 5, and 4 and 5, were copolymerized with 1 and 2 by AIBN according to the same experimental manner as that for the basic PIL film (Figure 13). The fluorescence spectra of the resulting films showed two kinds of emissions due to the incorporated dye moieties by excitation at 260-400 nm. These data suggested that the respective dye groups in the PIL films were individually emitted by direct excitation or FRET. The PIL film carrying three dye moieties was similarly prepared by copolymerization of 1, 2, with the three polymerizable dyes. The fluorescence spectra of the resulting film also showed three kinds of emissions excited at 260-400 nm. Thus, the resulting films exhibited yellow, magenta, cyan, and white fluorescent emissions, respectively, by UV light irradiation at 365 nm (Figure 13). These results indicated that the PIL films carrying proper fluorescent dye moieties emitted tunable multicolors by excitation at a sole wavelength.

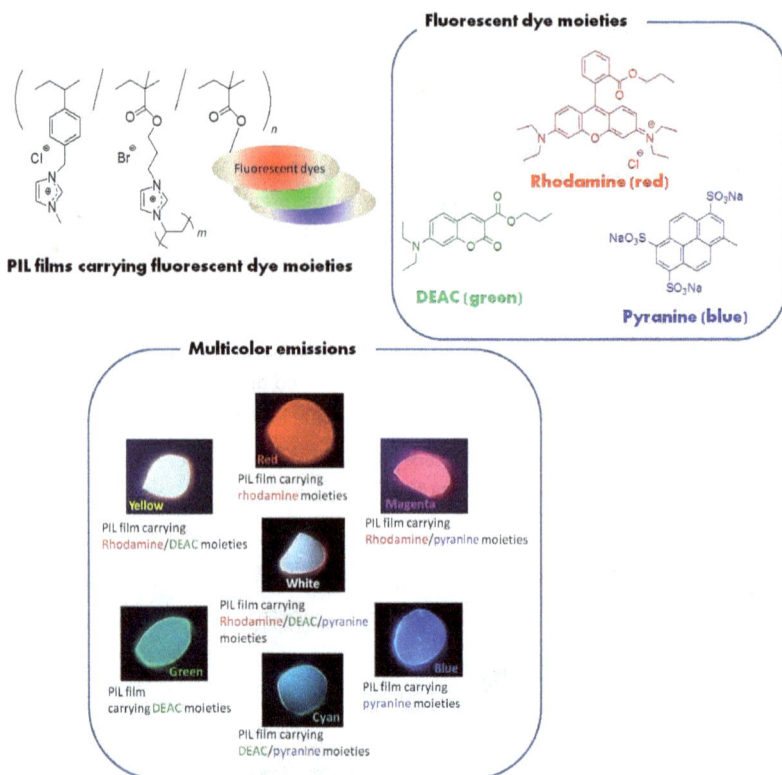

Figure 13. Multicolor emissions of PIL films carrying various combinations of fluorescent dye moieties.

4. Preparation of Photo Functional Ion Gels of Polysaccharides with an ionic liquid

4.1. Ion Gels of Polysaccharides with Ionic Liquids

Because ILs have been found to be used as good solvents for natural polysaccharides such as cellulose [26-29], and accordingly, can be considered to have a specific affinity for polysaccharides, efficient methods to produce new polysaccharide-based materials compatibilized with ILs have the potential to lead to the practical use of natural polysaccharides as the promising biomass resources [30,31]. On the basis of these viewpoints, the author has reported the facile preparation of gel materials of abundant polysaccharides such as cellulose, starch, and chitin, which include ILs as disperse media in the polysaccharide network matrixes, so-called ion gels [32-35]. Besides such abundant polysaccharides, many kinds of natural polysaccharides from various sources have been known [36]. For example, some polysaccharides such as guar gum and xanthan gum are used as hydrocolloid polysaccharides for a stabilizer, a viscous agent, and a structure provider in food industries [37]. Guar gum is a galactomannan extracted from the seed of the leguminous shrub *Cyamopsis tetragonoloba* and consists of a (1→ 4)-linked β-D-mannopyranose main-chain with a branched α-D-galactopyranose unit at 6 position (Figure 14). Xanthan gum produced by *Xanthomonas campestris* has a cellulose-type main-chain (β-(1→ 4)-glucan) with trisaccharide side chains attached to alternate glucose units in the main-chain (Figure 14). The author has reported that functional ion gels of hydrocolloid polysaccharides, e.g., guar gum and xanthan gum, with BMIMCl were obtained when the corresponding solutions of the polysaccharides in BMIMCl in appropriate concentrations were left standing at room temperature [38-42]. These ion gels have been applied to providing functional materials by means of the specific fluorescent behaviors of ILs.

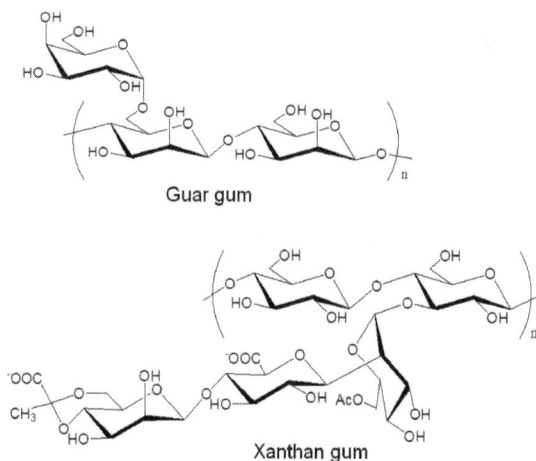

Figure 14. Structures of guar and xanthan gums.

4.2. FRET Function of Ion Gel of Guar Gum with an Ionic Liquid

For the preparation of gel materials exhibiting the aforementioned unique FRET function, the gelling system of BMIMCl using guar gum was employed. When the fluorescence spectra of the guar gum/BMIMCl ion gel was measured by excitation at 260-600 nm, the similar excitation-wavelength-dependent fluorescence behavior as that of a sole BMIMCl was appeared (Figure 15). Accordingly, the guar gum/BMIMCl ion gel containing rhodamine 6G (1.5 mmol/L) was prepared from the mixture of rhodamine 6G and guar gum with BMIMCl. The fluorescence spectra of the resulting ion gel exhibited emissions due to rhodamine 6G by excitation at 260-600 nm, whereas no emissions due to BMIMCl were observed (Figure 16). These results indicated the occurrence of FRET from BMIMCl to rhodamine 6G in the ion gel. Indeed, the gel showed the red emissions by photo irradiation at various wavelengths (Figure 17).

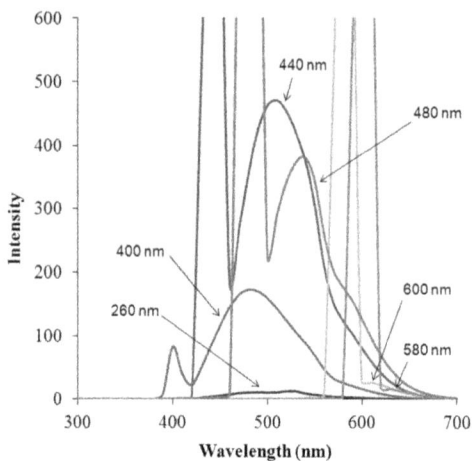

Figure 15. Fluorescence spectra of guar gum/BMIMCl ion gel by excitation at 260-600 nm.

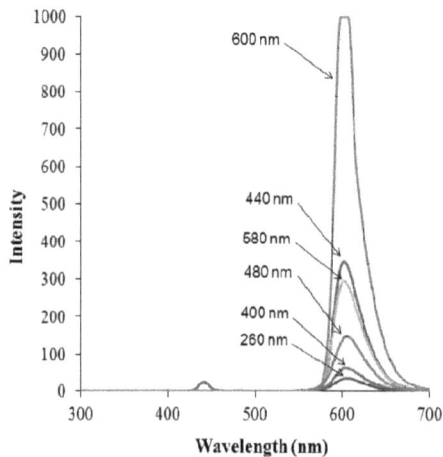

Figure 16. Fluorescence spectra of guar gum/rhodamine 6G/BMIMCl ion gel by excitation at 260-600 nm.

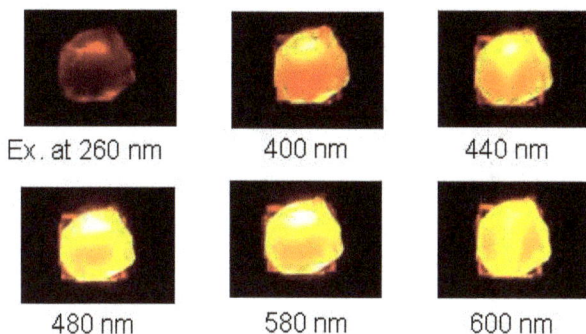

Ex. at 260 nm   400 nm   440 nm

480 nm   580 nm   600 nm

Figure 17. Photographs of guar gum/rodamine 6G/BMIMCl ion gel by excitation at 260-600 nm.

4.3. Fluorescent Behaviors of Ion Gel of Xanthan Gum with an Ionic Liquid

The author has been interested in the association state of BMIMCl in the xanthan gum/ BMIMCl ion gels because nano-ordered association of 1-butyl-3-methylimidazolium-type ionic liquids in the liquid state was suggested in previous report [43]. The UV-vis spectra of the ion gels were measured to evaluate the association states of BMIMCl [41]. A liquid BMIMCl showed significant absorptions at wavelengths below 250 nm besides very small absorptions at 250-450 nm (Figure 18(a)). However, the strong absorptions in a wide range from 200 to 450 nm were observed in the UV-vis spectra of the ion gels with different contents (10 and 30% (w/w), Figure 18(b) and (c)). Such strong absorption was not observed in the UV-vis spectrum of guar gum/BMIMCl ion gel. These results suggested the presence of the different association state of BMIMCl in the xanthan gum/BIMICl ion gel from that in the liquid and the guar gum/BIMICl ion gel. The presence of the specific association state of BMIMCl in the xanthan gum/BMIMCl ion gel was also confirmed by the ^{1}H NMR analysis.

On the basis of the above findings, the fluorescent behaviors of the xanthan gum/BMIMCl ion gels were investigated. Figure 19 shows the fluorescence spectra of the ion gels in the different xanthan gum contents (10, 20, 40, and 60% (w/w)) by excitation at 360-480 nm. Emission maxima were obviously shifted to the longer wavelength areas with increasing xanthan gum contents. Such red-shift was probably due to the presence of the specific association states of BMIMCl depending to the xanthan gum contents in the gels. Actually, the colors of the gels were changed from yellow to red-brown with increasing the xanthan gum contents (Figure 20). These results suggest that the present xanthan gum/BMIMCl ion gels can be applied to the new fluorescent gel materials in the future.

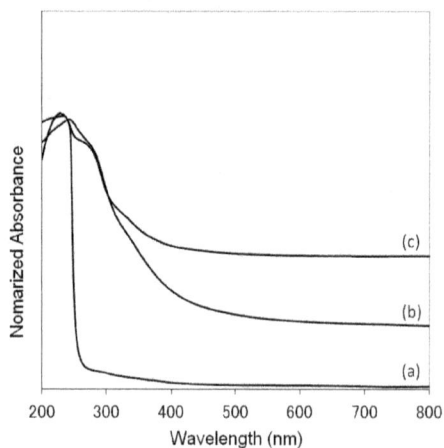

Figure 18. UV-vis spectra of a liquid BMIMCl (a), and xanthan gum/BMIMCl ion gels in 10 and 30% (w/w) contents ((b) and (c), respectively).

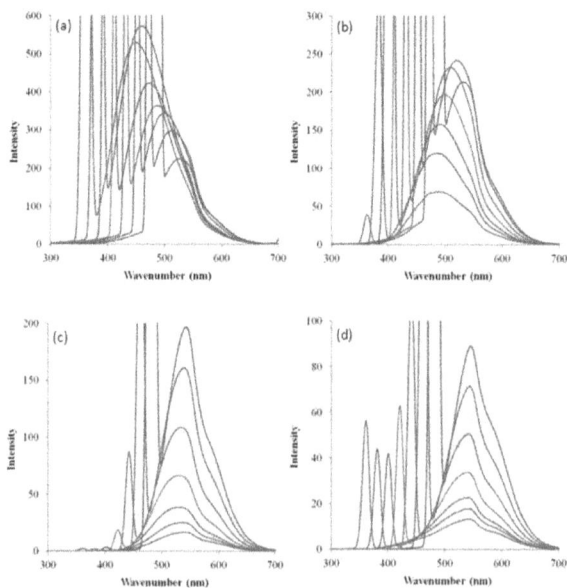

Figure 19. Fluorescence spectra of xanthan gum/BMIMCl ion gels in 10, 20, 40, and 60% (w/w) contents by excitation at 360-480 nm ((a) – (d), respectively).

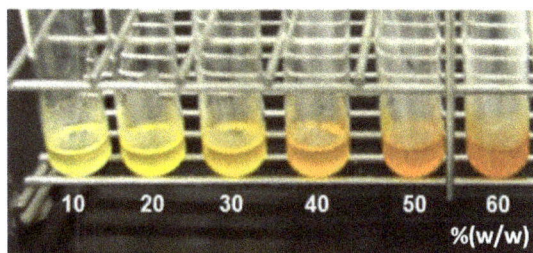

Figure 20. Photographs of xanthan gum/BMIMCl ion gels in 10-60% (w/w) contents.

5. Conclusion

This chapter overviewed the preparation of new fluorescent materials composed of the ILs as components which exhibited specific and unique photo functions. The unique FRET system using rhodamine 6G and the imidazolium-type IL, BMIMCl, was successfully appeared. The radical copolymerization of two PILs, which had one and two polymerizable groups, respectively, was carried out with AIBN as an initiator to give the transparent polymeric ionic liquid (PIL) film. The fluorescence spectra of the film exhibited excitation-wavelength-dependence fluorescent emission maxima at around 430 – 470 nm by excitation at 260 – 400 nm. On the basis of the above results, the PIL films carrying fluorescent dye moieties were prepared by radical copolymerization of polymerizable ionic liquids with appropriate polymerizable fluorescent dye derivatives. The films carrying rhodamine, 7-(diethylamino)coumarin-3-carboxylic acid (DEAC), and pyranine moieties exhibited the three primary color emissions, i.e., red, green, and blue, respectively, by excitation at wide wavelength areas. By incorporating possible combinations of the dye moieties in the PIL backbones, furthermore, the PIL films, which emitted tunable multicolors, were successfully obtained.

For the preparation of materials exhibiting the unique fluorescent behaviors, the gelling system of BMIMCl using guar gum of a natural polysaccharide containing rhodamine 6G was employed. The fluorescence spectra of the resulting ion gel showed the emissions due to rhodamine 6G by excitation at 260 – 600 nm, whereas no emissions due to BMIMCl were observed, indicating the occurrence of FRET from BMIMCl to rhodamine 6G in the gel. The fluorescent behaviors of xanthan gum/BMIMCl ion gels were also investigated. The gels exhibited the xnathan gum content-dependent emission changes, probably owing to the presence of specific association states of BMIMCl in the gels.

The specific fluorescent functions of the materials described in this chapter are realized by the unique photo properties of the ILs. The present materials have the potential for the practical applications in the various fields in the future.

Acknowledgements

The author is indebted to the co-workers, whose names are found in references from his papers, for their enthusiasistic collaborations.

Author details

Jun-ichi Kadokawa*

Address all correspondence to: kadokawa@eng.kagoshima-u.ac.jp

Graduate School of Science and Engineering, Kagoshima University, Japan

References

[1] Simmons, J., & Potter, K. S. (2000). Optical Materials. Waltham: Academic Press

[2] Kim, E., & Park, S. B. (2009). Chemistry as a Prism: A Review of Light-emitting Materials Having Tunable Emission Wavelengths. *Chemistry- A European Journal, 4,* 1646-1658.

[3] Obata, M., Morita, M., Nakase, K., Mitsuo, K., Asai, K., Hirohara, S., & Yano, S. (2007). Synthesis and Photophysical Properties of Rhodamine B Dye-bearing Poly(isobutyl Methacrylate-co-2, 2, 2 trifluoroethyl Methacrylate) as a Temperature-sensing Polymer Film. *Journal of Polymer Science, Part A: Polymer Chemistry,* 45, 2876, 2885.

[4] Shiraishi, Y., Miyamoto, R., & Hirai, T. (2008). Rhodamine-conjugated Acrylamide Polymers Exhibiting Selective Fluorescence Enhancement at Specific Temperature Ranges. *Journal of Photochemistry and Photobiology A: Chemistry,* 200, 432-437.

[5] Zhu, M., Zhou, C., Zhao, Y., Li, Y., Liu, H., & Li, Y. (2009). Synthesis of a Fluorescent Polymer Bearing Covalently Linked Thienylene Moieties and Rhodamine for Efficient Sensing. *Macromolecular Rapid Communications,* 30, 1339-1344.

[6] Welton, T. (1999). Room-temperature Ionic Liquids. Solvents for Synthesis and Catalysis. *Chemical Reviews,* 99, 2071-2083.

[7] Plechkova, N. V., & Seddon, K. R. (2008). Applications of Ionic liquids in the Chemical Industry. *Chemical Society Reviews,* 37, 123-150.

[8] Giernoth, R. (2010). Task-specific Ionic Liquids. Angewandte Chemie International Edition , 49, 2834-2839.

[9] Davis, J. H. (2004). Task-specific Ionic Liquids. *Chemistry Letters,* 33, 1072-1077.

[10] Lee, S. G. (2006). Functionalized Imidazolium Salts for Task-specific Ionic Liquids and Their Applications. *Chemical Communications*, 1049-1063.

[11] Paul, A., Mandal, P. K., & Samanta, A. (2005). How Transparent are The Imidazolium Ionic Liquids? A Case Study with 1 -Methyl-3-butylimidazolium Hexafluorophosphate, [bmim][PF_6]. *Chemical Physics Letters*, 402, 375-379.

[12] Paul, A., Mandal, P. K., & Samanta, A. (2005). On The Optical Properties of The Imidazolium Ionic Liquids. *The Journal of Physical Chemistry B*, 109, 9148-9153.

[13] Paul, A., & Samanta, A. (2006). Optical Absorption and Fluorescence Studies on Imidazolium Ionic Liquids Comprising The Bis(trifluoromethanesulphonyl)imide anion. *Journal of Chemical Sciences*, 118, 335-340.

[14] Mandal, P. K., Paul, A., & Samanta, A. (2006). Excitation Wavelength Dependent Fluorescence Behavior of The Room Temperature Ionic Liquids and Dissolved Dipolar Solutes. *Journal of Photochemistry and Photobiology A: Chemistry*, 182, 113-120.

[15] Miyawaki, A., Llopis, J., Heim, R., Mc Caffery, J. M., Adams, J. A., Ikura, M., & Tsien, R. Y. (1997). Fluorescent Indicators for Ca^{2+} Based on Green Fluorescent Proteins and Calmodulin. *Nature*, 388, 882-887.

[16] Lakowicz, J. R. (1999). Principle of Fluorescence Spectroscopy. , 2nd Edition. New York Plenum

[17] Kikuchi, K. (2010). Design, Synthesis, and Biological Application of Fluorescent Sensor Molecules for Cellular Imaging. In: Endo I, Nagamune T. (eds.) Nano/Micro Biotechnology- Advances in Biochemical Engineering-Biotechnology. Berlin Springer , 119, 63-78.

[18] Kikuchi, K. (2010). Design, Synthesis and Biological Application of Chemical Probes for Bio-imaging. *Chemcal Society Reviews*, 39, 2048-2053.

[19] Izawa, H., Wakizono, S., & Kadokawa, J. (2010). Fluorescence Resonance-energy-transfer in Systems of Rhodamine 6G with Ionic Liquid Showing Emissions by Excitation at Wide Wavelength Areas. *Chemical Communications*, 46, 6359-6363.

[20] Beija, M., Afonso, C. A. M., & Martinho, J. M. G. (2009). Synthesis and Applications of Rhodamine Derivatives as Fluorescent probes. *Chemical Society Reviews*, 38, 2410-2433.

[21] Bhattar, S. L., Kolekar, G. B., & Patil, S. R. (2008). Fluorescence Resonance Energy Transfer between Perylene and Riboflavin in Micellar Solution and Analytical Application on Determination of Vitamin B-2. *Journal of Luminescence*, 128, 306-310.

[22] Green, O., Grubjesic, S., Lee, S., & Firestone, M. A. (2009). The Design of Polymeric Ionic Liquids for The Preparation of Functional Materials. *Journal of Macromolecular Science: Part C, Polymer Reviews*, 49, 339-360.

[23] Mecerreyes, D. (2011). Polymeric Ionic Liquids: Broadening The Properties and Applications of Polyelectrolytes. *Progress in Polymer Science*, 36, 1629-1648.

[24] Wakizono, S., Yamamoto, K., & Kadokawa, J. (2011). FRET Function of Polymeric Ionic Liquid Film Containing Rhodamine Moieties for Exhibiting Emissions by Excitation at Wide Wavelength Areas. *Journal of Photochemistry and Photobiology A: Chemistry*, 222, 283-287.

[25] Wakizono, S., Yamamoto, K., & Kadokawa, J. (2012). Tunable Multicolor Emissions of Polymeric Ionic Liquid Films Carrying Proper Fluorescent Dye Moieties. Journal of Materials Chemistry , 22, 10619-10624.

[26] Seoud, O. A. E., Koschella, A., Fidale, L. C., Dorn, S., & Heinze, T. (2007). Applications of Ionic Liquids in Carbohydrate Chemistry: A Windows of Opportunities. *Biomacromolecules*, 8, 2629-2647.

[27] Liebert, T., & Heinze, T. (2008). Interaction of Ionic Liquids with Polysaccharides 5. Solvents and Reaction Media for The Modification of Cellulose. *BioResources*, 3, 576-601.

[28] Feng, L., & Chen, Z. I. (2008). Research Progress on Dissolution and Functional Modification of Cellulose in Ionic Liquids. *Journal of Molecular Liquids*, 142, 1-5.

[29] Pinkert, A., Marsh, K. N., Pang, S., & Staiger, M. P. (2009). Ionic Liquids and Their Interaction with Cellulose. *Chemical Reviews*, 109, 6712-6728.

[30] Kadokawa, J. (2011). Preparation of Polysaccharide-based Materials Compatibilized with Ionic Liquids. In: Kokorin A. (ed.) Ionic Liquids, Application and Perspectives. Rijeka InTech , 95-114.

[31] Kadokawa, J. (2012). Preparation of Functional Ion Gels of Polysaccharides with Ionic Liquids. In: Mun J, Sim H. (eds.) Handbook of Ionic Liquids: Properties, Applications and Hazards. Hauppauge Nova Science Publishers , 455-466.

[32] Kadokawa, J., Murakami, M., & Kaneko, Y. (2008). A Facile Preparation of Gel Materials from a Solution of Cellulose in Ionic Liquid. *Carbohydrate Research*, 343, 769-772.

[33] Kadokawa, J., Murakami, M., Takegawa, A., & Kaneko, Y. (2009). Preparation of Cellulose-starch Composite Gel and Fibrous Material from a Mixture of The Polysaccharides in Ionic Liquid. *Carbohydrate Polymers*, 75, 180-183.

[34] Prasad, K., Murakami, M., Kaneko, Y., Takada, A., Nakamura, Y., & Kadokawa, J. Weak. (2009). Gel of Chitin with Ionic Liquid, 1Allyl-3-methylimidazolium Bromide. *International Journal of Biological Macromolecules*, 45, 221-225.

[35] Takegawa, A., Murakami, M., Kaneko, Y., & Kadokawa, J. (2010). Preparation of Chitin/cellulose Composite Gels and Films with Ionic Liquids. *Carbohydrate Polymers*, 79, 85-90.

[36] Schuerch, C. (1986). Polysaccharides. In: Mark HF, Bilkales N, Overberger CG. (eds.) Encyclopedia of Polymer Science and Engineering, 2nd Edition. New York John Wiley & Sons , 13, 87-162.

[37] Stephen, A. M., Philips, G. O., & Williams, P. A. (1995). Food Polysaccharides and Their Applications. London, Taylor & Francis.

[38] Prasad, K., Kaneko, Y., & Kadokawa, J. (2009). Novel Gelling Systems of κ-, ι- and λ-Carrageenans and Their Composite Gels with Cellulose Using Ionic Liquid. *Macromolecular Bioscience*, 9, 376-382.

[39] Prasad, K., Izawa, H., Kaneko, Y., & Kadokawa, J. (2009). Preparation of Temperature-induced Shapeable Film Material from Guar Gum-based Gel with an Ionic Liquid. *Journal of Materials Chemistry*, 19, 4088-4090.

[40] Izawa, H., Kaneko, Y., & Kadokawa, J. (2009). Unique Gel of Xanthan Gum with Ionic Liquid and Its Conversion into High Performance Hydrogel. *Journal of Materials Chemistry*, 19, 6969-6972.

[41] Izawa, H., & Kadokawa, J. (2010). Preparation and Characterizations of Functional Ionic Liquid-gel and Hydrogel of Xanthan Gum. *Journal of Materials Chemistry*, 20, 5235-5241.

[42] Mine, S., Prasad, K., Izawa, H., Sonoda, K., & Kadokawa, J. (2010). Preparation of Guar Gum-based Functional Materials Using Ionic Liquid. *Journal of Materials Chemistry*, 20, 9220-9225.

[43] Romero, C., Moore, H. J., Lee, T. R., & Baldelli, S. (2007). Orientation of 1-Butyl-3-methylimidazolium Based Ionic Liquids at a Hydrophobic Quartz Interface Using Sum Frequency Generation Spectroscopy. *The Journal of Physical Chemistry C*, 111, 240-247.

Chapter 11

Ionic-Liquid-Assisted Synthesis of Hierarchical Ceramic Nanomaterials as Nanofillers for Electromagnetic-Absorbing Coatings

Elaheh Kowsari

Additional information is available at the end of the chapter

1. Introduction

The use of electronic devices is experiencing an exponential growth in all the fields of human life, and most of them (personal computers, communication, medical and analytic devices, a lot of domestic appliances) work in the microwave frequency range. This growth gives rise to an increase of Electro Magnetic Interference (EMI) so that it is mandatory to develop systems to protect electronic devices from external interferences.

Electromagnetic (EM) wave absorption materials have attracted much attention owing to the expanded EM interference problems. The EM absorbers are now requested to have not only strong absorption characteristics and wide absorption frequency, but also light weight and antioxidation. The electric permittivity (ε) and magnetic permeability (μ) are parameters related to the dielectric and magnetic properties of a material, and directly associated to their absorbing characteristics [1-5].

The relative permittivity and permeability are represented by Equations 1 and 2, respectively; the values of these parameters are calculated from the experimental values of the transmission and refection coeffcients of the material.

$$\varepsilon_r = \varepsilon' + i\varepsilon'' \tag{1}$$

$$\mu_r = \mu' + i\mu'' \tag{2}$$

In these equations, the primed and double-primed symbols denote real and imaginary components. When the material is lossy, the permittivity and permeability of are complex and some of the incident electromagnetic energy is dissipated [6-7].

In the case of a magnetic material, losses are produced by changes in the alignment and rotation of the magnetization spin [1,8, 9].

The traditional absorbers such as ferrite have strong absorption characteristics, but the thickness required is too large. Therefore, many nanostructures have recently been studied for attenuation of EM wave. These materials are involved with carbon nanotubes, iron and zinc oxide, etc [10-14]. EM wave absorption capability depends on the nature, shape, and size of an absorber.

The dramatic effect that shape anisotropy has on the electronic, optical, and catalytic properties of noble ceramic nanostructures makes the development of morphology-controlled synthesis strategies a main step toward the design of future nanodevices [15-19].

In this regard, the last years have been very prolific in the design of new procedures dealing with the synthesis of noble metal anisotropic structures (Au, Ag, Pt, Pd) such as nanowires, [20-22] nanoplates, [23-26] nanocubes, [27,28] and nanorods [29] with well-controlled size. Recently, some examples concerning the synthesis of branched nanostructures (monopod,bipod, tripod, tetrapod,multipod),[30-35] star polyhedral crystals [36], nanoflowers [37], and ringlike nanostructures[38] have also been reported.

Hierarchical nano-/micro-structures with specific morphology have fascinated scientists all over the world because of their sophisticated architectures which are expected to provide some unique and exciting properties. To date, many recent efforts have been devoted to the synthesis of inorganic materials with hierarchical shapes, including metal [39,40], metal oxide [41], sulfide [42], hydrate [43], and other minerals [44,45].

Recently, ionic liquids have aroused increasing interest because of their unique properties and the potential applications. Ionic liquid can act as a new reaction medium for reactants and morphology templates for the products at the same time, which enables the synthesis of hierarchical nano-/micro-structures with novel or improved properties [46]. Novel nanostructures can be produced by selecting suitable ionic liquids reaction systems. Various nano- or microstructured materials, such as Bi_2S_3 nanostructures [47, 48],Bi_2Se_3 nanosheets [49] and hollow TiO_2 microspheres [50] have been synthesized in ionic liquids.

In this chapter, an attempt has been made to develop hierarchical ceramic nanomaterials with a wide range of morphologies and sizes with improved reflection losses (RL). The duration of experimental processes parameters like amount of IL, pH and temperature have been extensively optimized to obtain morphologies and sizes.

1.1. Development of Innovative Synthesis Methods of Ceramic Nano-materials Using Ionic Liquid

Synthesis of nanomaterials is an increasing active area [51,52]. This interest arises from not only their unusual chemical and physical properties but also their potential application in many fields, which have stimulated the search for new synthetic methods for these materials. Size, morphology and dimensionality can strongly affect the properties of nanostructured materials. Recently, nanostructured metallic and semiconducting materials with various structures and morphologies have received muchattention due to their novel applications, intriguing properties, and quantum size effects [53]. Especially, a three-dimensional (3D) integrated platform of nanostructured materials is highly desirable for applications in advanced nanoelectronic, optoelectronic, solar cells, sensor, etc., [54–56].

Novel nanostructures can be produced by selecting suitable ILs reaction systems. Various nano- or microstructured materials, have been synthesized in ILs

Cupric oxide (CuO) with leaf-like, chrysanthemum-like and rod shapes have been synthesized by microwave-assisted approach using an ionic liquid 1-n-butyl-3-methyl imidazolium tetrafluoroborate ([BMIM]BF4) by Xu and coworkers [57]. By controlling the concentration of [BMIM]BF$_4$ and reaction time, shape transformation of CuO nanostructures could be achieved in a short period of time. The morphologies of the samples were shown in Figure. 1. Leaf-like CuO nanosheets with uniform shape and size were obtained on a large scale (Figure. 1(a)).

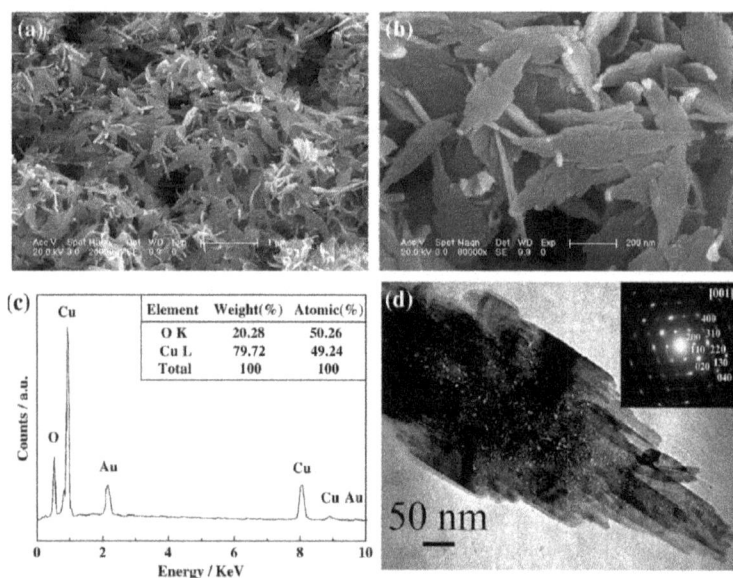

Figure 1. FESEM images of sample A (a–b), EDS results of sample A (c), and TEM image of sample A (d). The inset showed the SAED pattern taken from d.(Reproduced from X Xu, M Zhang, J Feng, M Zhang. Shape-controlled synthesis of single-crystalline cupric oxide by microwave heating using an ionic liquid. Mat Lett. 2008; 62:2787–2790, Copyright (2008), with permeation from Elsevier).

MoS_2 microspheres were successfully synthesized via a facile hydrothermal route assisted by an ionic liquid [BMIM][BF4] by Ma and coworkers [58]. SEM images showed that the MoS_2 microspheres had uniform sizes with mean diameter about 2.1 μm. The MoS$_2$ microspheres had rough surfaces and were constructed with sheetlike structures. Ionic liquid played a crucial role as a templating reagent in the formation of MoS$_2$ microspheres. A possible formation mechanism of MoS$_2$ microspheres was preliminarily presented. The size and morphology of the samples were examined by SEM. Figure. 2A shows that the as-synthesized MoS$_2$ products display a uniform spherical morphology with mean diameter of 2.1 μm.

Figure 2. SEM images of MoS2 microspheres prepared by IL-assisted hydrothermal process. (A) and (B) as-synthesized samples and (C) after annealing at 800 °C for 2 h. [Reproduced from Ma L, Chen WX, Li H, Zheng YF, Xu ZD. Ionic Liquid-Assisted Hydrothermal Synthesis of MoS2 Microspheres. Mat Lett. 2008; 62: 797–799, Copyright (2008), with permeation from Elsevier)].

Wurtzite CdSe nanoparticles-assembled microspheres with macropores have been successfully synthesized through a modified hydrothermal method with Cd(NO$_3$)$_2$ and Na$_2$SeO$_3$ as precursors and hydrazine hydrate as a reductant in the presence of 1-n-butyl-3-methyl-imidazolium bromide ([Bmim]Br) by Liu and coworkers [59]. The results indicated that the CdSe microspheres have an average size of about 3 μm and were assembled by CdSe nanoparticles with size ranging from 20 to 40 nm. It was found that the pH and [Bmim]Br have influence on the morphologies of the products. Figure. 3 shows Wurtzite CdSe nanoparticles-assembled microspheres.

ZnO/SnO$_2$ nanostructured have been successfully synthesized by a hydrothermal method in the presence of the chiral ionic liquid (CIL) ditetrabutylammonium tartrate, [TBA]2[L-Tar] by Kowsari and coworkers [60].The results revealed that using different ratios of

Zn^{2+}/Sn^{4+} affects the phase and morphology of the ZnO/SnO_2 nanocomposite materials. Figure. 4 shows ZnO/SnO_2 nanostructures.

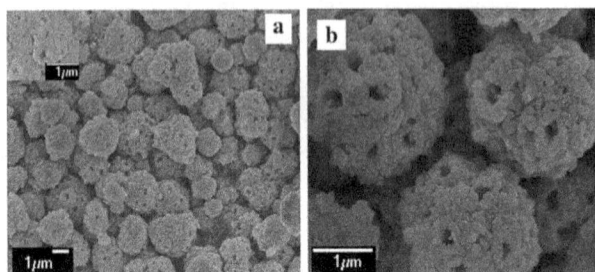

Figure 3. (a) Low magnification SEM image of the CdSe microspheres, the inset shows the structure on the surface of the microspheres; (b) Magnified SEM image of three individual CdSe microspheres. [Reproduced from Liu X, Peng P, Ma J, Zheng W. Preparation of novel CdSe microstructure by modified hydrothermal method. Mat Lett. 2009; 63: 673–675. ,Copyright (2009), with permeation from Elsevier)].

Figure 4. SEM images of products obtained at different Zn^{2+}/Sn^{4+} molar ratios: (a) 2:1, (b) 1:1, (c) 4:1, (d) 1:2, (e) 0:1, and (f) 1:0; the reaction time was kept constant at 24 h and the reaction temperature was 170 °C, CIL=0.05 g. [Reproduced from Kowsari E, Ghezelbash MR. Ionic liquid-assisted, Facile Synthesis of ZnO/SnO_2 Nanocomposites, and Investigation of Their Photocatalytic Activity. Mat Lett. 2012; 68: 17–20,Copyright (2009), with permeation from Elsevier)].

A hydrothermal method has been employed to prepare cactus-like zincoxysulfide ZnO_xS_{1-x} nanostructures with the assistance of a dicationic task-specific ionic liquid (TSIL), [mim] {$(CH)_2$}$_3$[imm]$(SCN)_2$ by kowsari and coworkers[61] To the best of our knowledge, this is the first time that this TSIL with the SCN anion has been used in place of conventional reagents as a source of S to prepare a ZnO_xS_{1-x} nanostructure. The effect of the TSIL concentration on the morphology of the products shows in Figure. 5.

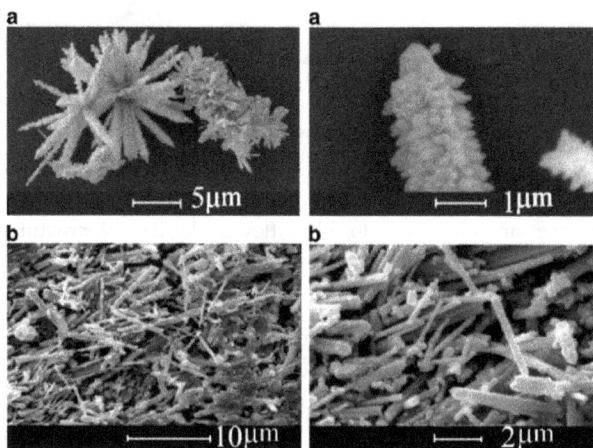

Figure 5. SEM images of products obtained with different amounts of FIL: (a) 0.22 g (S-1), (b) 0.1 g (S-2); the reaction time was constant at 24 h and Zn2+/OH–=1:20; the reaction temperature was 170 °C. [Reproduced from Kowsari E, Ghezelbash MR. Synthesis of Cactus-Like Zincoxysulfide (ZnO_xS_{1-x}) Nanostructures Assisted by a Task-Specific Ionic Liquid and Their Photocatalytic Activities. Mat Lett.2011; 65: 3371–3373.Copyright (2011), with permeation from Elsevier)].

2. Influence of ionic liquids on the growth of nanofillers for electromagnetic-absorbing coatings

2.1. The fabrication of $BaCO_3$ nanostructures as nanofillers for electromagnetic-absorbing coatings

An economical and efficient ionic liquid-assisted chemical method was demonstrated for the first time for the fabrication of $BaCO_3$ nanostructures. The shape of these $BaCO_3$ nanostructures could be readily controlled by changing the chemical conditions and the amount of the chiral ionic liquid (CIL), ditetrabutylammonium tartrate, [TBA]$_2$[L-Tar] by Kowsari and coworkers [62]. The CIL is a reagent and templating agent for the fabrication of $BaCO_3$ nanostructures. It was demonstrated that [TBA]$_2$[L-Tar] served as a modifier in the reactionsystem.Figure 6 shows typical SEM images of $BaCO_3$ nanostructures synthesized with 0.05 g CIL at 170 °C for 24 h. From the SEM image Figure 6, it is clear that the typical $BaCO_3$ dendritic nanostructures assembled were hyacinth-like with rod-like nanostructures with lengths of up to several micrometers.

Figure 6. Typical scanning electron micrographs of BaCO₃ nanostructures synthesized with 0.05 g CIL, NaOH = 0.04 M (2 ml) at 170 °C for 24 h, the Ba(NO₃)₂ = 0.53 g. [Reproduced from Kowsari E, Karimzadeh AH. Using a Chiral Ionic Liquid for Morphological Evolution of BaCO₃ and its Radar Absorbing Properties as a Dendritic Nanofiller. Mat Lett. 2012;74, 33-36, Copyright (2012), with permeation from Elsevier)].

With increase in the amount of CIL to 0.1 g, flower-like BaCO₃ structures composed of dendritic petals were produced, which appeared as a result of oriented attachment and self-assembly, as exhibited in Fig 7.

Figure 7. Typical scanning electron micrographs of BaCO₃ nanostructures synthesized with 0.1g CIL, NaOH = 0.04 M (2 ml) at 170 °C for 24 h, the Ba(NO₃)₂ = 0.53 g. [Reproduced from Kowsari E, Karimzadeh AH. Using a Chiral Ionic Liquid for Morphological Evolution of BaCO₃ and its Radar Absorbing Properties as a Dendritic Nanofiller. Mat Lett.2012;74, 33-36, Copyright (2012), with permeation from Elsevier)].

The effects of the synthetic parameters, such as the concentration of NaOH, reaction temperature, reaction time and [Ba²⁺], on the morphologies of the resulting products were investigated. Typical SEM images of the products with different morphologies brought about by varying the amount of NaOH are presented in Figure. 8.

Too much NaOH was found to be harmful to the formation of flower-like structures and resulted in atrophic flowers, which are analogs of a sphere as shown in Figure. 8. NaOH affected the reaction kinetics through tuning the dissolution-deposition equation of Ba(OH)₂, because soluble Ba(NO₃)₂ will first react with NaOH to form Ba(OH)₂ precipitate (Ksp) 1.09 × 10-15). The trend for Ba(OH)₂ to release Ba2+ decreased if the concentration of OH- is relatively high, leading to the thermodynamic change of nucleation and growth velocities, which is beneficial for the formation of spherical structures.

Figure 9 shows XRD patterns of the products obtained under different reaction conditions. It is clear that all of the peaks can be readily indexed to the pure orthorhombic phase of BaCO3 (JCPDS card no. 05-0378). The sharp diffraction peaks of the sample indicate that well-crystallized BaCO3 crystals can be easily obtained under the current synthetic conditions. On comparing the XRD patterns of the four products, It was shown that the relative intensity of the peaks varied slightly.

Figure 8. Typical scanning electron micrographs of BaCO$_3$ nanostructures synthesized with (a) 2.5 ml, CIL = 0.05 g at 170 °C for 24 h, Ba(NO$_3$)$_2$ = 0.53 g.] [Reproduced from Kowsari E, Karimzadeh AH. Using a Chiral Ionic Liquid for Morphological Evolution of BaCO$_3$ and its Radar Absorbing Properties as a Dendritic Nanofiller. Mat Lett.2012;74, 33-36, Copyright (2012), with permeation from Elsevier)].

Figure 9. Typical scanning electron micrographs of BaCO$_3$ nanostructures synthesized with CIL = 0.05 g at 170 °C for 24 h, Ba(NO$_3$)$_2$ = 0.53 g. [Reproduced from Kowsari E, Karimzadeh AH. Using a Chiral Ionic Liquid for Morphological Evolution of BaCO$_3$ and its Radar Absorbing Properties as a Dendritic Nanofiller. Mat Lett.2012;74, 33-36, Copyright (2012), with permeation from Elsevier)].

The value of the minimum reflection loss for the BaCO$_3$ dendritic nanostructures composite was -40 dB at 10.2 GHz (for a thickness of 4.0 mm) [Figure. 10]. According to the results shown above, BaCO$_3$ dendritic nanostructures showed very strong absorption of microwave compared with other samples. It is noteworthy that the BaCO$_3$ dendritic nanostructures have special geometrical morphology. Such isotropic crystal symmetry can form isotropic quasi antennas and some continuous networks in the composites. It is possible for the electromagnetic waves to penetrate the nanocomposites formed by the numerous antenna-like semiconducting BaCO$_3$ dendritic nanostructures and for the energy to be induced into a dissipative current; then the current will be consumed in the continuous networks, which leads to the energy attenuation (19-20). More importantly, the interfacial electric polarization should be considered. However, further experimental and theoretical work is needed to clarify this mechanism.

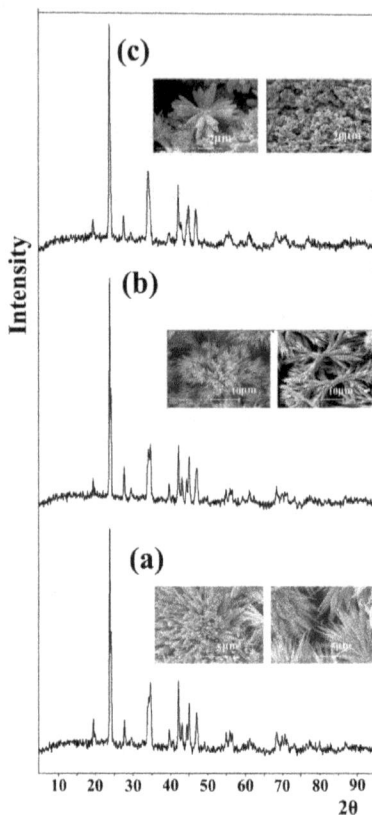

Figure 10. X-ray diffraction patterns of the $BaCO_3$ nanostructures synthesized with (a) 5 ml, (b) 2.5 ml, (NaOH (0.04 M) [CIL = 1.25 g /L at 170 °C for 24 h, $Ba(NO_3)_2$ = 13.25 g/L.], (c) 2.5 g/L CIL. [NaOH = 0.04 M (2 ml) at 170 °C for 24 h, the $Ba(NO_3)_2$ = 13.25 g/L]; [Reproduced from Kowsari E, Karimzadeh AH. Using a Chiral Ionic Liquid for Morphological Evolution of $BaCO_3$ and its Radar Absorbing Properties as a Dendritic Nanofiller. Mat Lett.2012;74, 33-36, Copyright (2012), with permeation from Elsevier)].

2.2. Fern-like, fish skeleton-like, bunched cubic, and butterfly-like BaO nanostructures as nanofillers for Radar-absorbing coatings

Fern-like, fish skeleton-like, bunched cubic, and butterfly-like BaO nanostructures have been synthesized by a hydrothermal method at 170 °C from $Ba(NO_3)_2$ •3H2O and NaOH in the presence of ammonium persulfate $(NH_4)_2S_2O_8$ (APS) by Kowsari and coworkers[63]. The effect of the chiral ionic liquid (CIL) ditetrabutylammonium tartrate, [TBA]$_2$[L-Tar], on the morphologies of the products has been investigated. It was demonstrated that [TBA]$_2$[L-Tar] served as a modifier in the reaction system. Furthermore, the BaO nanostructures have been used as fillers in high-performance microwave-absorbing coatings.

By controlling the concentrations of CIL, Ba^{2+}, and APS, shape transformations of the BaO could be achieved. When the amount of CIL was zero, rotor-like BaO was formed (Figure 11).

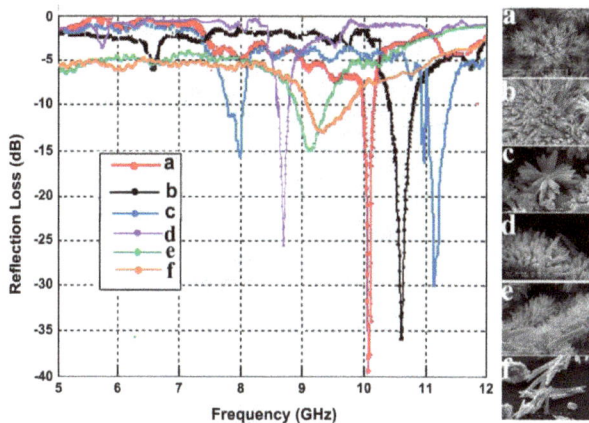

Figure 11. Reflection loss (dB) as a function of frequency (GHz) of $BaCO_3$ nanostructures synthesized with: (a) NaOH = 2.5 mL (0.04 M), CIL = 1.25 g /L for 24 h $Ba(NO_3)_2$ = 13.25 g/L; (b) NaOH = 5 mL (0.04 M) and CIL = 1.25 g /L for 24 h, $Ba(NO_3)_2$ = 13.25 g/L; (c) NaOH = 2 mL (0.04 M) and CIL = 2.5 g/L, for 24 h, amount of $Ba(NO_3)_2$ = 13.25 g/L; (d) NaOH = 2 mL (0.04 M), CIL = 1.25 g /L for 24 h, $Ba(NO_3)_2$ = 6.52 g/L; (e) NaOH = 2 mL (0.04 M), CIL = 1.25 g /L for 24 h, $Ba(NO_3)_2$ = 13.25 g/L; (f) NaOH- = 2 mL (0.04 M), CIL = 1.25 g /L for 48 h, $Ba(NO_3)_2$ = 13.25 g/L, (at 170 °C and with filler = 0.025 g in coating). [Reproduced from Kowsari E, Karimzadeh AH. Using a Chiral Ionic Liquid for Morphological Evolution of $BaCO_3$ and its Radar Absorbing Properties as a Dendritic Nanofiller. Mat Lett.2012;74, 33-36, Copyright (2012), with permeation from Elsevier)].

Figure 12. SEM images of products obtained with : 0.1 g CIL, keeping the reaction time fixed at 24 h; $Ba(NO_3)_2$ = 0.2 g, APS = 0.27 g, reaction temperature 170 °C. [Reproduced from Kowsari E, Karimzadeh AH. Fabrication of Fern-Like, Fish Skeleton-Like, and Butterfly-Like BaO Nanostructures as Nanofillers for Radar-Absorbing Nanocomposites. Mat Lett. 2012; 74:33–36, Copyright (2012), with permeation from Elsevier].

When the amount of CIL was increased gradually, while maintaining the same reaction time, rotor-like BaO changed to bunched cubic BaO, as shown in Figure 14. The cubes have a porous surface structure.

Figure 13. SEM images of products obtained with 0.1 g, CL= 0.15 g keeping the reaction time fixed at 24 h; Ba(NO$_3$)$_2$ = 0.2 g, APS = 0.27 g, reaction temperature 170 °C. [Reproduced from Kowsari E, Karimzadeh AH. Fabrication of Fern-Like, Fish Skeleton-Like, and Butterfly-Like BaO Nanostructures as Nanofillers for Radar-Absorbing Nanocomposites. Mat Lett.2012; 74:33–36, Copyright (2012), with permeation from Elsevier].

Figure 14. SEM images of products obtained with Ba(NO$_3$)$_2$ = 0.8 g, CIL = 0; APS = 0.27 g; (the reaction time was kept constant at 24 h and the reaction temperature was 170 °C). [Reproduced from Kowsari E, Karimzadeh AH. Fabrication of Fern-Like, Fish Skeleton-Like, and Butterfly-Like BaO Nanostructures as Nanofillers for Radar-Absorbing Nanocomposites. Mat Lett.2012; 74:33–36, Copyright (2012), with permeation from Elsevier].

The CIL clearly plays a key role in tailoring the form of the resultant BaO nanostructures. It is thought that hydrogen bonds formed between the hydrogen atom at the position-2 (connected to oxygen) of the CIL cation and the oxygen atoms of O–Ba crystal cores may act as effective bridges in connecting the produced BaO nuclei and CIL cations, playing a crucial role in the directional growth of the 2D nanocrystals.

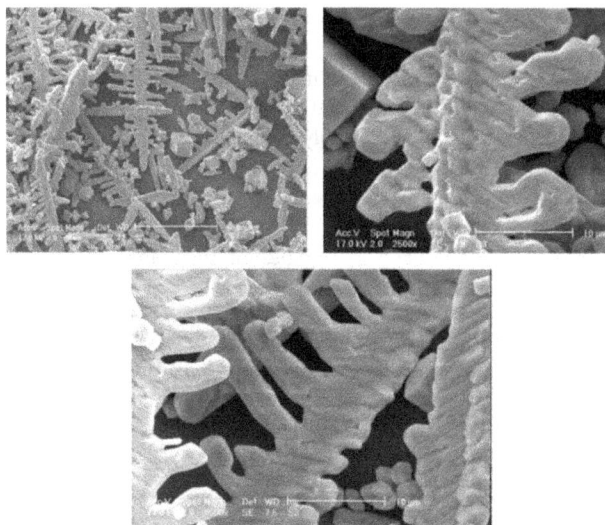

Figure 15. SEM images of products obtained with: Ba(NO$_3$)$_2$ = 0.4 g, CIL = 0, APS = 0.27 g; (the reaction time was kept constant at 24 h and the reaction temperature was 170 C). [Reproduced from Kowsari E, Karimzadeh AH. Fabrication of Fern-Like, Fish Skeleton-Like, and Butterfly-Like BaO Nanostructures as Nanofillers for Radar-Absorbing Nanocomposites. Mat Lett.2012; 74:33–36, Copyright (2012), with permeation from Elsevier].

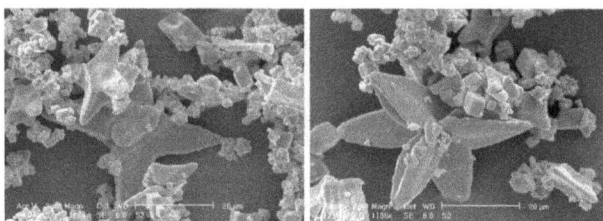

Figure 16. SEM images of products obtained with Ba(NO$_3$)$_2$ = 0.2 g, CIL = 0, APS = 0.27 g; (the reaction time was kept constant at 24 h and the reaction temperature was 170 C). [Reproduced from Kowsari E, Karimzadeh AH. Fabrication of Fern-Like, Fish Skeleton-Like, and Butterfly-Like BaO Nanostructures as Nanofillers for Radar-Absorbing Nanocomposites. Mat Lett.2012; 74:33–36, Copyright (2012), with permeation from Elsevier].

We also investigated the effect of the amount of Ba(NO$_3$)$_2$ on the morphology of the products. The respective products obtained from this series of experiments are depicted in Figure 13, 14, 15.

The value of the maximum reflection loss for the composite with BaO cubic nanostructures was measured as ⊘20 dB at 10.5 GHz for a thickness of 4.0 mm. According to the results shown in Figure 4, the BaO cubic nanostructures showed very strong absorption of micro-waves compared with the other samples. It may be noted that the BaO cubic nanostructures had a special geometrical morphology.

2.3. Morphology evolution of the ZnO/Zn(OH)2 nanofillers using ionic liquids

An efficient ionic-liquid-assisted chemical method for the fabrication of ZnO/Zn(OH)$_2$ nanoplates is demonstrated by kowsari and coworkers [64]. The shape of the resulting ZnO/Zn(OH)$_2$ nanostructures could be readily controlled by changing the chemical condi-tions and the amount of [Cn(mim)]$^+$ H2PO$_4$. as a task-specific ionic liquid (TSIL). The TSIL thus serves as a reagent and templating agent for the fabrication of ZnO/Zn(OH)$_2$ nano-plates. Furthermore, a possible growth mechanism of the ZnO/Zn(OH)$_2$ nanostructures is proposed. The effects of different morphologies of the nanofillers on electromagnetic properties have been investigated.

By controlling the concentration of the TSIL and the reaction time, shape transformations of the ZnO/Zn(OH)$_2$ nanostructures could be achieved. Our results show that the quantity of the TSIL affected the morphology of the Zn(OH)$_2$ and caused changes in the interplanar sep-aration and the arrangement of the nanoplates. Salient SEM images are compared in Fig. 1. In this figure, it can be seen that as the quantity of the TSIL was increased from 0.1 g to 0.2 g, the separation between the nanoplates decreased and the number of nanoplates increased.

The TSIL clearly played a key role in tailoring the form of the resultant ZnO/Zn(OH)$_2$ nano-composites.To analyze the effect of the reaction time on morphology, the reaction was car-ried out for 12, 24, and 48 h, while keeping the amounts of Zn2+ and TSIL constant. Fig. 2 shows the morphologies of the products. It can be seen that with increasing reaction time, the separation between the nanoplates increased and the number of nanoplates decreased

(a)   (b)

Figure 17. SEM images of products obtained with different time of reaction: (a) 12 h, (b) 48 h, Zn(NO$_3$)$_2$ = 0.8 g, OH = 0.27 g, reaction temperature 170 °C, TSIL =0.1 g.

Fig. 18 clearly shows the process of TSIL-assisted hydrothermal growth of ZnO and Zn(OH)2 crystals. The amount of the growth unit [Zn(OH)$_2$(H$_2$PO$_4$)2]$^{2-}$ is greatly increased, and further ZnO and Zn(OH)$_2$ nuclei directly conglomerate to form a two-dimensional nanosheet structure in order to lower the surface potential. Thereafter, the nanosheets self-assemble to produce a cabbage-like structure.

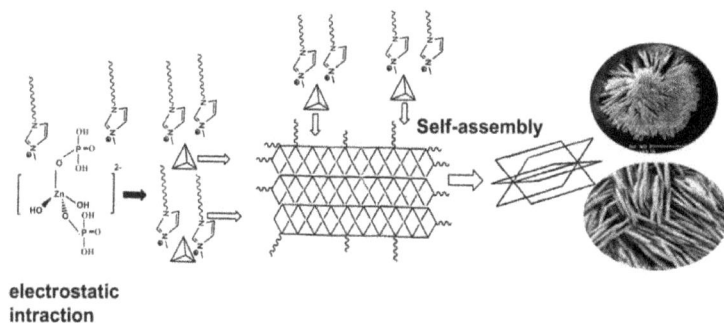

electrostatic intraction   **Self-assembly**

Figure 18. Schematic illustration of the processes of $Zn(OH)_2$/ZnO nanostructure assembly by IL-assisted hydrothermal growth.

The TSIL clearly played a key role in tailoring the form of the resultant $ZnO/Zn(OH)_2$ nanocomposites. In a general manner, the produced coatings showed broadband absorber behavior, which may be attributed to the dielectric properties and the particular shape of the $Zn(OH)_2$/ZnO nanofillers. When radar impinges on a radar -absorbing coating, the incident radiation is not totally absorbed immediately. The radar attenuation by the radar-absorbing coating is significantly complex and different attenuation mechanisms can occur. Equal amounts of three types of $Zn(OH)_2$/ZnO nanostructures with different morphologies and different separations between their nanoplates, as synthesized under different conditions, were used as fillers in coatings.

The value of the maximum reflection loss for the composite with $Zn(OH)_2$/ZnO nanostructures was measured ~20 dB at 8 GHz for a thickness of 4.0 mm. The samples incorporating $Zn(OH)_2$/ZnO nanostructures of a morphology with a shorter inter-plate separation displayed higher absorptions. The absorption was also shifted toward higher frequencies. Moreover, a higher bandwidth was observed. It may be noted that the $ZnO/Zn(OH)_2$ nanostructures had a special geometrical morphology, namely that of a multi-layer microwave absorber. More importantly, the interfacial electric polarization should be considered. It is well known that the permittivity mainly originates from electronic polarization, ion polarization, and intrinsic electric-dipole polarization, on which the crystal structure, size, and shape of nanomaterials may have important influences. On the basis of the electron microscopy characterization results, we can conclude that the radar-absorbing properties are associated with the crystal structure, the crystallization, and the degree of aggregation of the nanocrystal building blocks. This also explains why the $ZnO/Zn(OH)_2$ nanostructures studied here show different radar-absorption properties. However, further experimental and theoretical work is needed to clarify the mechanism.

3. Conclusion

Over the past years, there has been an increasing interest in developing hierarchically structural materials on a nanometer scale due to their novel or enhanced properties. By controlling the condition of reaction, shape transformations of the resulting hierarchical ceramic nanomaterials could be achieved. In this chapter, the ceramic nanomaterials have been synthesized in a single reaction system by simply adjusting the reaction conditions and used as fillers in high-performance microwave-absorbing coatings, with epoxy resin as the polymer matrix. The task specific ionic liquids thus serves as a reagent or templating agent for the fabrication of ceramic oxides. The effects of different morphologies of the nanofillers on electromagnetic properties have been investigated. The maximum reflection attenuation was measured as −20 dB in the frequency range 5–12 GHz.

Author details

Elaheh Kowsari*

Address all correspondence to: kowsarie@aut.ac.ir

Department of Chemistry, Amirkabir University of Technology, Tehran, Iran

References

[1] Balanis, C. A. (1989). *Advanced Engineering Electromagnetics*, New York, John Wiley and Sons.

[2] Lee, S. M. (1991). *International Encyclopedia of Composites*, New York, VCH Publishers.

[3] Clark, D. E., Diane, C. F., Stephen, J. O., & Richards, S. (1995). *Microwaves: Theory and Application in Materials Processing III*, Westerville, The American Ceramic Society.

[4] Hippel, A. (1954). *Dielectric Materials and Applications*, London, Artech House.

[5] Yusoffa, A. N., Abdullah, M. H., Ahmad, S. H., Jusoh, S. F., Mansor, A. A., & Hamid, S. A. A. (2002). Electromagnetic and absorption properties of some microwave absorbers. *J.Appl Phy*, 92(2), 876-882.

[6] Orfanidis, S. J. (2008). *Electromagnetic Waves and Antennas*, Available from, http://www.ece.rutgers.edu/~orfanidi/ewa/, accessed Oct.

[7] Sucher, M. ., & Fox, J. (1963). *Handbook of the microwave measurements* (3 ed.), New York, John Wiley and Sons.

[8] Thostenson, E. T., & Chou, T. W. (1999). Microwave processing: fundamentals and applications. *Composites: Part A: Applied Science and Manufacturing*, 30(9), 1055-1071.

[9] Jarem, L. M., Johnson, J. B. ., & Scott, W. (1995). Measuring the permittivity and per-
     meability of sample at Ka Band using a partially flled waveguide. *IEEE Transactions
     on Microwave and Techniques*, 43(12), 2654-2667.

[10] Watts, P. C. P., Hsu, P. C. P., Barnes, W. K., & Chambers, A. (2003). B High Permittiv-
     ity from Defective Multiwalled Carbon Nanotubes in the X-Band. *Adv Mater.*, 15(7-8),
     600-603.

[11] Wadhawan, A., Garrett, D., & Perez, J. M. (2003). Nanoparticle-assisted microwave
     absorption by single-wall carbon nanotubes. *Appl Phys Lett.*, 83(13), 2683-2685.

[12] Deng, L. J., & Han, M. (2007). Microwave Absorbing Performances of Multiwalled
     Carbon Nanotube Composites with Negative Permeability. *Appl Phys Lett.*, 91(2),
     023119-023121.

[13] Liu, J. R., Itoh, M., Terada, M., Horikawa, T., & Machida, K. I. (2007). Enhanced Elec-
     tromagnetic Wave Absorption Properties of Fe Nanowires in Gigaherz Range. *Appl
     Phys Lett*, 91(9), 093101-093103.

[14] Chen, Y. J., Cao, M. S., Wang, T. H., & Wan, Q. (2004). Microwave absorption proper-
     ties of the ZnO nanowire-polyester composites. *Appl. Phys. Lett.*, 84(17), 3367-3370.

[15] Tang, Z., & Kotov, N. A. (2005). One-dimensional assemblies of nanoparticles: Prepa-
     ration, properties, and promise. *AdV Mater.*, 17(8), 951-962.

[16] Murphy, C., Sau, T. K., Gole, A. M., Orendorff, C. J., Gao, J., Gou, L., Hunyadi, S., &
     Li, T. (2005). Anisotropic Metal Nanoparticles: Synthesis, Assembly, and Optical Ap-
     plications. *J. Phys. Chem. B*, 109(29), 13857-13870.

[17] Xia, Y., & Halas, N. (2005). Shape-Controlled Synthesis and Surface Plasmonic Prop-
     erties of Metallic Nanostructures. *J MRS Bull.*, 30(5), 338-348.

[18] Liz-Marza´n, L. M. (2006). Tailoring Surface Plasmons through the Morphology and
     Assembly of Metal Nanoparticles. *Langmuir*, 22(1), 32-41.

[19] Link, S., & El -Sayed, M. A. (1999). Spectral Properties and Relaxation Dynamics of
     Surface Plasmon Electronic Oscillations in Gold and Silver Nanodots and Nanorods.
     J. Phys. Chem. B, 103(40), 8410-8426.

[20] Vasilev, K., Zhu, T., Wilms, M., Gillies, G., Lieberwirth, I., Mittler, S., Knoll, W., &
     Kreiter, M. I. (2005). One-step Synthesis of Gold Nanowires in Aqueous Solution.
     Langmuir, 21(26), 12399-12403.

[21] Wiley, B., Sun, Y., Mayers, B., & Xia, Y. (2005). Shape-Controlled Synthesis of Metal
     Nanostructures: The Case of Silver. *Chem Eur J*, 11(2), 454-463.

[22] Giersig, M., Pastoriza-Santos, I., & Liz-Marza´n, L. M. (2004). Evidence of an Aggre-
     gative Mechanism During the Formation of Silver Nanowires in N,N-dimethylforma-
     mide. *J Mater Chem*, 14, 607-610.

[23] Umar, A. A., & Oyama, M. (2006). Formation of Gold Nanoplates on Indium Tin Ox-
     ide Surface: Two-Dimensional Crystal Growth from Gold Nanoseed Particles in the
     Presence of Poly(vinylpyrrolidone). *Cryst Growth Des.*, 6(4), 818-821.

[24] Shankar, S. S., Rai, A., Ankamwar, B., Singh, A., Ahmad, A., & Sastry, M. (2004). Bio-
     logical Synthesis of Triangular Gold Nanoprisms. *Nat Mater*, 3(7), 482-488.

[25] Xiong, Y., Mc Lellan, J. M., Chen, J., Yin, Y., Li, Z., & Xia, Y. (2005). Kinetically Con-
     trolled Synthesis of Triangular and Hexagonal Nanoplates of Palladium and Their
     SPR/SERS Properties. *J Am Chem Soc.*, 127(48), 17118-17127.

[26] Yang, J., Lu, L., Wang, H., Shi, W., & Zhang, H. (2006). Glycyl Glycine Templating
     Synthesis of Single-Crystal Silver Nanoplates. *Cryst Growth Des.*, 6(9), 2155-2158.

[27] Sun, Y., & Xia, Y. (2002). Shape-Controlled Synthesis of Gold and Silver Nanoparti-
     cles. *Science*, 298(5601), 2176-2179.

[28] Yu, D., & Yam, V. W. (2004). Controlled Synthesis of Monodisperse Silver Nanocubes
     in Water. *J. Am Chem Soc.*, 126(41), 13200-13201.

[29] Gou, L., & Murphy, C. J. (2005). Fine-Tuning the Shape of Gold Nanorods. *Chem. Ma-
     ter.*, 17(14), 3668-3672.

[30] Chen, S., Wang, Z. L., Ballato, J., Foulger, S. H., & Carrol, D. L. (2003). Monopod, Bi-
     pod, Tripod, and Tetrapod Gold Nanocrystals. *J Am ChemSoc.*, 125(52), 16186-16187.

[31] Sau, T. K., & Murphy, C. J. (2004). Room Temperature, High-Yield Synthesis of Mul-
     tiple Shapes of Gold Nanoparticles in Aqueous Solution. *J Am Chem Soc*, 126(28),
     8648-8649.

[32] Yamamoto, M., Kashiwagi, Y., Sakata, T., Mori, H., & Nakamoto, M. (2005). Synthesis
     and Morphology of Star-Shaped Gold Nanoplates Protected by Poly(N-vinyl-2-pyr-
     rolidone). *Chem Mater.*, 17(22), 5391.

[33] Kuo, C., & Huang, M. H. (2005). Synthesis of Branched Gold Nanocrystals by a Seed-
     ing Growth Approach. *Langmuir*, 21(5), 2012-2016.

[34] Hao, E., Bailey, R. C., Schatz, G. C., Hupp, J. T., & Li, S. (2004). Synthesis and Optical
     Properties of "Branched" Gold Nanocrystals. *Nano Lett.*, 4(2), 327-330.

[35] Bakr, O. M., Wunsch, B. H., & Stellacci, F. (2006). High-Yield Synthesis of Multi-
     Branched Urchin-Like Gold Nanoparticles. *Chem Mater.*, 18(14), 3297-3301.

[36] Burt, J. L., Elechiguerra, J. L., Reyes-Gasga, J., Montejano-Carrizales, J. M., & Yaca-
     man, M. J. (2005). Beyond Archimedean Solids: Star Polyhedral Gold Nanocrystals. *J
     Cryst Growth*, 285(4), 681-691.

[37] Wang, T., Hu, X., & Dong, S. (2006). Surfactantless Synthesis of Multiple Shapes of
     Gold Nanostructures and Their Shape-Dependent SERS Spectroscopy. *J Phys Chem.
     B*, 110(34), 16930-16936.

[38] Ma, H., Huang, S., Feng, X., Zhang, X., Tian, F., Yong, F., Pan, W., Wang, Y., & Chen, S. (2006). Electrochemical Synthesis and Fabrication of Gold Nanostructures Based on Poly(N-vinylpyrrolidone). *Chem Phys Chem;*, 7(2), 333-335.

[39] Shen, G. Z., Bando, Y., & Golberg, D. (2007). Self-Assembled Hierarchical Single-Crystalline β-SiC Nanoarchitectures. *Cryst Growth Des.*, 7(1), 35-38.

[40] Teng, X. W., & Yang, H. (2005). Synthesis of platinum multipods: an induced aniso-tropic growth. *Nano Lett.*, 5(5), 885-891.

[41] Liu, B., & Zeng, H. C. (2004). Fabrication of ZnO "Dandelions" via a Modified Kir-kendall Process. *J Am Chem Soc.*, 126(51), 16744-16746.

[42] Xie, S. H., & Zhao, D. Y. (2002). A Simple Route for the Synthesis of Multi-Armed CdS Nanorod-Based Materials. *Adv Mater.*, 14(21), 1537-1540.

[43] Zhang, Z., Sun, H., Shao, X., Li, D., Yu, H., & Han, M. (2005). Three-Dimensionally Oriented Aggregation of a Few Hundred Nanoparticles into Monocrystalline Archi-tectures. *Adv Mater.*, 17(1), 42-47.

[44] Kowsari, E., & Faraghi, G. (2010). Synthesis by an ionic liquid assisted method and optical properties of nanoflower Y_2O_3. *J Mater Res Bull*, 45, 939-945.

[45] Murray, C. B., Kagan, C. R., & Bawendi, M. G. (1995). Self-Organization of CdSe Nanocrystallites into Three-Dimensional Quantum Dot Superlattices. *Science*, 270(5240), 1335-1338.

[46] Andreas, T. (2004). CuCl Nanoplatelets from an Ionic Liquid-Crystal Precursor. *An-gew Chem Int. Ed. Engl.*, 43(40), 5380-5382.

[47] Jiang, Y., & Zhu, Y. J. (2005). Microwave-Assisted Synthesis of Sulfide M2S3 (M = Bi, Sb) Nanorods Using an Ionic Liquid. *J Phys Chem., B*, 109(10), 4361-4364.

[48] Jiang, J., Yu, S. H., Yao, W. T., Ge, H., & Zhang, G. Z. (2005). Morphogenesis and Crystallization of Bi_2S_3 Nanostructures by an Ionic Liquid-Assisted Templating Route: Synthesis, Formation Mechanism, and Properties. *Chem. Mater.*, 17(24), 6094-6100.

[49] Jiang, Y., Zhu, Y. J., & Cheng, G. F. (2006). Synthesis of Bi_2Se_3 Nanosheets by Micro-wave Heating Using an Ionic Liquid Cryst Growth Des. 9, 2174-2176.

[50] Nskashima, T., & Kimizuka, N. (2003). Interfacial Synthesis of Hollow TiO_2 Micro-spheres in Ionic Liquids. *J Am Chem Soc.*, 125(21), 6386-6387.

[51] Rai, P., Jo, J. N., Lee, I. H., & Yu, Y. T. (2010). Fabrication of 3D rotorlike ZnO nano-structure from 1D ZnO nanorods and their morphology dependent photolumines-cence property. *J Solid State Sci*, 12, 1703-1710.

[52] Patzke, G. R., Krumeich, F., & Nesper, R. (2002). Oxidic Nanotubes and Nanorods-Anisotropic Modules for a Future Nanotechnology Angew. *Chem., Int. Ed. Engl.*, 41(14), 2446-2461.

[53] Cui, Y., & Lieber, C. M. (2001). Functional Nanoscale Electronic Devices Assembled Using Silicon Nanowire Building Blocks. *Science*, 291(5505), 851-853.

[54] Chen, A. C., Peng, X. P., Koczkur, K., & Miller, B. (2004). Super-hydrophobic tin oxide nanoflowers. *Chem Commun.*, 1964-1965.

[55] Gur, I., Fromer, N. A., Geier, M. L., & Alivisatos, A. P. (2005). Air-stable all-inorganic nanocrystal solar cells processed from solution. *Science*, 310(5747), 462-465.

[56] Antonietti, M., Kuang, D. B., Smarsly, B., & Yong, Z. (2004). Ionic liquids for the convenient synthesis of functional nanoparticles and other inorganic nanostructures. *Angew ChemInt Ed*, 43(38), 4988-4992.

[57] Xu, X., Zhang, M., Feng, J., & Zhang, M. (2008). Shape-controlled Synthesis of Single-Crystalline Cupric Oxide by Microwave Heating Using an Ionic Liquid. *Mat Lett.*, 62(17-18), 2728-2790.

[58] Ma, L., Chen, W. X., Li, H., Zheng, Y. F., & Xu, Z. D. (2008). Ionic Liquid-Assisted Hydrothermal Synthesis of MoS_2 Microspheres. *Mat Lett.*, 62(6-7), 797-799.

[59] Liu, X., Peng, P., Ma, J., & Zheng, W. (2009). Preparation of Novel CdSe Microstructure by Modified Hydrothermal Method. *Mat Lett.*, 63(8), 673-675.

[60] Kowsari, E., & Ghezelbash, M. R. (2012). Ionic Liquid-Assisted, Facile Synthesis of ZnO/SnO_2 Nanocomposites, and Investigation of Their Photocatalytic Activity. *Mat Lett.*, 68, 17-20.

[61] Kowsari, E., & Ghezelbash, Mohammad Reza. (2011). Synthesis of cactus-like zincoxysulfide (ZnO_xS_{1-x}) nanostructures assisted by a task-specific ionic liquid and their photocatalytic activities. *Mat Lett.*, 65-3371.

[62] Kowsari, E., & Karimzadeh, A. H. (2012). Using a Chiral Ionic Liquid for Morphological Evolution of $BaCO_3$ and its Radar Absorbing Properties as a Dendritic Nanofiller. *Mat Lett.*, 74, 33-36.

[63] Kowsari, E., & Karimzadeh, A. H. (2012). Fabrication of Fern-Like, Fish Skeleton-Like, and Butterfly-Like BaO Nanostructures as Nanofillers for Radar-Absorbing Nanocomposites. *Mat Lett.*, 74, 33-36.

Chapter 12

Use of Ionic Liquid Under Vacuum Conditions

Susumu Kuwabata, Tsukasa Torimoto,
Akihito Imanishi and Tetsuya Tsuda

Additional information is available at the end of the chapter

1. Introduction

Ionic liquid (IL) is a kind of salt that can stay as a liquid phase even at room temperature. However, researchers sometimes prefer to call it room temperature ionic liquid (RTIL) to distinguish between liquid salt at around room temperature and that at high temperature. Fig. 1 shows structural formulas of well-known ILs with their names and abbreviations. The liquid possesses several attracting characteristics like high ionic conductivity, wide electrochemical windows, and negligible vapor pressure [1-5]. All features imply that IL has high stability and inertness, which have become very useful to utilize IL as electrolytes for Li-ion secondary batteries and PEM fuel cells, reaction solvents for organic synthesis and nanoparticle preparation, and lubricants usable in cosmic space [4].

The negligible vapor pressure of most ILs at room-temperature means that such the ILs can be put in a vacuum chamber without any vaporization. This fact invented a new technological concept because there are several instruments that require vacuum conditions for sample analyses and material manufacturing. Those instruments are basically designed for dealing with solid samples because it is quite common sense that the vacuum conditions should provide dry atmosphere. In other words, conventional procedures with such the instruments cannot be applied to any wet sample, although researchers frequently meet the cases where they would like to deal with wet samples and liquid itself in vacuum equipments. Possibility to introduce ILs to the vacuum instruments could innovatively change the techniques requiring vacuum conditions. Then, some researchers including our research group started to put ILs in vacuum chambers of several instruments.

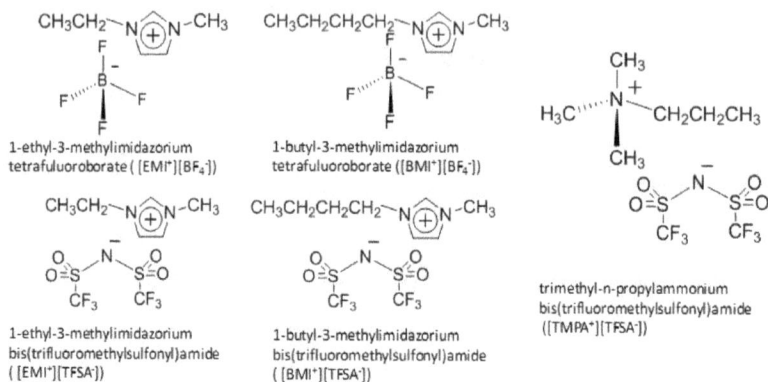

Figure 1. Structure formulas of typical ionic liquids with names and their abbreviations.

In this chapter, new techniques developed by putting ILs in the vacuum chambers of instruments will be introduced. Actually there are some other liquids having very low vapor pressure like silicon grease, which also can be introduced into the vacuum chamber. However, such oily medias cannot work as solvent for chemical and physical reaction because of their extremely high viscosity. On the other hand, ILs, which work well as solvent and electrolyte, make the dry vacuum conditions become wonderful wet world as will be introduced in this chapter.

2. XPS analysis

XPS is an instrument for analyzing the composition of solid materials and chemical state of each element. To detect generated photoelectrons with high sensitivity, vacuum condition is required. This condition is also effective for avoiding contamination of the sample surfaces. Although many researchers would like to put liquid into this instrument, vacuum condition in its sample chamber does not allow it. Nevertheless, some papers have reported the aggressive attempt to analyze the liquid surface by XPS with intricately designed sample stages [6-9]. In contrast, IL that is not vaporized under ultra-vacuum conditions can easily be put in the XPS chamber without any specific technique of modification.

$[EMI^+][C_2H_5SO_3^-]$ was the first IL that has been subjected to XPS analyses. It was found that IL emitted stable photoelectron flux, giving XPS spectra with high resolution [10-13]. Then, the obtained spectra enable a peak separation by the fitting calculations. Figure 2(a) shows the C1s spectrum of $[EMI^+][C_2H_5SO_3^-]$ together with the fitting curves based on C elements in five locations in the IL [14]. The XPS analysis of IL also allows detection of species dissolved in the IL, enabling in situ analysis of chemical reactions. An example is shown in Fig. 2(b) which were XPS spectra of $Pd(OAc)_2(PPh_3)$ dissolved in $[EMI^+][C_2H_5SO_3{}^-]$ [14]. Successive measurements showed that the intensity of spectra due to Pd(II) decreased, while the spectra due to Pd° increased, showing decomposition of Pd complex known as the Heck catalyst.

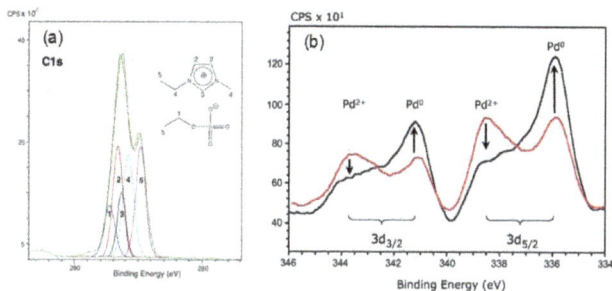

Figure 2. High resolution XPS spectra of $[EMI^+][C_2H_5OSO_3{}^-]$ detailing the C1s photoemission (a) and those of $Pd(Oac)_2$ in $[EMI^+][C_2H_5OSO_3{}^-]$. The red line shows data recorded at the start of the XPs experiment and the black line presents data recorded 6 h later.

3. MALDI Mass Spectroscopy

The matrix-assisted laser desorption/ionization mass spectrometry (MALDI-MS) is one of the groundbreaking analysis techniques that vaporize samples by laser irradiation with assistance of an appropriate matrix, as schematically illustrated in Fig. 3(a). This way allows to vaporize very large molecules like proteins that could not been analyzed without decomposition of the sample by the conventional mass spectrometry. The Matrix selection is quite essential, and an ideal matrix is a material possessing sufficient absorption coefficient for the laser beam, low vapor pressure, ability to dissolve or co-crystallize with sample, and ability to promote ionization of sample without its significant decomposition. The liquid matrices having low vapor pressure like glycerol and 3-nitrobenzyl alcohol have been utilized in the early research but their inherent volatility still causes some problems, such as decrease in their amounts with time. Another problem is that these liquids possess no UV absorbability, requiring addition of another photosensitive component.

Since IL possesses both no volatility and UV absorbability, it seems to be an ideal solution matrix for MALDI. However, usual ILs such as $[BMI^+][BF_4{}^-]$ and $[BMI^+][PF_6{}^-]$ were unfortunately unable to ionize samples dissolved in them [15]. Then, new ionic liquid family for the liquid matrixes were synthesized using solid acidic compounds of α-cyano-4-hydroxycinnamic acid (CHCA), sinapinic acid (SA), and 2,5-dihydroxybenzoic acid (DHB), which are widely used as solid matrixes for MALDI-MS [15,16]. Then, some of them were found to keep liquid state at room temperature and work as liquid matrices for detection of polymer and some biomolecules by MALDI-MS [15-18].

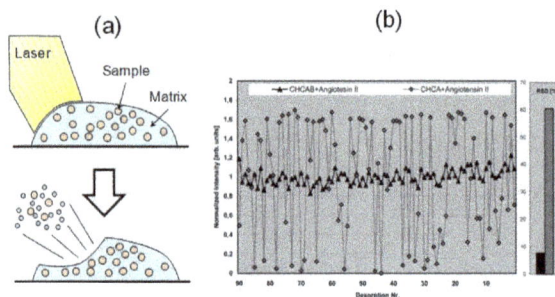

Figure 3. a) Schematic illustration of MALDI. (b) [M + H]+ ion intensities from 90 positions on a human angiotensin II preparation with ionic liquid matrix CHCAB (black triangles) and with traditional CHCA matrix (grey squares). The relative standard deviations (RSD) of the data series are given as bar graphs. Black bars indicate RSD values found using ionic liquid matrixes, and gray bars indicate RSD values of the data series yielded by the respective traditional MALDI matrixes.

Fig.3(b) shows change in signal intensities of [M + H]+ obtained at 90 different positions on a spot of sample-matrix mixture. As expected, the IL matrix of α-cyano-4-hydroxycinnamic acid butylamine (CHCAB) gave much narrow data dispersion than that obtained for the solid matrix of CHCA, indicating evidently usefulness of IL matrix for improvement of reproducibility. Another feature of the IL matrix is higher ability to suppress decomposition of sample than the conventional solid matrix. Use of CHCA-based guanidium salt and its analogous salts as IL matrixes enabled detection of oligosaccharides, which exhibit poor ionization efficiencies and tend to get thermal fragmentation through the loss of SO_3 groups, with suppression of loss of SO_3 [17,18].

4. Use of ILs for SEM observations

The first attempt was to observe ILs with a scanning electron microscope (SEM) [19]. The fact that ILs possess ionic conductivity but they do not possess electric conductivity gave an anticipation of charging of a IL drop during SEM observation. As a matter of fact, nonvolatile silicon oil, which can also be put in a vacuum chamber without vaporization, exhibited a white image with lots of noise because of charging behavior (Fig. 4(a)). Surprisingly, however, IL droplet gave a dark contrast images without any noise (Figure 4(b) -(d)), indicating that ILs are not charged by electron beam irradiation. Pulse radiolysis studies on ILs have revealed that electrons injected in ILs with high accelerated voltage are stabilized in condensed ions, allowing electrons to move in the liquid [20]. Consequently, ILs behave like electrically conducting materials for SEM observations. Based on this fact, several attempts have been done for SEM observations using ILs. The simplest way must be putting conductivity in place of metal or carbon deposition to insulating materials to observe them with a SEM. However, if neat IL was put onto the surface of the insulating material, existence of the liquid pools interfered observations of surface details of the abrasive paper, as shown in Fig. 5 (b), as compared with the case of the Au-deposited sample (Fig. 5 (a)). This troublesome can be resolved by dilution of IL with volatile solvent like alcohol. The abrasive paper was soaked in 2 mol dm^{-3} [BMI$^+$][TFSA$^-$] / ethanol solution in a couple of seconds and was taken out of the solution. Leaving it in air for several ten seconds allowed vaporization of ethanol, resulting in stay of a thin IL layer on the sample. In fact, its SEM image as shown in Fig. 5c was quite similar to that obtained for the Au-coated abrasive paper [21].

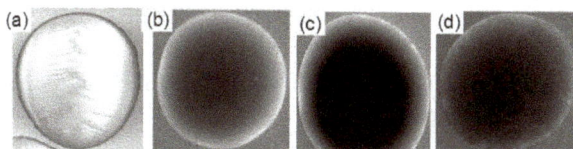

Figure 4. SEM images of droplets of silicon oil (a), [BMI$^+$][BF$_4^-$] (b), [EMI$^+$][BF$_4$] (c), and [EMI$^+$][TFSA$^-$]

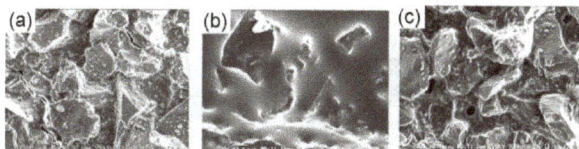

Figure 5. SEM images of surfaces of abrasive paper coated with gold (a), neat [BMI$^+$][TFSA$^-$] (b), and /ethanol solution (c).

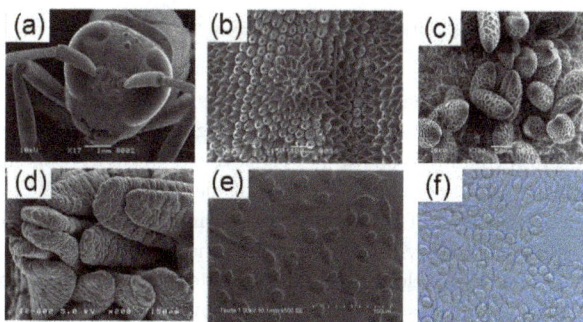

Figure 6. SEM images of insect, flower, tissue, and cell pretreated with IL; a) head of a yellow jacket, b) stamen of asteraceae, c) pollens of lily, d) villi of mouse small intestine, and mouse-derived fibroblast L929 cells (e). A picture (f) is optical microscope image of L929 cells for comparison.

Application of IL as an electric conducting material to an insulating sample gives another advantages, as compared with metal or carbon deposition. The liquid can keep the sample wet conditions even in vacuum chamber. This possibility has in particular a positive effect on observation of biological specimens [22-25]. Some examples are shown in Fig. 6 [22]. Since biological specimens have complex surface structures, metal or carbon deposition cannot perfectly deposit conducting films on the surfaces having dimples and indented places. However, liquid can reach anywhere on the complex surfaces, resulting in complete suppression of the charging behavior. Also replacement of water contained in the biological specimens with IL keeps the sample wet condition. As a result, SEM image of IL-treated fibrous blast cells, as shown in Fig. 6(e) was quite similar to shapes of cells containing water, which were observed by an optical microscope (Fig. 6(f)).

The IL treatment became now significant for biological and medical studies by electron microscope observations. An example is shown in Fig. 7 [23,24]. The metastasis is one of the most serious problems during cancer treatment. Such action has never been seen for the normal cells which prefer to adhere to each other during and after the cell fissions. This means that the normal cells must have some specific means to keep connections with neighboring cells. It is already known that something works well to make cell-to-cell junction, but direct observation of the something by an electron microscope has not yet been succeeded.

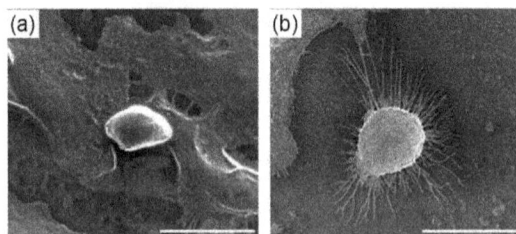

Figure 7. SEM images of A549 cells with (a) and without (b) pre-TGF-β1 treatment for 18 h. The IL treatment was conducted for the both samples before SEM observation.

The cells shown in Fig. 7 are human lung epithelial cells (A549). Before SEM observation of the cells, fixation treatment and metal or carbon coating are required to keep the cell morphology under vacuum conditions and to put electrical conductivity. Sometimes, however, such the treatments deform the delicate moieties of the cells. The A549 cells, which were subjected to dehydration and Pt-sputtering gave their SEM images showing relatively smooth surface with short microvilli. On the other hand, if IL was applied to the cells in place of the Pt-sputtering to put electrical conductivity to the cell, the SEM image shown in Fig. 7(b) was obtained. It indicates evidently long microvilli around the cell and ruffled cell surface. It is noteworthy that some of long microvilli made connections between separated cells, forming cellular bridges. The transforming growth factor (TFG)-β1 is a representative epthelial-mesenchymal transition (EMT), which is a key event in cancer metastasis. In fact, it is well known that the cells pre-treated with (TFG)-β1 lose their polarity and cell-to-cell contact. Fig. 7(a) shows A549 cells, which were subjected to the (TFG)-β1 treatment before SEM observation using IL. It is apparent by comparison of Fig. 7(a) with (b) that the A549 treated with (TFG)-β1 completely lost microvilli, implying that EMT-inducing TGF- β1 on the regulation of filopodia formation in mitotic cells. As a result the cells, which are free from other cells, enable their metastases.

5. Observation of electrochemical reactions

ILs work as a favorable electrolyte for several kinds of electrochemical reactions, implying that such the electrochemical reactions can be induced in a vacuum chamber, allowing observation of the reactions by an electron microscope [26,27]. As the first attempt, SEM observation of electrochemical Ag deposition has been conducted. Since the electrochemical reaction proceeds in IL as an electrolyte, IL may disturb the observation. However, the reaction occurring at several μm from the IL surface can be observed because the accelerated electron beam of SEM can penetrate such a thing IL layer.

In situ SEM observation of silver deposition was made by applying electrode potential to the working electrode, while observing the electrode surface from the top. The polarization potential chosen was -0.22 V, which was a little more negative than the onset potential of silver deposition (-0.15 V), and -1.14 V vs. Ag/Ag⁺ where the reaction rate is determined by diffusion of Ag^+. It is well known that silver deposition with nucleus growth is dominant when overpotential is small, whereas aciculate deposits becomes dominant at the potentials where diffusion determines the reaction rate. Such the natural rule was well represented by the SEM images taken with the reaction time, as shown in Fig. 8 [26].

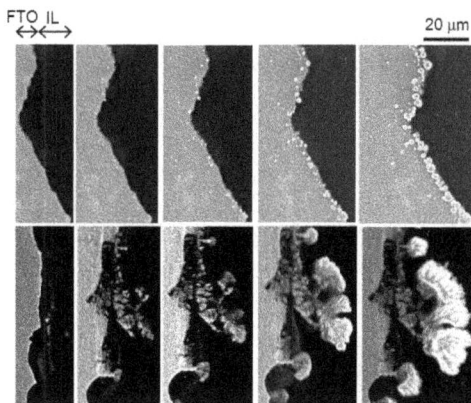

Figure 8. SEM images of gradual deposition of silver particles polarized at 0.22 V (upper) and -1.14 V (lower) vs. Ag/Ag⁺ for 0, 15, 30, 60 min, and180 min.

6. EDX analysis

Energy dispersive X-ray spectrometry (EDX) has become powerful for elemental analysis at small parts when it is combined with electron microscopes. Our previous studies revealed that EDX analysis is effective for detecting changes in components caused by electrochemical reactions in ILs. As a concrete application, we utilized this technique to reveal reaction mechanism of the electrochemical actuator.

The actuator device was prepared using a film of poly(vinylidene fluoride-co-hexafluoropropylene) (PVdF-HFP) containing IL. This composite film was sandwitched between two thin Au or Pt layers that were deposited by a metal sputtering. When [EMI$^+$][TFSA$^-$] was used to prepare the PVdF-HFP-IL composite, the resulting actuator bent toward the positive side (Fig. 10 (a)). The same tendency was observed for other polymer-IL composite actuators. Based on such the results, some researchers explained that bending is caused by difference in size between cation and anion of IL, the former is a little larger than the latter. Therefore, larger cations and smaller anions are attracted to the negative and the positive metal layer, respectively, resulting in expansion of the former side more than the latter.

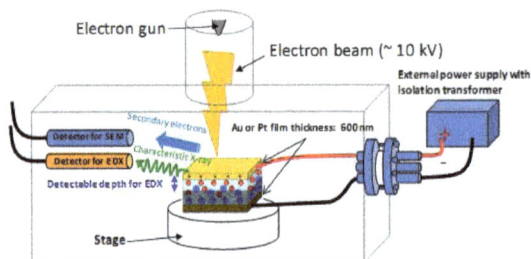

Figure 9. Schematic illustration of in situ SEM system for an actuator to investigate component changes by its reaction.

However, unexpected result was obtained when the fluorohydrogenate IL, [EMI$^+$][(FH)$_{2.3}$F$^-$] was used for the composite preparation. The actuator fabricated using this composite bent toward its negative side, as shown in Fig. 10 (b) although size of [(FH)$_{2.3}$F$^-$] is smaller than [TFSA$^-$], requiring another reaction mechanism that can explain reasonably the bending toward positive and negative sides. Then, we attempted in situ EDX analysis using the specifically modified SEM instrument, as shown in Fig. 9.

The prepared actuator was put on the sample stage of the SEM and DC voltage of + 3.5 V was applied to the Pt layers from a power supply outside of the SEM chamber. The one side of the actuator was observed by the SEM and change in amount of ions at vicinity of the Pt layer was detected by EDX while changing polarity of the DC supply. Typical change in the EDX spectra were shown in Fig. 10 (c) and (d). In case of the [EMI$^+$][TFSA$^-$] composite, the peak intensities for the anion components such as O, F, and S did not show significant change at all after changing polarity, while the intensity of carbon, which is mainly contained in the cation, markedly increased when the polarity was changed from plus to minus. On the contrary, in case of the $[EMI^+][(FH)_{2.3}F^-]$ composite, the F intensity decreased by changing voltage from +3.5 to -3.5 V, while almost no change was observed for the carbon intensity. Such the results indicated that cations and anions moved dominantly by changing the voltage polarity. Those behavior may be comprehensible because the transport numbers of cations and anions are larger for $[EMI^+][TFSA^-]$ and $[EMI^+][(FH)_{2.3}F^-]$, respectively. Based on those results and information, schematic illustrations shown in Fig. 10 (e) and (f) can be depicted to explain the actuator's bending. If only cations move in the polymer-IL composite, population of ions existing in the vicinity of the positive side decreases and that in the vicinity of the negative side increases, resulting in bending toward the positive side. It is, therefore, explainable that the bending toward the negative side in case of the IL, anion of which has larger transport number than cation, as shown in Fig. 10 (f) [28].

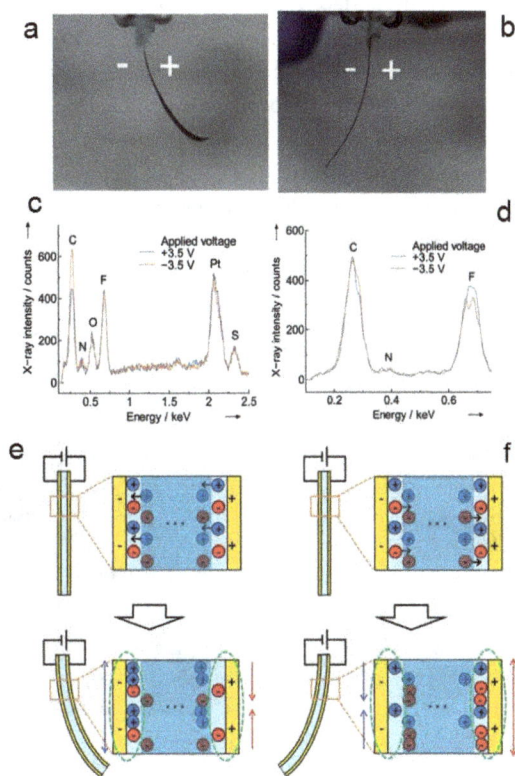

Figure 10. Motion of the electrochemical actuators prepared using ILs of $[EMI^+][TFSA^-]$ (a) and $[EMI^+][(FH)_{2.3}F^-]$ (b), their in situ EDX results (c, d), and plausible reaction mechanisms (e, f).

7. Preparation of metal nanoparticles by plasma deposition method

Ag, Cu, and Al nanoparticles have been successfully synthesized by plasma deposition method [29-31]. This approach called glow discharge electrolysis, is based on historical articles reported about 100 years ago. Schematic illustration of the system for the plasma deposition is illustrated in Fig. 11 (a). The plasma generation does not need vacuum conditions, but appropriate gas of low pressure is required to generate a stable plasma. However, the use of ILs is also essential in this case because the presence of vapor of a volatile liquid in the gas phase would inhibit the plasma generation. A typical plasma experiment is shown in Fig. 11 (b) [31]. The reaction media was [EMI$^+$][TFSA$^-$] with 62 mmol L^{-1}Cu(I). A dark brown layer that appeared at the interphase between the IL and plasma phase was growing with plasma irradiation time, indicating that Cu(I) was reduced to Cu nanoparticles with an average size of ~11 nm. However, the surface was covered with a copper oxide layer. If the metal nanoparticles are yielded under vacuum or inert gas condition, this is a common issue for which it is very difficult to collect metallic state nanoparticles, especially base metal nanoparticles that are oxidized readily under atmospheric condition. One of solution methodologies about this will be introduced in a later section. Very recently, the plasma deposition method was adopted for preparation of Au [32,33] and Pt [33] nanoparticles. The relationship between deposition conditions and the characteristics of the prepared nanoparticles was studied in detail. One of interesting findings is that nanoparticles were prepared even if the cathode was placed in the IL phase and not the gas phase.

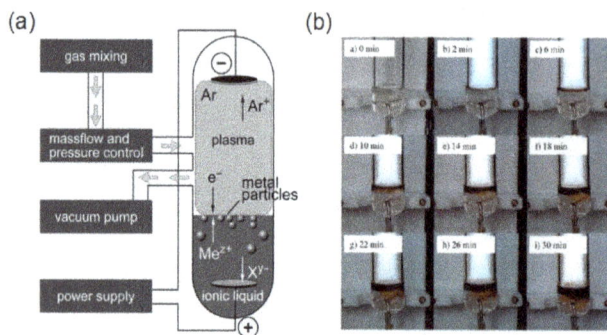

Figure 11. Schematic illustration of the experimental setup for plasma electrochemical reduction of metal ions dissolved in ILs (a) and photographs of the plasma electrochemical reduction experiment of Cu(I) dissolved in a IL at different reduction times (b).

8. Nanoparticle synthesis by sputtering

It is well known that metal vapor-deposition is a method to prepare ultrapure metal nanoparticles or films on solid substrates. Several research groups have developed the sputter deposition of metal nanoparticle onto pure ILs, as schematically illustrated in Fig. 12 (a), in order to prepare ultrapure metal nanoparticles [32-38]. The simple sputter deposition of Au onto ILs resulted in a solution containing highly dispersed Au nanoparticles whose size was dependent on the IL used [32]. In situ TEM observations of Au-deposited ILs revealed that highly-dispersed nanoparticles with no aggregation were seen in ILs, as shown in Fig. 12 (b), (c). Sputter deposition onto [EMI$^+$][BF$_4$ $^-$] gave spherical Au nanoparticles having an average diameter of 5.5 nm, while much smaller nanoparticles with size of 1.9 nm were obtained for the experiment using [Me$_3$PrN$^+$][TFSA$^-$]. Although sputtered species were assumed not to considerably suffer gas-phase collisions in the space between Au foil and IL solution because of low gas pressure, their injection into IL solution could make high concentration enough to coalesce with each other. Their coalescence would proceed until Au nanoparticles were stabilized by the adsorption of IL ions, the degree being dependent on IL. Since it is well-known that TFSA$^-$ anion makes a coordination bond with metal ions, it is expected that strong adsorption of TFSA$^-$ suppresses the growth and/or coalescence of particles.

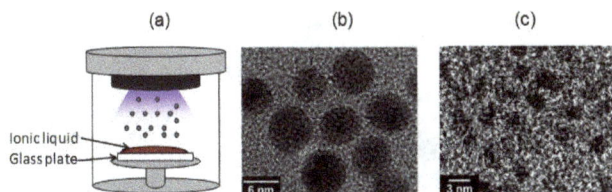

Figure 12. Schematic illustration of the metal nanoparticle formation by sputtering (a) and TEM image of Au nanoparticles sputter-deposited in EtMeImBF$_4$.

This method has achieved the preparation of various pure metal nanoparticles, such as Au [32-36], Ag [37,38], Pt [39,40] and so forth, possessing particle sizes less than 10 nm in diameter without any specific stabilizing agent. A small-angle X-ray scatting study revealed the initial formation mechanism of the gold nanoparticles during the sputtering process onto several [1,3-dialkylimidazolium][BF$_4$] [36]. The proposed formation mechanism is divided into two phases where it was concluded that both surface tension and viscosity of the IL are important factors for the Au nanoparticle growth and its stabilization.

The simultaneous sputter deposition of different pure metals on IL is a facile synthetic method to prepare bimetallic alloy nanoparticles [37]. The use of metal targets composed of radially-arranged Au and Ag foils allowed simultaneous sputter deposition of Au and Ag onto [BMI$^+$][PF$_6$ $^-$] in argon atmosphere at ca. 20 Pa. The color of [BMI$^+$][PF$_6$ $^-$] subjected to the sputter deposition was varied (Fig. 11), depending on a fraction of gold foils on targets. Each absorption spectrum of the resulting solution exhibited a single peak or shoulder assigned to the surface plasmon resonance (SPR) band of the metal particles and the peak wavelength increased almost linearly with the Au fraction. These results indicated that the simultaneous injection of sputtered species of Au and Ag into an IL caused coalescence with each other in the solution, resulting in the formation of AuAg alloy nanoparticles, as shown in Fig. 13. The chemical composition and optical properties of the alloy nanoparticles are easily controlled just by varying the area ratio of the individual pure metal foils in a sputtering target.

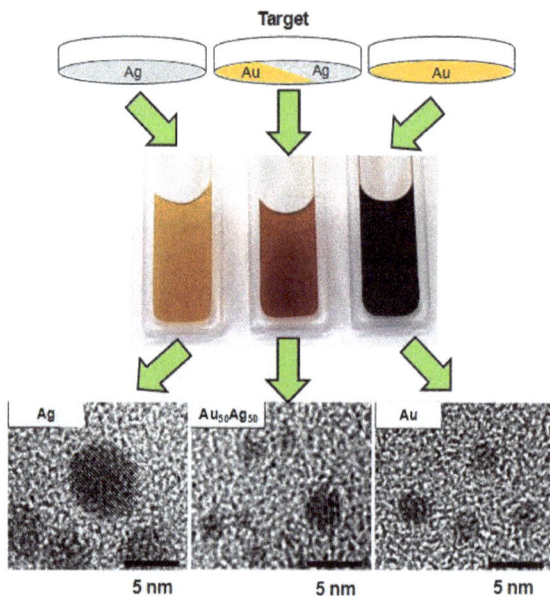

Figure 13. Photographs of [BuMeIm][PF$_6$] ILs after sputtering experiments at Au-Ag targets having different surface area ratios and TEM images of the resulting nanoparticles obtained at each Au-Ag target.

Furthermore, it was discovered that hollow nanoparticles can be synthesized by some modification of the sputtering method [39]. This fact was found when we attempted to produce indium metal nanoparticles by sputtering of In onto [BMI$^+$][BF$_4$$^-$]. SEM observation and some analyses of the obtained nanoparticles revealed that they had In/In$_2$O$_3$ core/shell configuration, as shown in Fig. 14(a). Since the melting point of In is 156.6 °C, we attempted to heat the resulting In/In$_2$O$_3$-dispersed IL at 250 °C, giving the In$_2$O$_3$ hollow nanoparticles, a TEM image of which is shown in Fig. 14(b). The plausible reaction mechanisms for synthesis of In/In$_2$O$_3$ and In$_2$O$_3$ hollow nanoparticles are schematically illustrated in Fig. 12(c) and (d), respectively. The oxidation of the In nanoparticle surfaces and the melted In by heating might be caused by oxygen that was dissolved in the IL when the IL was taken out from the sputtering instrument.

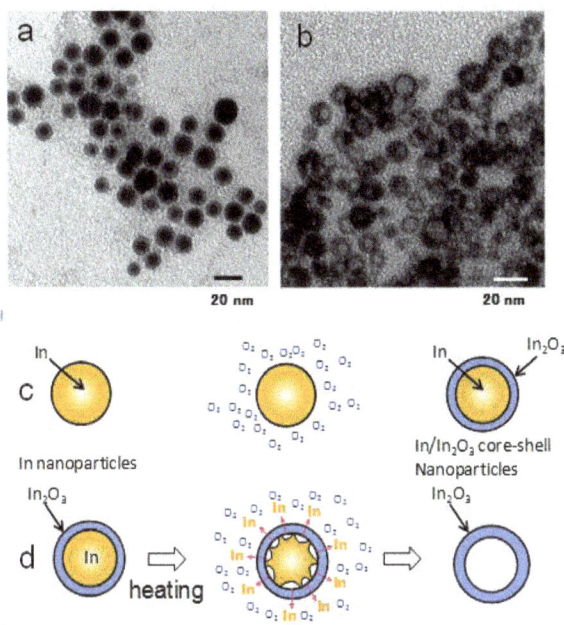

Figure 14. Nanoparticles synthesized by sputtering of In onto [EMI⁺][BF₄⁻] (a), those after heating at 250 °C, and the plausible reaction mechanisms for the In sputtering (c) and the heating (d) procedures.

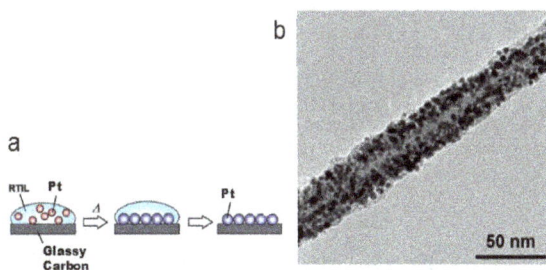

Figure 15. A way to immobilize Pt nanoparticles on surface of carbon substance (a) and Pt nanoparticles-immobilized carbon nanotube (b) prepared by this way.

The produced metal nanoparticles are stably dispersed in ILs for long time without any specific stabilizing agent. However, the nanoparticles can be immobilized onto carbon substances by putting the nanoparticles-dispersed IL on a carbon substrate followed by heating and then removal of IL by washing with acetonitrile (Fig. 15(a)) [34]. When the resulting Pt nanoparticles-immobilized carbon substance was used as an electrode, it exhibited high electro-catalytic activities toward O_2 reduction, indicating that the immobilized Pt nanoparticles kept their catalytic acitvities [40,41]. Similar method was found to be useful to immobilize Pt nanoparticles onto surfaces of carbon nanotubes, as shown in Fig. 15 (b) [42]. In this case, Pt-dispersed IL and carbon nanotubes are vigorously mixed and the resulting mixture was heated, followed by washing with acetonitrile. The Pt nanoparticles on the carbon nano-tubes also exhibited high electrocatalytic activities, as shown in Fig. 14.

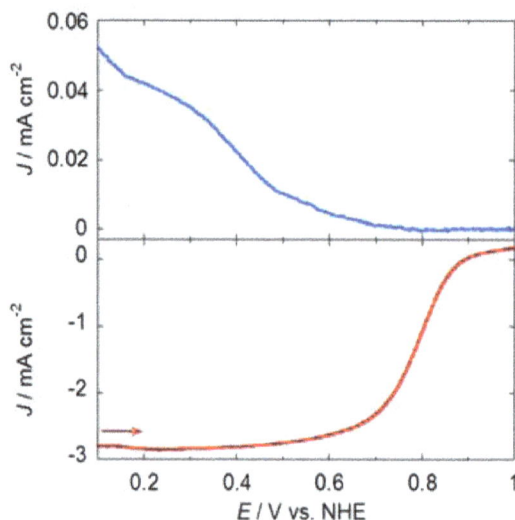

Figure 16. Hydrodynamic voltammograms for O_2 reduction at a Pt–SWCNT modified rotating ring disk electrode in O_2-saturated 0.1 M $HClO_4$ aqueous solution at 298 K. The electrodes were (top) Pt ring and (bottom) Pt–SWCNT modi-fied GC disk. The IL used for Pt nanoparticle preparation was [Me₃PrN⁺][TFSA⁻]. The potential for the ring electrode was 1.20 V. The scan rate was 10 mV s⁻¹; the rotation rate was 1200 rpm.

9. Nanoparticle preparation by quantum Beam

Irradiation of electron beam to IL is another way to synthesize metal particles. This fact was found first when we observed [BMI⁺][TFSA⁻] containing 0.1 mol dm⁻³ NaAuCl₄ [43]. As shown in Fig. 17(a)-(c), many bright lines appeared in the IL droplet on a FTO with observa-tion time. When the bright line was observed with higher magnification, a SEM image of Fig. 17 (d) was obtained, showing Several experiments and analyses have revealed that Au particles are produced by reduction of Au^{3+} ions by electron beam, and that particle size de-pends on the experimental conditions [44].

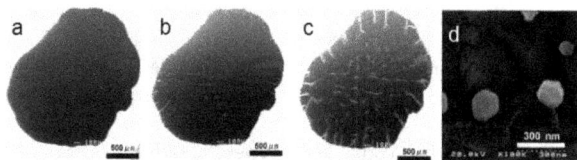

Figure 17. SEM images of the ionic liquid irradiated with an electron beam for 0 s (a), 90 s (b), and 300 s (c), and generated Au particles (d).

Accelerated electron beam and γ-ray generated by industrial plants, which are usually utilized for sterilizing medical kits, are also available for reduction of metal ions to metal nanoparticles [45]. In this case, since irradiation is made by conveying IL containing metal ions, which is sealed in a sample vial, through the generator, mass production of metal nanoparticles is possible. This technique was able to synthesize several kinds of metal nanoparticles, such as Au, Ag, Cu, Ni, Pd, Pt, Mg, Fe, Zn, Al, Sn, and FePt alloy [46].

10. Conclusion Remarks

Chemists prefer wet conditions than dry conditions because wet conditions are more desired for inducing chemical reactions. Since wet conditions are also required for all living things, investigation of biomaterials under wet conditions are much better than that under dry conditions. Unfortunately, however, many instruments for precise analyses and those for material production with micro or nano scales require vacuum conditions because the air intererferes precise proving and controlling. So far, vacuum and wet are contradictory words because there is no liquid which can stand in vacuum without vaporization. IL is the first liquid that can set the relationship between vacuum and wet on the right footing. There are more instruments requiring vacuum conditions than those introduced in this article. We would like to apply IL to more instruments to make many scientists know that IL is the key material to solve the mysteries of our wet world.

Author details

Susumu Kuwabata[1,2], Tsukasa Torimoto[2,3], Akihito Imanishi[2,4] and Tetsuya Tsuda[1,4]

*Address all correspondence to: kuwabata@chem.eng.osaka-u.ac.jp

1 Department of Applied Chemistry, Graduate School of Engineering, Osaka University, Japan

2 Japan Science and Technology Agency, CREST, Japan

3 Department of Crystalline Material Sciences, Graduate School of Engineering, Nagoya University, Japan

4 Department of Chemistry, Graduate School of Engineering Science, Osaka University, Japan

Frontier Research Base for Global Young Researchers, Graduate School of Engineering, Japan

References

[1]   Ohno, H. (2005). Electrochemical Aspects of Ionic Liquids. *Wiley-Interscience, New Jersey*.

[2]   Wasserscheid, P., & Welton, T. (2007). Ionic Liquids in Synthesis. *Wiley-VCH, Weinheim, Germany*.

[3]   Koel, M. (2009). Ionic Liquids in Chemical Analysis. *CRC Press, Boca Raton*.

[4]   Torimoto, T., Tsuda, T., Okazaki, K. I., & Kuwabata, S. (2010). New Frontiers in Materials Science Opened by Ionic Liquids. *Adv. Mater*, 22(5), 1196-1221.

[5]   Kuwabata, S., Tsuda, T., & Torimoto, T. (2010). Room-Temperature Ionic Liquid. A New Medium for Material Production and Analyses under Vacuum Conditions. *J. Phys. Chem. Lett.*, 1(21), 3177-3188.

[6]   Oliveira, F. C. C., Rossi, L. M., Jardim, R. F., & Rubim, J. C. (2009). Magnetic Fluids Based on γ-Fe_2O_3 and $CoFe_2O_4$ Nanoparticles Dispersed in Ionic Liquids. *J. Phys. Chem. C*, 113(20), 8566-8572.

[7]   Swatloski, R. P., Spear, S. K., Holbrey, JD, & Rogers, R. D. (2002). Dissolution of cellose with ionic liquids. *J. Am. Chem. Soc.*, 124(18), 4974-4975.

[8]   Zhu, S., Wu, Y., Chen, Q., Yu, Z., Wang, C., Jin, S., Ding, Y., & Wu, G. (2006). Dissolution of cellulose with ionic liquids and its application: a mini-review. *Green Chem*, 8(4), 325-327.

[9]   Feng, L., & Chen, Z. (2008). Research progress on dissolution and functional modification of cellulose in ionic liquids. *J. Mol. Liq*, 142(1-3), 1-5.

[10] Smith, E. F., Rutten, F. J. M., Villar-Garcia, I. J., Briggs, D., & Licence, P. (2006). Ionic Liquids in Vacuo: Analysis of Liquid Surfaces Using Ultra-High-Vacuum Techniques. *Langmuir*, 22(22), 9386-9392.

[11] Hofft, O., Bahr, S., Himmerlich, M., Krischok, S., Schaefer, J. A., & Kempter, V. (2006). Electronic Structure of the Surface of the Ionic Liquid [EMIM][Tf_2N] Studied by Metastable Impact Electron Spectroscopy (MIES), UPS, and XPS. *Langmuir*, 22(17), 7120-7123.

[12] Caporali, S., Bardi, U., & Lavacchi, A. (2006). X-ray Photoelectron Spectroscopy and Low Energy Ion Scattering Studies on 1-Butyl-3-methyl-imidazolium Bis(Trifluoro-methane) Sulfonamide. *J. Electron Spectrosc. Relat. Phenom*, 151(1), 4-8.

[13] Lovelock, K. R. J., Kolbec, C., Cremer, K. T., Paape, N., Schulz, P. S., Wasserscheid, P., Maier, F., & Steinruck, H. P. (2009). Influence of Different Substituents on the Surface Composition of Ionic Liquids Studied Using ARXPS. *J. Phys. Chem. B*, 113(9), 2854-2864.

[14] Smith, E. F., Villar, Garcia. I. J., Briggs, D., & Licence, P. (2005). Ionic Liquids in Va-cuo; Solution-Phase X-ray Photoelectron Spectroscopy. *Chem. Commun*, 5633-5635.

[15] Armstrong, D. W., Zhang, L. K., He, L., & Gross, M. L. (2001). Ionic Liquids as Ma-trixes for Matrix-Assisted Laser Desorption/Ionization Mass Spectrometry. *Anal. Chem*, 73(15), 3679-86.

[16] Mank, M., Stahl, B., & Boehm, G. (2004). Dihydroxybenzoic Acid Butylamine and Other Ionic Liquid Matrixes for Enhanced MALDI-MS Analysis of Biomolecules. *Anal. Chem*, 76(10), 2938-2950.

[17] Laremore, T. N., Zhang, F., & Linhardt, R. J. (2007). Ionic Liquid Matrix for Direc UV-MALDI-TOF-MS Analysis of Dermatan Sulfate and Chondroitin Sulfate Oligosac-charides. *Anal. Chem.*, 79(4), 1604-1610.

[18] Fukuyama, Y., Nakaya, S., Yamazaki, Y., & Tanaka, K. (2008). Ionic Liquid Matrixes Optimized for MALDI-MS of Sulfated/Sialylated/Neutral Oligosaccharides and Gly-copeptides. *Anal. Chem*, 80(6), 2171-2179.

[19] Kuwabata, S., Kongkanand, A., Oyamatsu, D., & Torimoto, T. (2006). Observation of Ionic Liquid by Scanning Electron Microsocope. *Chem. Lett*, 35(6), 600-601.

[20] Wishart, J. F., & Neta, P. (2003). Spectrum and Reactivity of the Solvated Electron in the Ionic Liquid Methyltributylammonium Bis(trifluoromethylsulfonyl)imide. *J. Phys. Chem. B.*, 107(30), 7261-7267.

[21] Arimoto, S., Sugimura, M., Kageyama, H., Torimoto, T., & Kuwabata, S. (2008). De-velopment of New Techniques for Scanning Electron Microscope Observation Using Ionic Liquid. *Electrochim. Acta*, 53(21), 6228-6234.

[22] Tsuda, T., Nemoto, N., Kawakami, K., Mochizuki, E., Kishida, S., Tajiri, T., Kushibiki, T., & Kuwabata, S. (2011). SEM Observation of Wet Biological Specimens Pretreated with Room-Temperature Ionic Liquid. *ChemBioChem*, 12(17), 2547-2550.

[23] Ishigaki, Y., Nakamura, Y., Takehara, T., Nemoto, N., Kurihara, T., Koga, H., Naka-gawa, H., Takegami, T., Tomosugi, N., Miyazawa, S., & Kuwabata, S. (2011). Ionic liquid enables simple and rapid sample preparation of human culturing cells for scanning electron microscope analysis. *Microsc. Res. Tech*, 74(5), 415-420.

[24] Ishigaki, Y., Nakamura, Y., Takehara, T., Shimasaki, T., Tatsuno, T., Takano, F., Ueda, Y., Motoo, Y., Takegami, T., Nakagawa, H., Kuwabata, S., Nemoto, N., Tomosugi, N.,

& Miyazawa, S. (2011). Scanning electron microscopy with an ionic liquid reveals the loss of mitotic protrusions of cells during the epithelial-mesenchymal transition. *Microsc. Res. Tech*, 74(11), 1024-1031.

[25] Yanaga, K., Maekawa, N., Shimomura, N., Ishigaki, Y., Nakamura, Y., Takegami, T., Tomosugi, N., Miyazawa, S., & Kuwabata, S. (2012). Use of ionic liquid in fungal taxonomic study of ultrastructure of basidiospore ornamentation. *Mycolog. Prog.*, 11(1), 343-347.

[26] Arimoto, S., Kageyama, H., Torimoto, T., & Kuwabata, S. (2008). Development of In Situ Scanning Electron Microscope System for Real Time Observation of Metal Deposition from Ionic Liquid. *Electrochem. Commun*, 10(12), 1901-1904.

[27] Arimoto, S., Oyamatsu, D., Torimoto, T., & Kuwabata, S. (2008). Development of in situ electrochemical scanning electron microscopy with ionic liquids as electrolytes. *Chem Phys Chem*, 9(5), 763-767.

[28] Tsuda, T., Baba, M., Sato, T., Sakao, R., Matsumoto, K., Hagiwara, R., & Kuwabata, S. (2011). Nonvolatile IL-based artificial muscle: Actuation mechanism identified by in situ EDX analysis. *Chem. Europ. J.*, 17(40), 11122-11126.

[29] Meiss, S. A., Rohnke, M., Kienle, L., Abedin, S. E., Endres, F., & Janek, J. (2007). Employing Plasmas as Gaseous Electrodes at the Free Surface of Ionic Liquids: Deposition of Nanocrystalline Silver Particles. *ChemPhysChem*, 8(1), 50-3.

[30] Abedin, S. Z. E., Pölleth, M., Meiss, S. A., Janek, J., & Endres, F. (2007). Ionic Liquids as Green Electrolytes for the Electrodeposition Nanomaterials. *Green Chem.*, 9(6), 549-553.

[31] Brettholle, M., Höfft, O., Klarhöfer, L., Mathes, S., Friedrichs, M., Abedin, S. Z. E., Krischok, S., Janek, J., & Endres, F. (2010). Plasma electrochemistry in ionic liquids: deposition of copper nanoparticles. *Phys. Chem. Chem. Phys.*, 12(8), 1750-1755.

[32] Torimoto, T., Okazaki, K., Kiyama, T., Hirahara, K., Tanaka, N., & Kuwabata, S. (2006). Sputter Deposition onto Ionic Liquids: Simple and Clean Synthesis of Highly Dispersed Ultrafine Metal Nanoparticles. *Appl. Phys. Lett*, 89(24), 243117/1-243117/3.

[33] Khatri, O. P., Adachi, K., Murase, K., Okazaki, K., Torimoto, T., Tanaka, N., Kuwabata, S., & Sugimura, H. (2008). Self-Assembly of Ionic Liquid (BMI-PF$_6$)-Stabilized Gold Nanoparticles on a Silicon Surface: Chemical and Structural Aspects. *Langmuir*, 24(15), 7785-7792.

[34] Okazaki, K., Kiyama, T., Suzuki, T., Kuwabata, S., & Torimoto, T. (2009). Thermally Induced Self-assembly of Gold Nanoparticles Sputterdeposited in Ionic Liquids on Highly Ordered Pyrolytic Graphite Surfaces. *Chem. Lett.*, 38(4), 330-331.

[35] Kameyama, T., Ohno, Y., Kurimoto, T., Okazaki, K., Uematsu, T., Kuwabata, S., & Torimoto, T. (2010). Size Control and Immobilization of Gold Nanoparticles Stabilized in an Ionic Liquid on Glass Substrates for Plasmonic Applications. *Phys. Chem. Chem. Phys.*, 12(8), 804-811.

[36] Hatakeyama, Y., Okamoto, M., Torimoto, T., Kuwabata, S., & Nishikawa, K. (2009). Small-Angle X-ray Scattering Study of Au Nanoparticles Dispersed in the Ionic Liquids 1-Alkyl-3-methylimidazolium Tetrafluoroborate. *J. Phys. Chem. C,* 113(10), 3917-3922.

[37] Okazaki, K., Kiyama, T., Hirahara, K., Tanaka, N., Kuwabata, S., & Torimoto, T. (2008). Single-Step Synthesis of Gold-Silver Alloy Nanoparticles in Ionic Liquids by a Sputter Deposition Technique. *Chem. Commun,* 691-693.

[38] Suzuki, T., Okazaki, K., Kiyama, T., Kuwabata, S., & Torimoto, T. (2009). A Facile Synthesis of AuAg Alloy Nanoparticles Using a Chemical Reaction Induced by Sputter Deposition of Metal onto Ionic Liquids. *Electrochemistry,* 77(8), 636-638.

[39] Suzuki, T., Okazaki-I, K., Suzuki, S., Shibayama, T., Kuwabata, S., & Torimoto, T. (2010). Nanosize-controlled syntheses of indium metal particles and hollow indium oxide particles via the sputter deposition technique in ionic liquids. *Chem. Mater.,* 22(18), 5209-5215.

[40] Tsuda, T., Kurihara, T., Hoshino, Y., Kiyama, T., Okazaki-I, K., Torimoto, T., & Kuwabata, S. (2009). Electrocatalytic activity of platinum nanoparticles synthesized by room-temperature ionic liquid-sputtering method. *Electrochemistry,* 77(8), 693-695.

[41] Tsuda, T., Yoshii, K., Torimoto, T., & Kuwabata, S. (2010). J. Power Sources , 195, 5980-5985.

[42] Yoshii, K., Tsuda, T., Arimura, T., Imanishi, A., Torimoto, T., & Kuwabata, S. (2012). Platinum nanoparticle immobilization onto carbon nanotubes using Pt-sputtered room-temperature ionic liquid. *RSC Adv,* 10.1039/C2RA21243A.

[43] Imanishi, A., Tamura, M., & Kuwabata, S. (2009). Formation of Au nanoparticles in an ionic liquid by electron beam irradiation. *Chem. Commun,* 1775-1777.

[44] Imanishi, A., Gonsui, S., Tsuda, T., Kuwabata, S., & Fukui-I, K. (2011). Size and shape of Au nanoparticles formed in ionic liquids by electron beam irradiation. *Phys. Chem. Chem. Phys.,* 13(33), 14823-14830.

[45] Tsuda, T., Seino, S., & Kuwabata, S. (2009). Gold nanoparticles prepared with a room-temperature ionic liquid-radiation irradiation method. *Chem. Comm,* 6792-6794.

[46] Tsuda, T., Sakamoto, T., Nishimura, Y., Seino, S., Imanishi, A., & Kuwabata, S. (2012). Various metal nanoparticles produced by accelerated electron beam irradiation of room-temperature ionic liquid. *Chem. Commun.,* 925-927.

Chapter 13

Plasma Process on Ionic Liquid Substrate for Morphology Controlled Nanoparticles

Toshiro Kaneko, Shohei Takahashi and
Rikizo Hatakeyama

Additional information is available at the end of the chapter

1. Introduction

The interaction of discharge plasmas with liquids [1-5] is one of the active topics in the realm of recent plasma science and science technology. It has pioneered new channels relating to nano material creation based on their distinct properties such as ultra-high density, high reactivity, high process rate, and so on. Especially, the boundary between the plasmas and the liquids, which activates physical processes and chemical reactions, has attracted much attention as a novel field in the nano-bio material creation. For example, the nanoparticle synthesis using the plasma-liquid interfaces [6-10] is especially advantageous in that a reducing agent is the plasma itself, and then, toxic stabilizers and reducing agents are unnecessary and the synthesis is continuous during the plasma irradiation. In these methods, although it has been reported that the metal salt is reduced by an electron or an active hydrogen, the precise control of the synthesis in terms of the synthesis rate, morphology (size, shape, structure, and so on) control remain unclear because the inevitable high voltage discharge in the atmospheric pressure and the consequential dynamic behavior of the gas-liquid interface prevent us from analyzing the precise properties of the plasmas in the interfacial region.

In this sense, for the purpose of the generation of the static and stable plasma contacting with the liquid, we adopt ionic liquids [11,12] which have the interesting characteristics such as their composition consisting of only positive and negative ions, i.e., no neutral solvent, extremely low vapor pressure, high heat capacity, and nonflammability. These characteristics enable us to introduce the ionic liquids to the vacuum system and the discharge plasma. Therefore, the ionic liquids are the most suitable liquid for the formation of nano-composite materials using the discharge plasmas in contact with the liquids [13-21].

On the other hand, recently, highly-ordered periodic structures of metal nanoparticles have attracted much attention due to their high catalytic activity, unique photosensitive reactivity, bio sensitivity, and so on [22-26]. One possibility is use of nano-carbons such as carbon nanotubes or graphenes as template for synthesis of the nanoparticles [27-32]. However, the structures of the nanoparticles are decided by the chemical properties of the nano-carbons and are difficult to be freely controlled by the external parameters. To realize the easy and flexible control of the periodic structure of the nanoparticles, we adopt a novel plasma technique combined with introduction of ionic liquids under strong magnetic fields up to several tesla (T), whose concept is schematically shown in Figure 1 [33].

Since the plasma generated under the strong magnetic field keeps its structure due to confinement along the magnetic field lines, the plasma structure can be transcribed to the liquid surface, resulting in the synthesis of the structured nanoparticles at the gas-liquid interface when the plasma reduces the metal chlorides in the liquid. This method could contribute to supplying a considerable amount of spatially-periodic nanoparticles available for the development of unique optoelectronic devices [34].

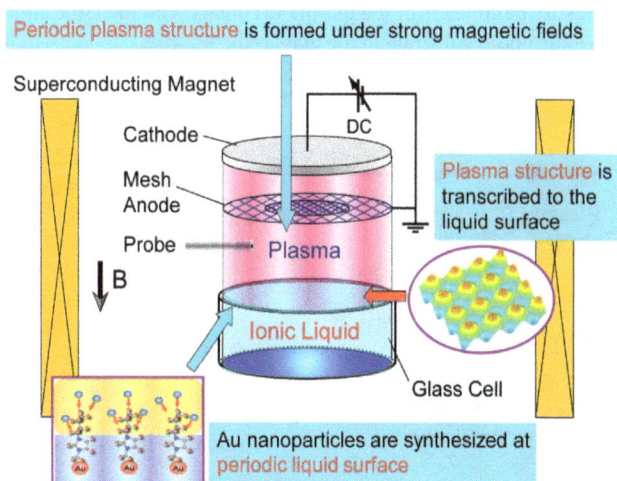

Figure 1. Concept of synthesis of periodic metal nanoparticles using discharge plasmas in contact with ionic liquid substrate.

2. Gas-liquid interfacial plasma process using ionic liquid substrate

Figure 2 shows schematic model of the gas-liquid interfacial plasmas for Au nanoparticle synthesis. An electrode which is made of a platinum (Pt) plate is located inside the glass cell, and a popular ionic liquid (1-buthl-3-methyl-imidazolium tetrafluoroborate: $[C_8H_{15}N_2]^+[BF_4]^-$) is introduced on the Pt electrode as cathode electrodes for the purpose of investigating the effects of the ionic liquid on the discharge. On the other hand, a grounded anode electrode which is made of the SUS plate is set in a gas phase (plasma) region at a distance of 60 mm from the cathode electrodes. This discharge configuration, in which the ionic liquid cathode electrode is in the glass cell, is defined as "ion irradiation mode", because the positive ions in the plasma are accelerated by the electric field formed on the ionic liquid as shown in Figure 2(a).

In order to examine the effects of the ion irradiation to the ionic liquid on discharge-related phenomena, the cathode electrode is switched to the SUS plate located at the top of the gas plasma region, which is defined as "electron irradiation mode" and the anode electrode consisting of the ionic liquid in the glass cell is grounded as shown in Figure 2(b). Removal of the water dissolved in the ionic liquid is performed under the vacuum condition for 2 hours after introducing the ionic liquid into the glass chamber. A negative direct current (DC) voltage is supplied to the cathode electrode, where typical discharge voltage V_D, discharge current I_D, plasma irradiation time t_{pi} are $V_D = 500 \sim 1500$ V, $I_D = 1 \sim 5$ mA, $t_{pi} = 1 \sim 40$ min, respectively. The argon gas is adopted as a discharge medium, and the gas pressure P_{gas} is varied from 20 Pa to 200 Pa approximately. A Langmuir probe is inserted at the position of $z = 0 \sim 60$ mm to measure parameters of the plasma in contact with the ionic liquid substrate ($z = 0$: surface of the ionic liquid substrate in the glass cell).

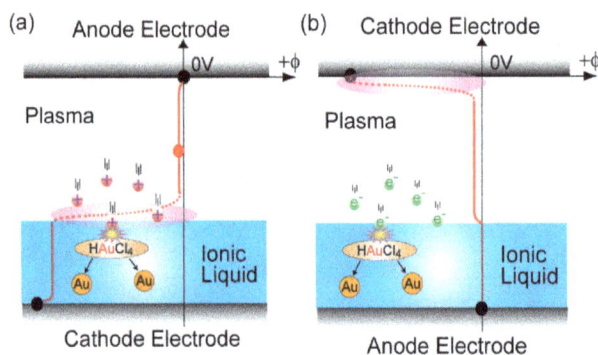

Figure 2. Schematic model of nanoparticle synthesis in the gas-liquid interface on the ionic liquid substrate. (a) ion irradiation mode, (b) electron irradiation mode.

Using this ion or electron irradiation, gold nanoparticles are synthesized in the ionic liquid by the reduction of Au chloride such as $HAuCl_4$ dissolved in the ionic liquid. Figure 3 shows transmission electron microscopy (TEM) images of the Au nanoparticles synthesized in (a) the ion irradiation mode and (b) the electron irradiation mode for $P_{gas} = 60$ Pa, $I_D = 1$ mA, and $t_{pi} = 40$ min. In both cases, the Au nanoparticles can be formed, however, it is found that, in ion irradiation mode, the average diameter of the Au nanoparticles is smaller and the particle number is larger than that in electron irradiation mode. The reduction reaction of the Au ions is believed to be caused by electrons injected from the plasma in electron irradiation mode, while in ion irradiation mode, the reduction may be caused by the hydrogen radical H*, which is generated by the dissociation of the ionic liquid. Based on this mechanism, the hydrogen radical is considered to be more effective for the reduction of Au ions than electrons, and efficient Au nanoparticle synthesis is realized using ion irradiation.

Since Au nanoparticles with diameter less than 100 nm are known to exhibit localized surface plasmon resonance, visible absorption spectra are obtained for a quantitative observation of the Au nanoparticle concentration. Figure 3(c) shows visible absorption spectra of the Au nanoparticles synthesized by an Ar plasma in ion and electron irradiation mode. The absorption peak appears around 550 nm, corresponding to the Au plasmon resonance, and the absorption-peak intensity in ion irradiation mode is obviously larger than that in electron irradiation mode. Ar ions with high energy can penetrate deep into the ionic liquid, promoting the generation of hydrogen radicals. The increased concentration of hydrogen radicals may reduce Au ions more effectively in ion irradiation mode than electron irradiation mode. The rate of Au nanoparticle synthesis could be controlled by the irradiation energy of inert gas ions such as Ar.

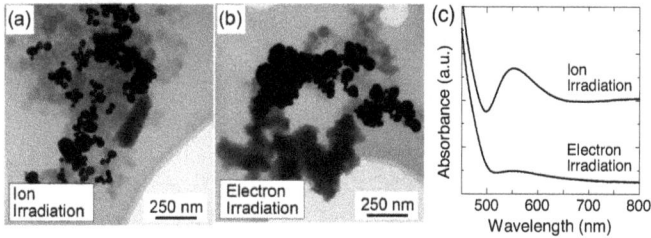

Figure 3. TEM images of Au nanoparticles synthesized in (a) the ion irradiation mode and (b) the electron irradiation mode. (c) UV-Vis absorption spectra of Au nanoparticles. P_{gas} = 60 Pa, I_D = 1 mA, t_{pi} = 40 min.

3. Control of nanoparticle morphology by gas-liquid interfacial plasmas

3.1. Periodic nanoparticle structure formed by periodic plasma

Figure 4(a) shows the schematic of an experimental setup for nanoparticle structure formation using the gas-liquid interfacial discharge plasma under strong magnetic fields, which has a glass cell with 15 mm inner diameter and 10 mm depth in a cylindrical glass chamber with 75 mm diameter and 200 mm length. A DC voltage V_D is supplied to an upper cathode electrode composed of a stainless steel (SUS) plate and a SUS mesh grid (10 meshes/inch) is used as an anode electrode to promote a spatial diffusion of the plasma. Typical discharge current is I_D=3 mA. Nitrogen gas is adopted as a discharge gas, and the gas pressure P_{gas} is varied from 20 to 100 Pa.

The new kind of the ionic liquid (N.N.N.-Trimethyl-N-propyl-ammonium Bis (trifluoro methane sulfonyl) imide) put in the glass cell is placed on a peltier element which is located at a distance of 50 mm from the anode electrode. Since this ionic liquid does not become supercooled state, we can make the ionic liquid solid state by cooling the ionic liquid using the peltier element located under the glass cell. When the strong magnetic fields are applied along the machine axis, the generated plasma is strongly magnetized, and then, the periodic plasma structure formed by the mesh anode is maintained just above the ionic liquid as shown in Figure 4(b).

The Au nanoparticles are synthesized in the ionic liquid by the plasma reduction of $HAuCl_4$. The ionic liquid can be cooled by the peltier element and becomes the solid state as mentioned above, with keeping the structure of Au nanoparticles synthesized by the plasma irradiation at the liquid interface.

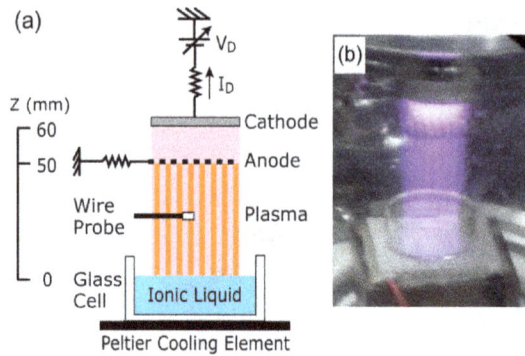

Figure 4. a) Schematic of the experimental setup and (b) photo of the synthesized periodic plasma structure for $B = 1$ T, $I_D = 3$ mA, $P_{gas} = 20$ Pa.

Figure 5 shows the radial profiles of the electron density n_e of the plasma as functions of (a) magnetic fields B and (b) gas pressure P_{gas}. When the strong mantic fields is applied (closed circles), the electron density has the depression periodically in the radial direction. The electron density in the high density region is about 10^8 cm^{-3}, and that in the depression region becomes one order smaller. The interval of the depression is about 2.5 mm, which is corresponding to the distance between the SUS wire of the mesh anode electrode. Therefore, these density depressions are caused by shielding of the plasma under the wire of the mesh anode, where the generated plasma cannot pass through toward the ionic liquid substrate. In the absence of the magnetic field, on the other hand, the radial profile of the electron density is almost flat even using the mesh anode, because the plasma diffuses in the radial direction and becomes uniform.

When the gas pressure is changed from P_{gas}=20 Pa to 40 Pa, the radial density profile drastically changes as shown in Figure 5(b). Since the collision between the ions and neutral particles becomes frequent with an increase in the gas pressure, the formed periodic structure of the electron density collapses and becomes relatively flat.

Based on these results, it is found that the structure of the plasma is sensitive to the magnetic field and gas pressure, which are necessary to be carefully adjusted to obtain the desired plasma structure.

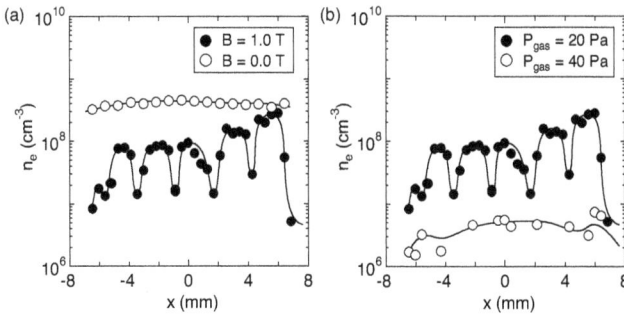

Figure 5. Radial profiles of the electron density of the plasmas (a) as a function of magnetic fields B for P_{gas} = 20 Pa and (b) as a function of gas pressure P_{gas} for B = 1 T. I_D = 3 mA.

Using this periodic plasma structure, we attempt to synthesize the Au nanoparticles with periodic morphology. Figure 6 shows photos of the temporal evolution of the periodic Au nanoparticle structure which is formed at the gas-liquid interface in accordance with the periodic plasma structure under the condition of the strong magnetic field of B = 1 T. The density of the periodic structured Au nanoparticles increases with plasma irradiation time t_{pi}, and the obvious structure is formed typically within t_{pi} = 5 min. The interval of the nanoparticle structure is about 2.5 mm, which is same as the distance between the wires of the mesh anode. As mentioned above, the electron density decreases under the shadow region of the mesh anode. Therefore, the nanoparticles are considered to be synthesized in the electron density depression region. This synthesis mechanism is discussed later.

In order to analyze the properties of the periodic structured Au nanoparticles, the ionic liquid is cooled using the peltier element located under the glass cell and is changed to the solid phase. The solid phase ionic liquid containing the Au nanoparticles can be extracted from the vacuum chamber with keeping its structure. Figure 7 shows the picture of the periodic structure of the Au nanoparticles which are synthesized on the ionic liquid substrate at room temperature, and (a) kept at room temperature and (b) cooled under $T_{sub} \sim 0$ °C after the synthesis. The ionic liquid becomes solid state under $T_{sub} = 0$ °C and the Au nanoparticles are fixed in the solid state of the ionic liquid.

However, when the plasma is irradiated for longer than 5 min, the Au nanoparticles diffuse and the structure is broadened as shown in Figure 7(b). In order to suppress the diffusion of the Au nanoparticles, the temperature of the ionic liquid is reduced during the plasma process using the peltier element, resulting in the increase in viscosity of the ionic liquid. Although the diffusion of the Au nanoparticles is suppressed, the Au nanoparticle synthesis rate becomes low. Therefore, it is necessary to precisely control the temperature of the ionic liquid for the fine periodic Au nanoparticle structure.

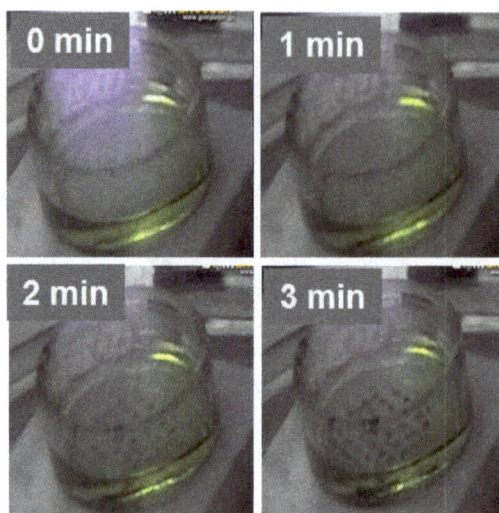

Figure 6. Temporal evolution of the periodic nanoparticle structure formed by controlled gas-liquid interfacial plasmas. $P_{gas} = 20$ Pa, $I_D = 3$ mA, $B = 1$ T.

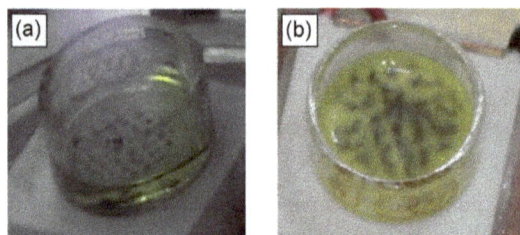

Figure 7. Pictures of the periodic nanoparticle structures which are formed at room temperature and (a) kept at room temperature and (b) cooled under $T_{sub} = 0$ °C. $P_{gas} = 20$ Pa, $I_D = 3$ mA, $B = 1$ T.

3.2. Ring-shaped nanoparticle structure formed by structure-controlled plasma

As the next step, we attempt to form more finely periodic structures of the Au nanoparticles based on the self-organizing behavior of turbulent plasmas generated by the nonlinear development of plasma fluctuations. For this purpose, a ring electrode is inserted in the plasma column as shown in Figure 8, and a positive DC bias voltage V_{RG} is applied to the electrode. It is found that the high frequency fluctuation (100 kHz – 1 MHz) is excited by the positive bias voltage, however the self-organized plasma structure is not observed at present. Therefore, in this experiment, the ring electrode is used for the plasma structure control without bias voltage.

Using this configuration, the Au nanoparticles are synthesized by reducing $HAuCl_4$. It is found that the ring shaped Au nanoparticle structure is formed corresponding to the shape of the inserted ring electrode as shown in Figure 8(b). This result means that the Au nanoparticles are synthesized at the region without plasma irradiation due to the shielding by the ring electrode. Although the Au nanoparticles are usually synthesized by the reduction effect of the electrons in the plasma [14-16], the electron is absent in the shadow region of the ring electrode in this experiment. Therefore, the reducing agent is not electrons in this case.

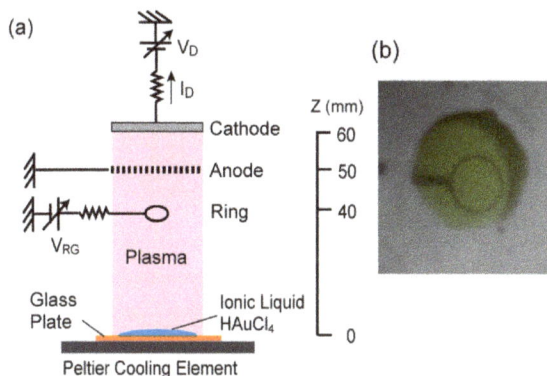

Figure 8. a) Experimental apparatus using ring electrode for nanoparticle structure control and (b) the synthesized ring shaped nanoparticle structure $B = 1$ T, $I_D = 2$ mA, $P_{gas} = 20$ Pa.

To explain the phenomena, we use the model of the Au nanoparticle synthesis as shown in Figure 9(a). As mentioned above, the charged particles such as the electrons and the positive ions cannot reach to the shadow region of the ring electrode, namely, only neutral radicals can arrive at the shadow region. The ionic liquid used in this experiment is described in Figure 9(b), which has C-H bond in cation (positive ion) and C-F bond in anion (negative ion). When the radicals in the plasma are irradiated to the ionic liquid, the C-H bond of the ionic liquid is considered to be dissociated, and the generated hydrogen radicals reduce the Au ions, resulting in the synthesis of Au nanoparticles in the shadow region of the ring electrode. On the other hand, in the plasma irradiation region, relatively high-energy ions are irradiated to the ionic liquid, and the C-F bond whose dissociation energy (D=5.07 eV) is larger than that of the C-H bond (D=4.29 eV), can be dissociated by the high-energy ions. Therefore, the Au nanoparticles are destroyed by the oxidation effect of the fluorine radicals which come from the dissociation of the ionic liquid by the collision with the high energy charged particle.

To understand the mechanism of this ring-shaped nanoparticle structure formation, the ring electrode is changed to disk electrode as shown in Figure 10, where the bias voltage of the disk electrode V_{DK} is floated in this experiment. In this case, since the shadow region of the disk electrode is wide and clear compare with the ring electrode, it is possible to clarify the species in the plasma, which are necessary for the synthesis of the Au nanoparticles. To change the plasma characteristics irradiated to the ionic liquid, we change the discharge current I_D.

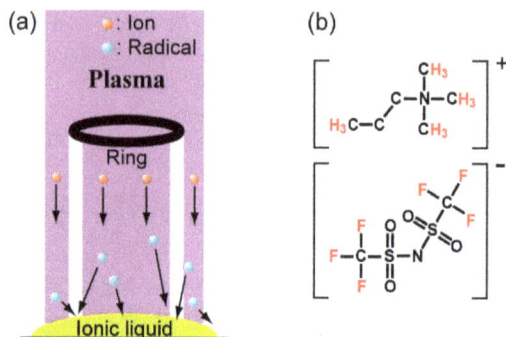

Figure 9. a) Model of the synthesis mechanism of the ring shaped Au nanoparticles and (b) chemical formula of the ionic liquid used in this experiment.

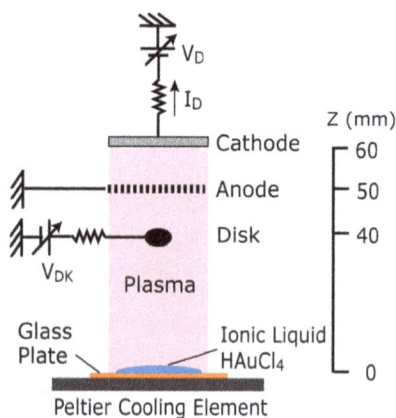

Figure 10. Experimental apparatus using disk electrode for understanding the mechanism of nanoparticle structure formation.

Figure 11 shows snapshots of the temporal evolution of the Au nanoparticle synthesis for (a) I_D= 1 mA and (b) I_D=3 mA. In the case for I_D=1 mA, the Au nanoparticles are synthesized from the outside of the ionic liquid in the plasma irradiation region, and are absent in the shadow region of the disk electrode for the plasma irradiation time t_{pi}=5 min. In the case of I_D=3 mA, on the other hand, the Au nanoparticles are synthesized not only from the outside of the ionic liquid but also the edge region of the disk electrode for t_{pi}=40 sec. Furthermore, the Au nanoparticles are not synthesized in the plasma irradiation region.

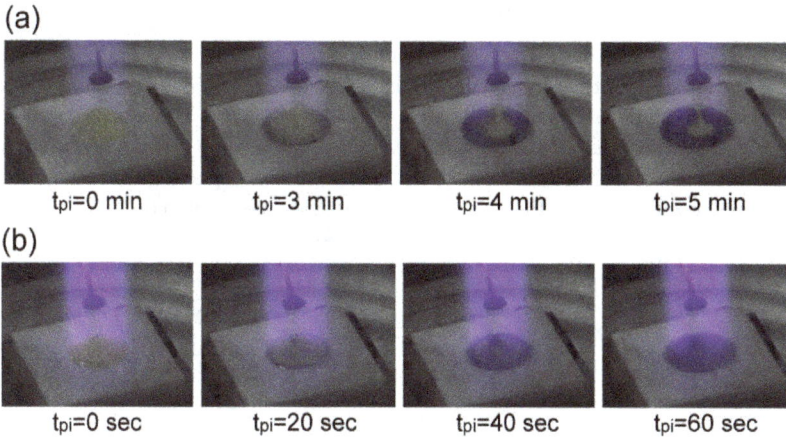

(a)

| t_{pi}=0 min | t_{pi}=3 min | t_{pi}=4 min | t_{pi}=5 min |

(b)

| t_{pi}=0 sec | t_{pi}=20 sec | t_{pi}=40 sec | t_{pi}=60 sec |

Figure 11. Photos of the Au nanoparticle structure as a function of plasma irradiation time t_{pi} for (a) I_D = 1 mA and (b) I_D = 3 mA. B = 1 T, P_{gas} = 20 Pa.

Since the electron density for I_D=1 mA is lower than that for I_D=3 mA, the sheath electric field formed above the ionic liquid is small for I_D=1 mA. Therefore, the ion irradiation energy to the ionic liquid is small, and as a result, the ion cannot dissociate the C-F bond of the ionic liquid. However, the C-H bond which has relatively low dissociation energy (4.29 eV) is dissociated by the low-energy ions or neutral nitrogen radicals. The dissociated hydrogen radicals can reduce the Au ions, resulting in the formation of the Au nanoparticles in the plasma irradiation region for I_D=1 mA.

For I_D=3 mA, on the other hand, the sheath electric field becomes large in the plasma irradiation region, and then, the high-energy ion irradiation can dissociate the C-F bond of the ionic liquid. Therefore, the Au ions are difficult to be reduced by the oxidation effect of the fluorine radical in the plasma irradiation region. However, in the shadow region of the disk electrode, the C-H bond of the ionic liquid is dissociated by the nitrogen radical of the plasma, and the dissociated hydrogen radical can synthesize the Au nanoparticles. In the center of the shadow region of the disk electrode, the Au nanoparticles are not synthesized because the nitrogen radicals cannot reach to the center of the shadow region.

Based on these results, it is found that the Au nanoparticles are synthesized by the reduction effect via the neutral radical irradiation, and are destroyed by the oxidation effect via the high-energy ion irradiation in the plasma irradiation region.

3.3. Size control of ring-shaped nanoparticle structure under inhomogeneous conversing magnetic fields

For the purpose of size control of the ring-shaped nanoparticle structure, we use the plasma - ionic liquid interface under the inhomogeneous converging magnetic fields (B) to shrink the size of the nanoparticle structure [33].

Figure 12 shows the schematic of a new experimental setup for the discharge plasma in contact with the ionic liquid containing a gold chloride ($HAuCl_4$), where the distance between a ring electrode and the ionic liquid on the glass plate is defined as z_r. The plasma is generated between a plate cathode electrode and a mesh anode electrode, and is irradiated to the ionic liquid. Here, a mirror ratio R_m is defined as the ratio of the magnetic field strength at the ring electrode position (B_{RG}) to that at the ionic liquid substrate position (B_{IL}).

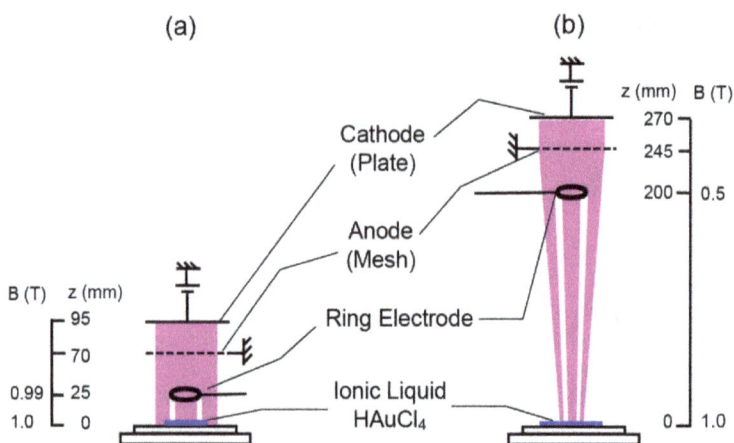

Figure 12. Experimental apparatus for size control synthesis Au nanoparticles under the inhomogeneous converging magnetic fields for mirror ratio (a) $R_m\sim1$ and (b) $R_m=2$.

Figure 13(a) shows pictures of the Au nanoparticle structure which is formed at the plasma-liquid interface in accordance with the plasma structure for $z_r = 25$ mm, i.e. $R_m\sim1$. The Au nanoparticles are synthesized in the shadow region under the ring electrode with the diameter of about 7 mm, where there are a large amount of neutral radicals and few charged ions. The Au nanoparticles are synthesized by the reduction effect of the radical in the plasma, while the charged ions inhibit the synthesis of the Au nanoparticles by the oxidation effect.

It is possible to control the Au nanoparticle structure by using shrinkage of the plasma diameter under the converging magnetic field. Figures 13(b) and 13(c) show pictures of the Au nanoparticle structure for $z_r = 200$ mm ($R_m=2$) and 550 mm ($R_m=10$), respectively.

The diameters of the Au nanoparticle structures are observed to be 5 mm and 2.2 mm for $R_m=2$ and $R_m=10$, respectively. Since the diameter of the plasma is proportional to the square root of the mirror ratio, the diameter of the ring shaped nanoparticles is also changed corresponding to the plasma diameter. Therefore, the diameter of the ring-shaped Au nanoparticle structure for $z_r = 200$ mm and 550 mm should be $1/\sqrt{2}$ and $1/\sqrt{10}$, respectively. We can easily control the size of the nanoparticle structure by simply change the magnetic field configuration, which is useful for the future nanoparticle nano-electronics devices.

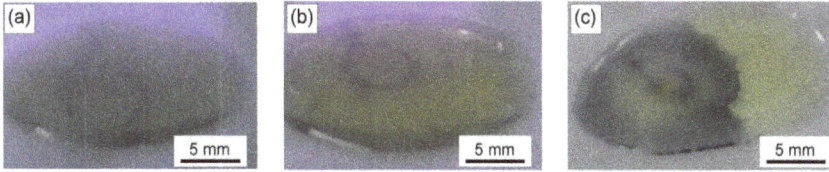

Figure 13. Picture of nanoparticle structures for (a) $R_m = 1$, (b) $R_m = 2$, and (c) $R_m = 10$. $I_0=2$ mA, $P_{gas}=20$ Pa, $B_{iL}=1$ T.

4. Conclusion

A direct current (DC) discharge plasma has stably been generated just above the ionic liquid by applying the DC voltage to an electrode immersed in the ionic liquid. The precise potential structure and the resultant plasma ion or electron irradiation to the ionic liquid are controlled. This ion irradiation is found to be effective for the synthesis of gold nanoparticles in comparison with the conventional electron irradiation system, and the control of the plasma-ion irradiation to the ionic liquid has the possibility of application to the synthesis of the various kinds of size- and yield-controlled nanoparticles.

Furthermore, the periodic and ring shaped Au nanoparticle structures are formed, which correspond to the shape of the strongly-magnetized plasmas generated using the mesh anode electrode or the ring/disk electrode inserted in the plasma.

It is very interesting that the structure of the Au nanoparticles depends on the discharge current, namely, the Au nanoparticles are synthesized from the periphery and absence in the shadow region of the disk electrode for small discharge current, while the Au nanoparticles are synthesized at the boundary of the disk electrode for relatively large discharge current. These phenomena are well explained by the reduction and oxidation effects of the radicals which are generated by the plasma irradiation to the ionic liquid and resultant dissociation of the ionic liquid.

Finally, the size of the Au nanoparticle structure can be controlled by using shrinkage of the plasma diameter under the converging magnetic field, which enables us to freely form the micro- or nano-sized nanoparticle structures.

Acknowledgements

The authors thank Prof. K. Tohji, K. Motomiya, T. Miyazaki, and H. Ishida for their technical assistance. We express our gratitude to Dr. K. Baba, Dr. Q. Chen, T. Harada, and T. Okuno for their collaboration. This work was supported by a Grant-in-Aid for Scientific Research from the Ministry of Education, Culture, Sports, Science and Technology, Japan.

Author details

Toshiro Kaneko, Shohei Takahashi and Rikizo Hatakeyama

Department of Electronic Engineering, Tohoku University, Sendai, Japan

References

[1] Gubkin J. Electrolytische Metallabscheidung an der freien Oberfläche einer Salzlösung. Annalen der Physik 1887; 268(9), 114-115.

[2] Kanzaki Y., Hirabe M., Matsumoto O. Glow Discharge Electrolysis of Aqueous Sulfuric Acid Solution in Various Atmosphere. Journal of The Electrochemical Society 1986; 133(11), 2267-2270.

[3] Baba K., Kaneko T., Hatakeyama R. Ion Irradiation Effects on Ionic Liquids Interfaced with RF Discharge Plasmas. Applied Physics Letters 2007; 90(20), 201501-1-3.

[4] Kaneko T., Baba K., Hatakeyama R. Static Gas-Liquid Interfacial Direct Current Discharge Plasmas Using Ionic Liquid Cathode. Journal of Applied Physics 2009; 105(10), 103306-1-5.

[5] Bruggeman P., Leys C. Non-Thermal Plasmas in and in Contact with Liquids. Journal of Physics D: Applied Physics 2009; 42(5), 053001-1-28.

[6] Koo I. G., Lee M. S., Shim J. H., Ahn J. H., Lee W. M. Platinum Nanoparticles Prepared by a Plasma-Chemical Reduction Method. Journal of Materials Chemistry 2005; 15(38), 4125-4128.

[7] Hieda J., Saito N., Takai O. Exotic Shapes of Gold Nanoparticles Synthesized Using Plasma in Aqueous Solution. Journal of Vacuum Science and Technology A 2008; 26(4), 854-856.

[8] Richmonds C., Sankaran R. M. Plasma-Liquid Electrochemistry: Rapid Synthesis of Colloidal Metal Nanoparticles by Microplasma Reduction of Aqueous Cations. Applied Physics Letters 2008; 93(13), 131501-1-3.

[9] Sato S., Mori K., Ariyada O., Atsushi H., Yonezawa T. Synthesis of Nanoparticles of Silver and Platinum by Microwave-Induced Plasma in Liquid. Surface and Coatings Technology 2011; 206(5), 955-958.

[10] Chen L., Iwamoto C., Omurzak E., Takebe S., Okudera H., Yoshiasa A., Sulaimanku-lova S., Mashimo, T. Synthesis of Zirconium Carbide (ZrC) Nanoparticles Covered with Graphitic "Windows" by Pulsed Plasma in Liquid. RSC Advances 2011; 1(6), 1083–1088.

[11] Seddon K. R. Ionic Liquids: A Taste of the Future. Nature Materials 2003; 2(6), 363-365.

[12] Rogers R. D., Seddon K. R. Ionic Liquids - Solvents of the Future?. Science 2003; 302(5646), 792-793.

[13] Baba K., Kaneko T., Hatakeyama R. Efficient Synthesis of Gold Nanoparticles Using Ion Irradiation in Gas-Liquid Interfacial Plasmas. Applied Physics Express 2009; 2(3), 035006-1-3.

[14] Kaneko T., Baba K., Harada T., Hatakeyama R. Novel Gas-Liquid Interfacial Plasmas for Synthesis of Metal Nanoparticles. Plasma Processes and Polymers 2009; 6(11), 713-718.

[15] Kaneko T., Chen Q., Harada T., Hatakeyama R. Structural and Reactive Kinetics in Gas-Liquid Interfacial Plasmas. Plasma Sources Science and Technology 2011; 20(3), 034014-1-8.

[16] Meiss S. A., Rohnke M., Kienle L., Zein El Abedin S., Endres F., Janek J. Employing Plasmas as Gaseous Electrodes at the Free Surface of Ionic Liquids: Deposition of Nanocrystalline Silver Particles. ChemPhysChem 2007; 8(1), 50-53.

[17] Torimoto T., Okazaki K., Kiyama T., Hirahara K., Tanaka N., Kuwabata S. Sputter Deposition onto Ionic Liquids: Simple and Clean Synthesis of Highly Dispersed Ultrafine Metal Nanoparticles. Applied Physics Letters 2006; 89(24), 243117-1-3.

[18] Xie Y. B., Liu C. J. Stability of Ionic Liquids under the Influence of Glow Discharge Plasmas. Plasma Processes and Polymers 2008; 5(3), 239-245.

[19] Kuwabata S., Tsuda T., Torimoto T. Room-Temperature Ionic Liquid. A New Medium for Material Production and Analyses under Vacuum Conditions. The Journal of Physical Chemistry Letters 2010; 1(21), 3177–3188.

[20] Wei Z., Liu C. J. Synthesis of Monodisperse Gold Nanoparticles in Ionic Liquid by Applying Room Temperature Plasma. Materials Letters 2011; 65(2), 353–355.

[21] Kareem T. A., Kaliani A. A. I-V Characteristics and the Synthesis of ZnS Nanoparticles by Glow Discharge at the Metal–Ionic Liquid Interface. Journal of Plasma Physics 2012; 78(2), 189-197.

[22] Shaw C. P., Fernig D. G., Levy R. Gold Nanoparticles as Advanced Building Blocks for Nanoscale Self-Assembled Systems. Journal of Materials Chemistry 2011; 21(33), 12181-12187.

[23] Huang L., Tu C. C., Lin L. Y. Colloidal Quantum Dot Photodetectors Enhanced by Self-Assembled Plasmonic Nanoparticles. Applied Physics Letters 2011; 98(11), 113110-1-3.

[24] Maye M. M., Nykypanchuk D., van der Lelie D., Gang O. DNA-Regulated Micro- and Nanoparticle Assembly. Small 2007; 3(10), 1678-1682.

[25] Chen Q., Kaneko T., Hatakeyama R. Rapid Synthesis of Water-Soluble Gold Nano- particles with Control of Size and Assembly Using Gas-Liquid Interfacial Discharge Plasma. Chemical Physics Letters 2012; 521, 113–117.

[26] Yu J., Rance G. A., Khlobystov A. N. Electrostatic Interactions for Directed Assembly of Nanostructured Materials: Composites of Titanium Dioxide Nanotubes with Gold Nanoparticles. Journal of Materials Chemistry 2009; 19(47), 8928-8935.

[27] Kaneko T., R. Hatakeyama R. Creation of Nanoparticle-Nanotube Conjugates for Life-Science Application Using Gas-Liquid Interfacial Plasmas. Japanese Journal of Applied Physics 2012; in press.

[28] Baba K., Kaneko T., Hatakeyama R., Motomiya K., Tohji K. Synthesis of Monodis- persed Nanoparticles Functionalizing Carbon Nanotubes in Plasma-Ionic Liquid In- terfacial Fields. Chemical Communications 2010; 46(2), 255-257.

[29] Georgakilas V., Gournis D., Tzitzios V., Pasquato L., Guldie D. M., Prato M. Decorat- ing Carbon Nanotubes with Metal or Semiconductor Nanoparticles. Journal of Mate- rials Chemistry 2007; 17(26), 2679–2694.

[30] Wildgoose G. G., Banks C. E., Compton R. G. Metal Nanoparticles and Related Mate- rials Supported on Carbon Nanotubes: Methods and Applications. Small 2006; 2(2), 182-193.

[31] Ye X., Lin Y., Wang C., Engelhard M. H., Wang Y., Wai C. M. Supercritical Fluid Syn- thesis and Characterization of Catalytic Metal Nanoparticles on Carbon Nanotubes. Journal of Materials Chemistry 2004; 14(5), 908-913.

[32] Han L., Wu W., Kirk F. L., Luo J., Maye M. M., Kariuki N. N., Lin Y., Wang C., Zhong C.-J. A Direct Route toward Assembly of Nanoparticle–Carbon Nanotube Composite Materials. Langmuir 2004; 20(14), 6019-6025.

[33] Kaneko T., Takahashi S., Hatakeyama R. Control of Nanoparticle Synthesis Using Physical and Chemical Dynamics of Gas-Liquid Interfacial Non-Equilibrium Plas- mas. Plasma Physics and Controlled Fusion 2012; in press.

[34] Krenn J. R. Nanoparticle Waveguides: Watching Energy Transfer. Nature Materials 2003; 2(4), 210-211.

Chapter 14

Preparation, Physicochemical Properties and Battery Applications of a Novel Poly(Ionic Liquid)

Takaya Sato, Takashi Morinaga and Takeo Ishizuka

Additional information is available at the end of the chapter

1. Introduction

Ionic liquids (ILs) are generally defined as a salt with a melting point lower than 100 °C whose properties include non-volatility, non-flammability, and a relatively high ionic conductivity [1]. Recently, therefore, interest has increased in the possible use of this type of liquid as an electrolyte in energy storage devices, for example, a lithium rechargeable battery [2–9] and an electric double layer capacitor [10–14].

At the same time, to further the development of large, thin, prismatic electrochemical devices for which there is a high market demand, solid electrolytes are generally preferred over liquid electrolytes from the view point of ease of production and long device lifetime. In the last decade, many researchers and battery companies have been developing a "solid state polymer electrolyte" or a "gel polymer battery" with a film-like shape. More recently, growing attention has been paid to poly(ionic liquids) as a class of polymeric materials that are highly non-flammable. Examples of new polyelectrolytes poly(ILs) have been produced from polymerizable ionic liquid monomers by several polymerization processes. Research into the application of poly(ILs) for polymer electrolytes were under intense study by Ono et al around at 2005 [15-17]. The ensuing intensive studies on poly(ILs) in the last five years significantly expanded the research scope of this new type of polymer, and some valuable review [18-19] and feature articles have been published [20].

Amid such developments, we also developed a polymerizable ionic liquid, N, N-diethyl- N-(2-methacryloylethyl)-N-methylammonium bis(trifluoromethylsulfonyl)imide (DEMM-TFSI, whose molecular structure is shown in Fig.1). From many previous studies, it was known that aliphatic quaternary ammonium based ionic liquids had an obviously higher cathodic stability than the aromatic type ionic liquids. Therefore, polymer materials made from these polymerizable ILs could have high durability in use as an electrolyte in various energy devices. However, because there have been only a few reports on this topic, we describe in this chapter the preparation, polymerization, physicochemical properties, and the application of quaternary ammonium type polymerizable ionic liquids and related polymers.

On the other hand, the use of ILs as the solvent in free radical polymerization media instead of a conventional organic solvent markedly affects the rate and degree of polymerization. It has been reported that poly(methylmethacrylate) (PMMA) reactions are much more rapid in an ionic liquid than in a nonpolar solvents, and that a PMMA prepared in an ionic liquid has a molecular weight approximately five times higher than in benzene and toluene [21]. The ionic liquid monomer, DEMM-TFSI, also gives the ultra high molecular weight poly(DEMM-TFSI) by bulk polymerization with a conventional 2-2'-azoisobutyronitrile (AIBN) initiator [22]. However, control of the molecular weight was difficult due to the strong enhancement of the propagation rate coefficients (k_p) in ILs. Although molecular weight control in the polymerization of polymerizable ILs is more successful using an atom transfer radical polymerization (ATRP) [20, 23, 24], there are few investigation of length control by other polymerization methods, such as conventional chain transfer radical (CTR) polymerization, adding chain transfer agents, or reversible addition fragmentation chain transfer (RAFT) polymerization. In this chapter, we describe the molecular weight-controlled polymerization of DEMM-TFSI by CTR polymerization and RAFT polymerization. We also detail the physicochemical properties of the resulting molecular weight-controlled ionic liquid polymer, poly(DEMM-TFSI), especially the thermal properties (glass transition temperature; T_g), ionic conductivity and the self-diffusion coefficient (D) of poly-cation and anion in solution using pulsed-gradient spin echo NMR (PGSE-NMR) spectroscopy.

For electrochemical device applications, a good method to produce a flexible polymer electrolyte membrane with high conductivity and non-flammability is to use poly(ILs) as a host polymer for the gelation of ILs [22, 25]. Poly(ILs) have a higher affinity for ionic liquids than that of conventional polymers such as PMMA, poly(ethylene oxide) (PEO) and poly(vinylidenefluoride) (PVdF). In fact, the ionic liquid gel composite material including poly(ILs) showed higher ionic conductivity than that of the materials including conventional polymer such as a PMMA. A high degree of compatibility with the ionic liquid results in a high ionic conductivity and good physical properties even when only a small amount of polymer material is added. This allows the IL to be completely encapsulated, thereby avoiding liquid leakage accidents. Work in this area will be also described in this chapter.

2. Preparation and physicochemical properties of DEMM-TFSI

We developed a new type of polymerizable IL, N, N-diethyl- N- (2-methoxyethyl)- N-methylammonium bis(trifluoromethylsulfonyl)imide (DEMM-TFSI in Fig. 1), with electrochemical stability in a wide range of potential. The preparation route of the DEMM-TFSI, was given elsewhere [22]. In brief, 2-(diethylamino)ethylmethacrylate was treated with 1.2 equiv. of methyliodide in tetrahydrofuran at 0 °C and stirred overnight, giving N,N-diethyl-N-(2-

methacryloylethyl)-*N*-methylammonium iodide as a precipitate, which was filtered off and recrystallized in tetrahydrofuran-ethanol solvent. The recrystallized product was treated with exactly 1.0 equiv. of lithium bis(trifluoromethylsulfonyl)imide in deionized water for 5 h. After the reaction, the mixture was separated into two phases, the bottom phase being *N*,*N*-diethyl-*N*-(2-methacryloylethyl)-*N*-   methylammonium   bis(trifluoromethylsulfon-yl)imide.

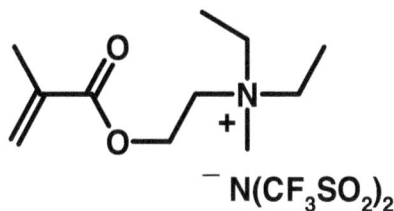

Figure 1. Molecular structure of DEMM-TFSI

The ionic conductivity is 0.6 mScm^{-1} at 25 °C measured by a conductivity meter (HM-30R, DKK-TOA Corporation).

Fig.2 depicts the differential scanning calorimetry (DSC) and thermal gravimetric analysis (TGA) curves of the DEMM-TFSI. This polymerizable IL did not exhibit a clear melting or freezing temperature in a typical DSC measurement; however, the glass transition tempera-ture T_g was observed at −68 °C. The decomposition temperature corresponding to a 10% weight loss according to TGA measurements occurred at 329 °C. DEMM-TFSI exists as an ionic liquid over a very wide temperature range, approximately 400 °C.

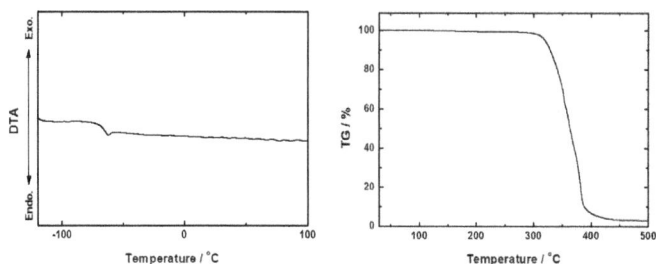

Figure 2. (a) Differential scanning calorimetry (DSC) and (b) thermal gravimetric analysis (TGA) results for the ionic liquid monomer, DEMM-TFSI at a heating rate of 5 °C min^{-1}.

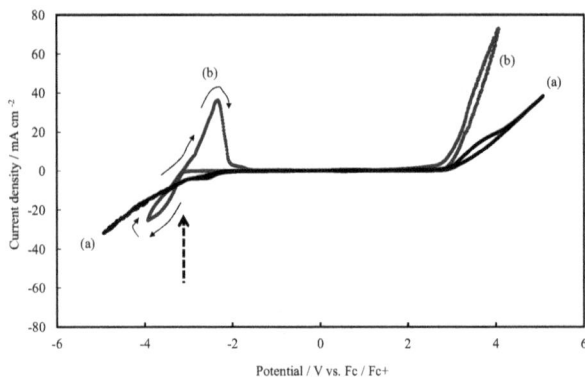

Figure 3. Cyclic voltammogram of an ionic liquid monomer, an organic electrolyte at 25 °C. Scan rate: 10mVs⁻¹; platinum working and counter electrodes; Ag/Ag⁺ reference electrode. The potential value (V) was referenced to the ferrocene (Fc)/ferrocenium (Fc⁺) redox couple in each salt. (a) 0.1 mol kg⁻¹ of DEMM-TFSI in propylene carbonate (PC) solution, (b) 0.1 mol kg⁻¹ of LiTFSI in PC solution. The concentration is defined in terms of molality = (mol solute/kg solvent). The dashed line arrow indicates the potential of the Li/Li⁺ couple.

Fig. 3 shows both the limiting reduction potentials ($E_{red.}$) on platinum of the ionic liquid monomer as measured by cyclic voltammetry at room temperature, and a voltammogram of an ordinary organic electrolyte LiTFSI in PC. The sharp peaks around −3.0V are probably due to the deposition and dissolution of Li metal, since a color change of the working electrode by the metal deposition appears with a current loop at that potential. The $E_{red.}$ and the limiting oxidation potentials ($E_{oxd.}$) were defined as the potential where the limiting current density reached 1mAcm⁻². The $E_{red.}$ of the DEMM-TFSI ionic liquid monomer was positioned about 0.7V positive against the Li/Li⁺. We confidently expected this result, because the monomer molecule had an easily reduced double bond.

3. Preparation of poly(DEMM-TFSI)

3.1. Radical polymerization of ionic liquid monomer; DEMM-TFSI

The poly(IL), poly(DEMM-TFSI), was synthesized by the bulk polymerization method. First, the monomer was dissolved in acetonitrile, and the solution was treated with activated carbon; the resultant acetonitrile solution was evaporated and the purified monomer was dried in vacuum at 25 °C. The ionic liquid monomer and 2,2'-Azoisobutyronitrile (AIBN), at a ratio of 1.0 mol% to the amount of methacryl groups present in the monomer, were mixed until they became homogeneous. The mixture was degassed in vacuum at 50 °C, and kept standing at 70 °C for 15 h. After polymerization, the product polymer was dissolved in acetonitrile and precipitated into ethanol and water, before a final drying in vacuum at 70 °C. The preparation scheme is shown in Fig. 4.

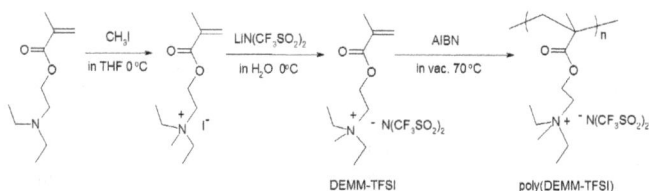

Figure 4. Synthesis of ionic liquid monomer (DEMM-TFSI) and poly(DEMM-TFSI).

It has been reported that poly(methyl methacrylate) (PMMA) reactions are much more rapid in an ionic liquid than in a nonpolar solvent such as benzene, and that a PMMA prepared in an ionic liquid has a molecular weight approximately five times higher than in benzene [21]. The ionic liquid monomer DEMM-TFSI also gives the ultra high molecular weight poly(DEMM-TFSI) by bulk polymerization with an AIBN initiator. Fig. 5 illustrates the GPC trace of the poly(DEMM-TFSI) product polymerized using a mol ratio of [monomer]/[AIBN] = 1:0.01. The resulting polymer was a rubbery solid, with a weight average molecular weight (M_w) of 1,084,000 and a polydispersity index by GPC analysis of 2.95 [PDI = (M_w)/number-average molecular weight (M_n)].

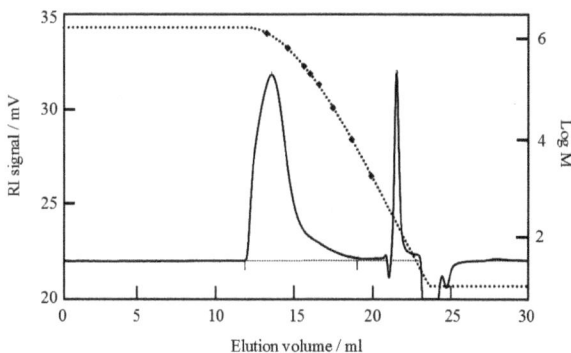

Figure 5. GPC trace of poly(DEMM-TFSI). M_n = 368,000; M_w = 1,084,000; M_w / M_n = 2.95. GPC analyses were performed at 40 °C, with a Shodex GPC-101 equipped with two, series-connected OHpak SB-806M HQ columns with a solution of 0.5 M acetic acid and 0.2 M sodium nitrate in acetonitrile and water (1/1 v/v) as the eluent. The weight- and number-average molecular weight were estimated on the basis of the calibration curve established with standard poly(ethylene oxide)s with the Shodex 480-II data station.

3.2. Molecular weight control of poly(DEMM-TFSI)

It is well known when the radical polymerizations are conducted in an ionic liquid, a significant increase of the k_p / k_t ratio is normally observed in comparison to those carried out in
other polar solvents [18, 26]. Not only the rates of polymerization, but also the molecular
weights of the polymers produced were considerably higher for several kinds of monomers.
Also, the polymerizable ionic liquids, in our case DEMM-TFSI, give considerably higher molecular weight polymers in bulk radical polymerizations. The difficulty in molecular weight
control might become disadvantageous in material development. In this section, we show
how we succeeded in the molecular weight control of the radical polymerization process using DEMM-TFSI by simply adding a chain transfer reagent (CTA).

Fig. 6 illustrates the relationships between the number average molecular weights and the
concentration of the charge transfer reagent, 3-mercapto-1-hexanol in bulk radical polymerization media using AIBN as a initiator at 50 °C. We were able to synthesize poly (DEMM-
TFSI) over M_n range from 5000 to 50,000, indicating that 3-mercapto-1-hexanol is an effective
CTA under these conditions. The polydispersity index (PDI = M_w / M_n) was approximately
1.8 and 2.7 for polymer with a M_n of ten thousand or less and several tens of thousands or
more, respectively; these PDI values are not very different from those found with a conventional free radical polymerization.

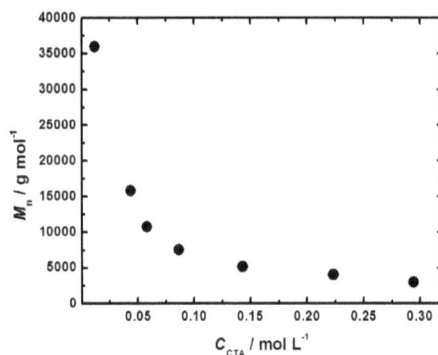

Figure 6. The relationship between the number average molecular weight of polymerized ionic liquid, DEMM-TFSI
and the concentration of the charge transfer reagent 3-mercapto-1-hexanol, using 2,2'-azobis(isobutyronitrile) (AIBN)
as initiator at 50 °C. Molar ratio: [DEMM-TFSI] / [AIBN] = 1 / 0.015. Number-average molecular weights were as determined by GPC.

Using the controlled/living radical polymerization techniques that have shown remarkable
progress recently is perhaps the best choice for preparation of the poly(ILs) with controlled
molecular weight and a narrow PDI. Two such possibilities are atom transfer radical polymerization (ATRP) and reversible addition-fragmentation chain transfer (RAFT) polymerization [27]. We have already achieved exceptionally dense grafting of well-defined
pol(DEMM-TFSI) on a solid surface, known as a concentrated polymer brush prepared by
surface initiated ATRP. Such a polymer brush surface showed unique properties, including

high modulus, super lubrication, and unique size exclusion that were quite different and even unpredictable from those of the previously studied semidilute polymer brushes [24]. The polymerization of DEMM-TFSI was well controlled and exhibited living characteristics when a Cu(I)Cl and Cu(II)Cl$_2$ mixture and 2,2'-bipyridine complex was used as the catalyst system and ethyl-2-bromoisobutyrate as initiator. However, if such synthesized poly(ILs) are used for the electrochemical devices, some residue of the metallic catalyst may ruin the reliability of the devices. An apparent disadvantage of applying ATRP for the synthesis of poly(ILs) is the unavoidable complexation of polymers with the catalytic copper ions. However, RAFT polymerization is free of this problem, as no metal source is involved.

Compared to ATRP, there are not many examples of research into RAFT polymerization. Because papers concerning the RAFT polymerization of polymerizable quaternary ammonium type ionic liquids are very few in number, we investigated the kinetics of the RAFT polymerization of DEMM-TFSI. Figure 7(a) shows the variation in $\ln([M]_0 / [M]_t)$ versus polymerization time for the polymerization of DEMM-TFSI in acetonitrile at 50 °C with AIBN in the presence of 2-cyano-2-propyl benzodithioate (CPBT). The reaction was conducted at a ratio of $[DEMM\text{-}TFSI]_0 / [CPBT]_0 = 200 / 1$, the concentration of DEMM-TFSI and AIBN being 75 wt.% and 0.5 wt.% in acetonitrile, respectively. Almost full conversion was reached after 5 hours and an almost linear first-order kinetic plot is seen until almost 100% conversion. Nevertheless, a linear increase in the number-average molecular weight determined by GPC spectroscopy with conversion is observed, indicating a constant number of propagating chains throughout the polymerization (Figure 7(b)). The PDI value is consistently small ($M_w / M_n = 1.45\text{-}1.24$) from the first stage of the polymerization to the end. This is an indication that the polymer chain end is capped with the fragments of CPBT, as expected according to the general mechanism of the RAFT process. These data indicated the molecular weight controlled synthesis of poly(DEMM-TFSI) with a narrow PDI can be successfully performed by RAFT polymerization of DEMM-TFSI.

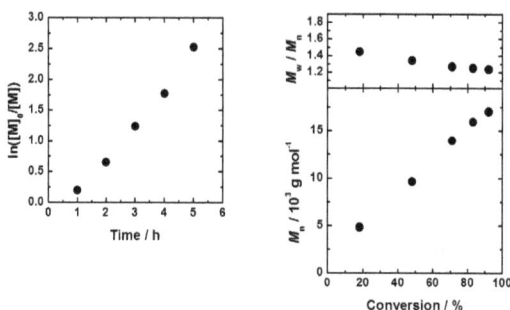

Figure 7. (a) Plot of $\ln([M]_0/[M]_t)$ versus time and (b) evolution of number average molecular weight (M_n) and polydispersity index (M_w/M_n) for the RAFT polymerization of DEMM-TFSI with 2,2'-azobis(isobutyronitrile) (AIBN) in the presence of 2-cyano-2-propyl benzodithioate (CPBT) in acetonitrile at 50 °C: $[DEMM\text{-}TFSI]_0 / [CPBT]_0 / [AIBN]_0 = 200 / 1 / 3.91$. The number average molecular weights and conversion were determined by GPC and ^{1}H-NMR.

4. Physicochemical properties of poly(DEMM-TFSI)

In Fig. 8 are shown the T_g values of poly(DEMM-TFSI) with various M_n values and relatively small polydispersity index values ($M_w / M_n < 1.2$) measured by differential scanning calorimetry (DSC) analysis during heating at 10 deg/min. versus the number average molecular weight (M_n). The polymer was prepared by atom transfer radical polymerization (ATRP) using a complex catalyst consisting of copper chloride and 2,2'-bipyridine in acetonitrile [24]. ATRP represents one of the branches of living radical polymerization (LRP). The M_n value was estimated as an absolute value, assuming a 100% initiation, from the monomer-to-initiator molar ratio and the conversion determined by ¹H-NMR using a JEOL JEM-ECX400 spectrometer. The M_w / M_n value was determined by poly(ethyleneglycol)-calibrated gel permeation chromatography (GPC) using a Shodex GPC-101 high-speed liquid chromatography system.

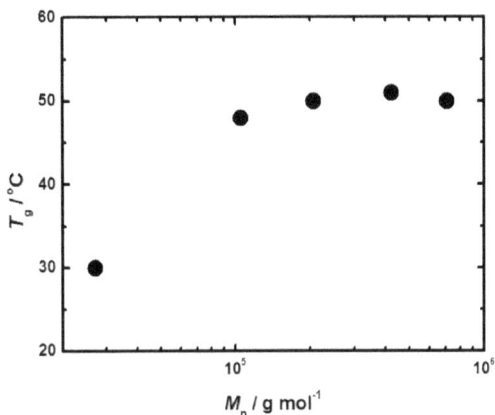

Figure 8. Glass transition temperatures, measured calorimetrically during heating at 10 deg. / min, versus number average molecular weight.

The glass transition temperature is related to the start of the segmental motion of polymers. In general, polymers with more free volume have a lower T_g value and that temperature is essentially independent of molecular weight because the free volume ratio per segment of polymer materials is the same even if the length of a polymer main chain changes. However, it is known that a lower molecular weight polymer usually has lower T_g value. Because the chain ends have high mobility and more free volume than the middle part of polymer, the free volume increases with the numbers of terminals, and T_g decreases as a result.

A poly-cation type material such as poly(DEMM-TFSI) has bulky tetra-alkyl ammonium functional groups in each unit and has a relatively low T_g value due to its large free volume due to the effect of the electrostatic repulsions of cationic groups. A polymer with a molecular weight that exceeds 100,000 has roughly a T_g of 50 °C while, on the other hand, a 20 degrees lower value of T_g of 30 °C was found for a polymer with a M_n about 10,000, presumably due to the chain ends effect. The relationships between M_n and T_g of this poly(IL) had a similar trend to that of polystyrene [28, 29]. We realized that the T_g rose greatly on polymerization only up to about 50 of degree of polymerization, though the T_g of the monomer had shown an extremely low value of -68 °C.

It was thought that a excellent solid polymer electrolyte with high physical strength and high ionic conductivity could be developed by the polymerization of ionic liquid monomers, so much attention has been directed toward developing poly(ILs) in order to avoid some disadvantages of liquid electrolytes, such as leakage and flammability, in energy device applications, including the lithium ion rechargeable battery, the electric double layer capacitor and dye-sensitized solar cell. However, the ionic conductivity of the polymerized ionic liquids was generally considerably lowered due to both the considerable elevation of the glass transition temperature and a reduced number of mobile ions, as one of ions is fixed to the polymer chain and so cannot move.

Fig. 9 shows the temperature dependence of the ionic conductivity for a series of poly(DEMM-TFSI)s with different M_n values. In the case of our polymer, the ionic conductivity of poly(DEMM-TFSI) was four digits or more lower than that of an ionic liquid such as DEME-TFSI (molecular structure is indicated in Fig. 11.) in the room temperature region. Unfortunately, this level of conductivity is not practicable for devices. However, an interesting point is that neither the value of the ionic conductivity nor the temperature dependency of the conductivity do not depend on the molecular weight of the polymer, and exhibit an almost constant value. Thus, polymers with M_n of 27,000, 250,000 and 710,000 have T_g values of 40, 50 and 50 °C, respectively. The ionic conductivity of these polymers continuously changes without an inflection point in the area before and after the T_g. There seems to be no correlation between the ionic conductivity value and the glass transition temperature. It is widely accepted that ion conduction in an amorphous polymer matrix should occur above the glass transition temperature as it is coupled with a segmental motion of polymer chain [30]. The matrix polymer solvated the mobile ions and created a liquid-like environment around the ions. However, in the poly(DEMM-TFSI) matrix, it seems the anion moves independently of the polymer chain which stops its segmental motion below the glass transition temperature. The mobility of the anion in this poly(IL) seems to be different from that of the solvated ions in the polymer without any dissociable ionic substituent, such as polyethylene oxide (PEO) that contains a lithium salt. Ionic conduction does not appear in a PEO matrix below the glass transition temperature. In the poly(DEMM-TFSI), an alternative anion conduction mechanism that does not involve polymer motion might exist. There are only a few previous studies related to the ionic conduction mechanism in poly-cation and poly-anion systems below T_g. We are planning to combine DSC and ionic conductivity measurements using a dielectric relaxation spectroscopy measurement in order to improve our understanding of this interesting phenomenon.

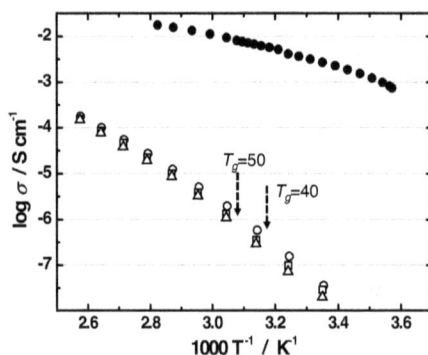

Figure 9. Temperature dependence of the ionic conductivity for poly(DEMM-TFSI)s with different M_n values and an ionic liquid by the complex impedance method [24]. Open circle, Poly(DEMM-TFSI) with M_n=27,000; open triangle, M_n=250,000; open square, M_n=710,000; filled circle, ionic liquid (DEME-TFSI). Ionic conductivity measurements were made by an AC-impedance method using a multi-frequency LCR meter (Agilent Technology E4980A Precision LCR). The sample was loaded at 100 °C between two polished stainless-steel discs acting as ion-blocking electrodes with ceramic-sheet spacer with a 100 µm thickness. The measurement cell was set in a thermostat oven chamber and collected at various frequencies ranging from 20 Hz to 2 MHz and various temperatures ranging from –5 to 125 °C. The conductivity of the ionic liquid was measured at various temperatures by a conductivity meter (HM-30R, DKK-TOA Corporation).

Hayamizu et al. have succeeded in measuring the self diffusion coefficients (D) values of the solvent molecule, the cation, the anion, and the polymer molecule in a PEO-type gel electrolyte system by the method of pulsed gradient spin-echo multinuclear NMR (PGSE-NMR) [31, 32, 33]. We also carried out PGSE-NMR measurements in order to evaluate the D value of the poly-cation and counter anion (TFSI) in a relatively high viscosity poly(DEMM-TFSI) solution in acetonitrile. The polymer concentration of the solution used for the measurement is 20 percent by weight, a value that far exceeded the overlap concentration, C*, where the polycation chains can contact and entangle with each other. So, the movement of the polymer chain was suppressed in this concentration region, hence it can be said that the environment was in a gel-like state.

Fig. 10 represents the NMR echo signal attenuation of a poly(DEMM-TFSI) solution in acetonitrile at 25 °C probed by ^{1}H and ^{19}F nuclear signals attributed to the methyl proton on the ammonium cation and acetonitrile and the TFSI anion. The attenuation due to free diffusion in the Stejskal and Tanner sequence using half-sine-shaped gradient pulses is given by

$$E = S / S_0 = \exp(-\gamma^2 g^2 \delta^2 D(4\Delta - \delta) / \pi^2) \qquad (1)$$

where γ is the gyromagnetic ratio, S is the amplitude of the echo signal and S_0 is the amplitude where $g = 0$, g is the amplitude of the gradient pulse, δ is the duration of the gradient pulse, and Δ is the interval between the gradient pulses. Thus, D could be determined from the slope of a plot of ln E against varying g. In the present experiments, the maximum g value was 13.5 T/m, the Δ was set to 10 ms, and the δ values varied in the range of 0.1 to 3 ms. The self-diffusion coefficient D of the poly-cation, anion, and solvent molecules are 1.6×10^{-11}, 4.5×10^{-10} and 2.8×10^{-9} m^2/s, respectively. The D value of the anion molecule was about 1/6 of that of the solvent molecule. However, one might think that the anion has the same degree of mobility as the solvent molecule because it has a several times larger molecular volume than the solvent molecule. That is to say, the anion can move like a solvent molecule in a very high concentration solution of poly-cation without suffering a strong restraint due to electrostatic effects. The anion seems to be able to escape from the restraint of the poly-cation even in a very high concentration poly-cation solution.

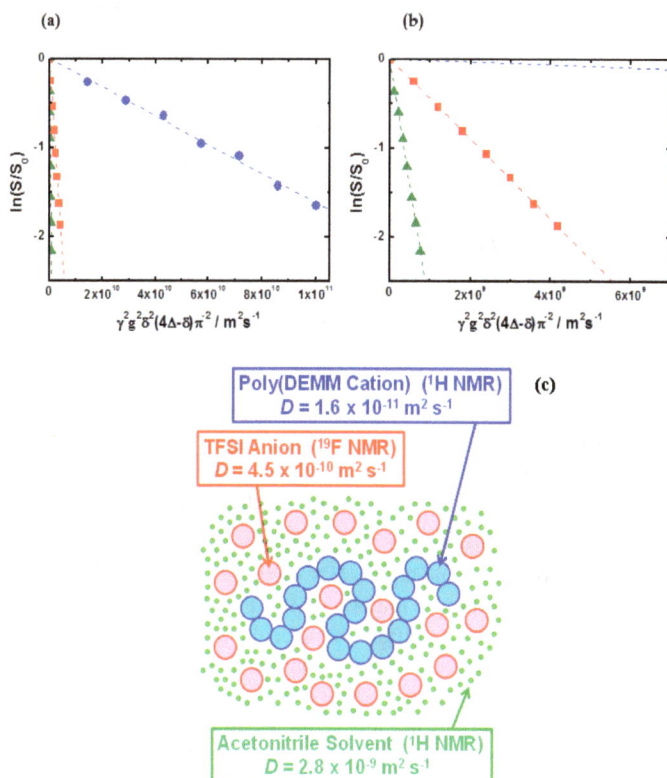

Figure 10. (a) PGSE attenuation plots for the single N-CH$_3$ of [Poly(DEMM-TFSI)] and the CH$_3$ signal of acetonitrile probed by the ^{1}H nucleus and the TFSI anion probed by ^{19}F at 25 °C obtained by varying δ at different g values for Δ = 10 ms. (b) Maginified plots area. (c) Self-diffusion coefficient values in high concentration poly(DEMM-TFSI) in acetonitorile solution.

5. High rate performance of a lithium polymer battery using an ionic liquid polymer composite

Some investigators have attempted to develop a non-flammable polymer electrolyte system; we have developed a polymer-gel electrolyte system consisting of a lithium salt in an ionic liquid and poly(ILs), which we have called a *LILP* composite system (Li-salt + ionic liquids + poly(ILs)). Generally, a binary Li-IL, specifically, a lithium salt dissolved in an ionic liquid having the same anion are used for *LILP* system. Several *LILP* systems with conductivities over 10^{-3} S cm^{-1} at room temperature have been developed [34, 25, 17]. However, since the binary Li-IL have a considerably high viscosity, the cell containing such liquids has a poor charge / discharge performance at a relatively large current, namely a lower power density, compared to conventional cells using the flammable organic solvent. And, there have been few reports of the performance of Li ion cells incorporating a *LILP* system. We are aiming to develop a truly safe Li ion polymer cell with a good charge and discharge performance at a large current, we discovered that the choices of $LiMn_2O_4$ and $Li_4Ti_5O_{12}$ [36] as, respectively, the cathode and anode active material, produced a faster charge / discharge reaction than conventional $LiCoO_2$ and graphite systems, and we combined these active materials and poly(DEMM-TFSI), which is very compatible with ionic liquids, as a novel *LILP* system. Our novel Li polymer cell has the following structure: negative electrode: $Li_4Ti_5O_{12}$/*LILP*-including ultrahigh molecular weight ionic liquid polymer / positive electrode: $LiMn_2O_4$. In this section, we will discuss the performance of this cell.

Poly (DEMM-TFSI) has the capability to dissolve a lithium salt independent of the presence of a liquid electrolyte. In a polarizing microscope analysis of Poly(DEMM-TFSI) which contained 1M concentration of dissolved lithium salt $LiClO_4$, we did not observe the birefringence indicating the existence of crystals. Thus, there must have been a complete dissociation of the lithium salt in the polymer matrix as the polymer itself has a non-crystalline nature. DSC measurements indicate that prepared poly(DEMM-TFSI) has a glass transition temperature (T_g) of approximately 50 °C. Thus, a poly(DEMM-TFSI) / lithium salt composite could potentially serve as an all-polymer electrolyte at temperatures over 50 °C.

Moreover, poly(DEMM-TFSI) dissolves in a variety of quaternary ammonium ionic liquids to make a gel. For example, the ionic liquid, *N*,*N*-diethyl-*N*-(2-methoxyethyl)-*N*-methylammonium bis(trifluoro- methylsulfonyl) imide (DEME-TFSI, molecular structure is indicated in Fig. 11.) containing only 5% of the ultrahigh molecular weight poly(DEMM-TFSI) lost its liquid characteristics and became a gel. It seems that the strong cohesiveness and loss of liquidity appear because of the entanglement effect of long polymer chains. When we adjusted the solution to a suitable viscosity by adding a supplementary solvent, in our case a propylene carbonate (PC) and vinylene carbonate (VC) mixture, it filled the pore space in the electrode and the separator.

The ionic liquid/poly(ILs) composite from which the solvent is removed by vacuum evaporation at a relatively high temperature has no components that leak out of the electrode. However, the high-polarity *LILP* matrix interacted with the supplementary solvent (PC + VC) and probably obstructed perfect evaporation of added solvents. The weight percentage of remained solvent in the composite was 1.7 wt%. We reported that the vinylene carbonate (10 wt%) in the electrolyte Li-DEME-TFSI, composed of an ionic liquid, DEME-TFSI and Li-TFSI was effective as solid electrolyte interface (SEI) forming additives on the carbon materials such as graphite used as an active material in the anode of a lithium ion cell [9]. Holzapfel and co-workers reported that 2% VC to ionic liquid, EMI-TFSI, contributed to the SEI formation, although the effect was not perfect [37]. Also in this case, there will be a possibility that the remained VC in composite contributes to the SEI formation. However, we realized that the effective SEI to prevent capacitance deterioration with charge/discharge cycles was not formed by the too small amounts (1.7 wt%) of carbonate solvents in *LILP* composite from the cycling behavior of the cell. A highly reliable polymer battery is not easily obtained when the polymerization is carried out in the battery bag. In such an "in-situ radical polymerization", some initiator and unreacted monomer may remain in the polymer matrix. However, our process makes possible the preparation of an electrolyte with few impurities through the use of the purified polymer combined with the binary Li-IL.

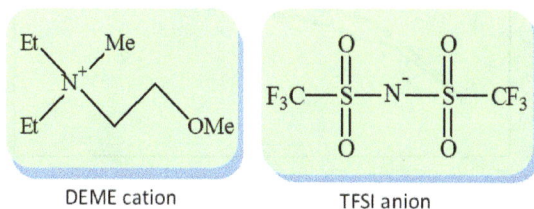

DEME cation   TFSI anion

Figure 11. Molecular structure of ionic liquid, DEME-TFSI

Fig. 12 shows both the limiting reduction ($E_{red.}$) and oxidation ($E_{oxd.}$) potentials of the poly(DEMM-TFSI) in PC as measured by cyclic voltammetry at room temperature. The $E_{red.}$ of the poly(DEMM-TFSI) can definitely be seen at around 2.0–2.5V positive relative to Li/Li +. However, its presence is not clear, and the current density is small because, we carried out the measurement in rather dilute conditions to avoid turbulence caused by an increased viscosity in the more concentrated polymer solution. In addition, the $E_{red.}$ of the DEME based ionic liquids was merely somewhat positive against the Li/Li+ [9]. Thus, we realized the need either to select an electrode that would avoid cathodic decomposition during the charge–discharge cycling, or to form a more effective protective layer, such as a solid electrolyte interface (SEI), on the negative electrode material.

We selected high power active electrode materials, and combined them with a *LILP* system and prepared two type vapor-free lithium ion polymer cells for demonstration purposes. The negative $Li_4Ti_5O_{12}$ electrodes (AKO-6), had a charge capacity of 0.42 mAh cm^{-2}, an area density of 3.00 mg cm^{-2}, and had an active electrode layer 25 micron-m thick on a copper foil. The other negative electrode used was a hard carbon electrode (AKT-2) with a charge capacity of 1.29 mAh cm^{-2}, an area density of 2.96 mg cm^{-2}, and a 33 micron-m in thick electrode active layer on a copper foil. The first positive electrode, a LiMn$_2$O$_4$ electrode (CKT-22) that was paired with the AKO-6 negative electrode, had a charge capacity of 0.49 mAh cm^{-2}, an area density of 6.60 mg cm^{-2} and an electrode active layer 36 -37 micron-m in thickness on an aluminum foil. The second LiMn$_2$O$_4$ electrode (CKT-9), paired with the AKT-2 had a charge capacity of 1.075 mAh cm^{-2}, an area density of 12.51 mg cm^{-2}, and a 65 micron-m thick electrode active layer on an aluminum foil. The specification of the electrodes are summarized in Table 1. The details of the preparation method have been described in our previous paper [9].

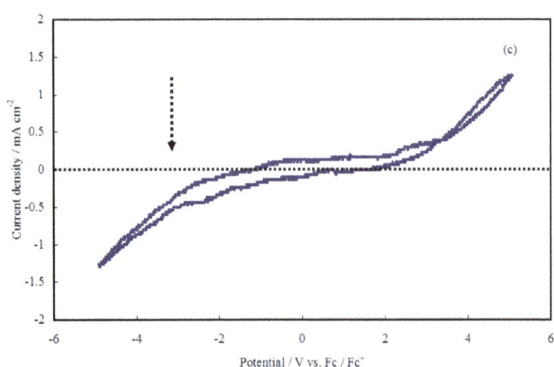

Figure 12. Cyclic voltammogram of poly(DEMM-TFSI) in PC at 25 °C. Scan rate: 10mVs^{-1}; platinum working and counter electrodes; Ag/Ag$^+$ reference electrode. The potential value (V) was referenced to the ferrocene (Fc)/ferrocenium (Fc$^+$) redox couple in each salt. Polymer concentration, 0.002 mol kg^{-1} poly(DEMMTFSI) in PC solution. The concentration is defined in terms of molality = (mol solute/kg solvent). The dashed line arrow indicates the potential of the Li/Li$^+$ couple.

Code	Active material	Polarity	Charge capacity (mAh cm^{-2})	Area density (mg cm^{-2})	Active layer thickness (μm)	Current collector/(μm)
AKO-6	Li^{4}Ti^{5}O12	Negative	0.42	3.00	25	Copper/13
CKT-22	LiMn$_2$O4	Positive	0.49	6.60	36-37	Aluminium/20
AKT-2	Hard carbon	Negative	1.29	2.96	33	Copper/13
CKT-9	LiMn$_2$O^4	Positive	1.075	12.51	65	Aluminium/20

Table 1. Electrode specification.

In Fig. 13, we show the first and second charge–discharge potential curves at 40 °C of the demonstration cells, consisting of hard carbon/*LILP* (1.3 mol of LiTFSI dissolved in poly(DEMM-TFSI)/ DEME-TFSI composite, giving a polymer concentration of 5.4 wt%)/

$LiMn_2O_4$. About 26% of the charge capacity was lost in the first cycle; however, from that point on, the cell exhibited an efficiency of 96% or more.

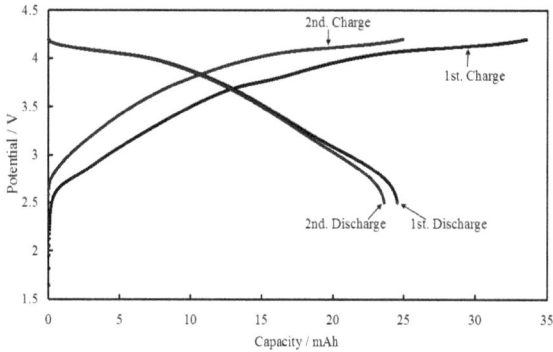

Figure 13. First and second charge/discharge curves of lithium polymer cell at 0.05C current at 40 °C. Positive electrode = $LiMn_2O_4$; negative electrode = hard carbon; electrolyte = 1.30 mol kg^{-1} of LiTFSI in a DEME-TFSI and poly(DEMM-TFSI) composite. The concentration is defined in terms of molality = (mol solute/kg polymer-ionic liquid composite). The polymer concentration for the composite electrolyte was 5.4 wt%.

The rate capability of this cell at 40 °C appears in Fig. 14. As the discharge current increased, the discharge capacity of this cell decreased significantly faster than that of a cell using conventional materials. The capacity at 1C discharge was approximately 63% of that at 0.1C discharge. Most likely, the greater decrease at large discharge currents in the capacity of the cell using the *LILP* electrolyte resulted from a large internal resistance in the cell. The cycling behavior of this cell, plotted in Fig. 15, was not suitable for practical use. When we set the upper limit voltage at 4.2V, the cell deteriorated sooner than with a voltage of 4.0V. The CV measurements suggested that degradation of the polymer or ionic liquid was occurring during the charge–discharge cycles. It is necessary to establish a method of forming an effective SEI if a practicable cycle performance is to be achieved.

We then prepared another type of lithium polymer cell, comprising $Li_4Ti_5O_{12}$ / *LILP* (1.3 mol of LiTFSI dissolved in poly(DEMM-TFSI) / DEME-TFSI composite, 5.4 wt% polymer concentration) / $LiMn_2O_4$. Because the intercalation-deintercalation potential of $Li_4Ti_5O_{12}$ is around 1.5V vs. Li / Li^+ potential, our prepared cell had about 3.0 V of charging potential with $LiMn_2O_4$. The overall cell reactions can be described as follows:

negative electrode,

$$Li[Li_{1/3}Ti_{5/3}]O_4 + xLi^+ + xe^- \Leftrightarrow Li_{1+x}[Li_{1/3}Ti_{5/3}]O_4 \tag{2}$$

positive electrode,

$$LiMn_2O_4 \Leftrightarrow Li_{1-x}Mn_2O_4 + xLi^+ + xe^- \qquad (3)$$

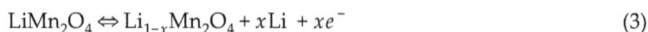

overall reaction,

$$Li[Li_{1/3}Ti_{5/3}]O_4 + LiMn_2O_4 \Leftrightarrow Li_{1+x}[Li_{1/3}Ti_{5/3}]O_4 + Li_{1-x}Mn_2O_4 \qquad (4)$$

The theoretical capacity of $Li_4Ti_5O_{12}$ was expected to be approximately 175 mAh g^{-1}. In our electrode, the discharge capacity was 171 mAh g^{-1}. The performances of the prepared cell are shown in Figs. 16-18.

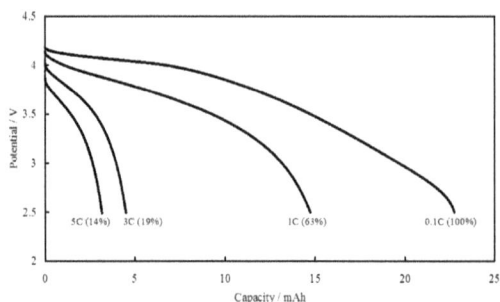

Figure 14. Discharge curves of the lithium polymer cell at various current densities at 40 °C. The positive electrode = LiMn$_2$O$_4$; negative electrode = hard carbon; electrolyte = 1.30 mol kg^{-1} of LiTFSI in DEME-TFSI and poly(DEMM-TFSI) composite. The concentration is defined in terms of molality = (mol solute/kg polymer-ionic liquid composite). The polymer concentration for the composite electrolyte was 5.4 wt%.

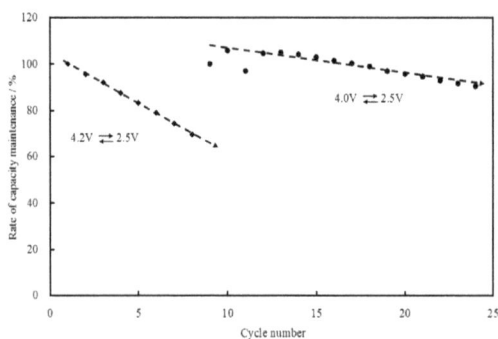

Figure 15. Cycle life of lithium polymer cell including hard carbon as negative active material. The charge–discharge process was performed at 0.1C at 40 °C. The cut-off voltages were 4.2V and 2.5V for nine cycles, and 4.0V and 2.5V from 10th to 24th cycle.

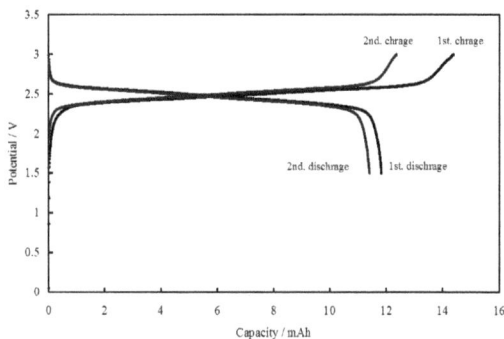

Figure 16. First and second charge/discharge curves of lithium polymer cell at 0.05 C current at 40 °C. The positive electrode = $LiMn_2O_4$; negative electrode = $Li_4Ti_5O_{12}$; electrolyte = 1.30 mol kg^{-1} of LiTFSI in DEME-TFSI and poly(DEMM-TFSI) composite. The concentration is defined in terms of molality = (mol solute/kg polymer-ionic liquid composite). The polymer concentration for the composite electrolyte was 5.4 wt%.

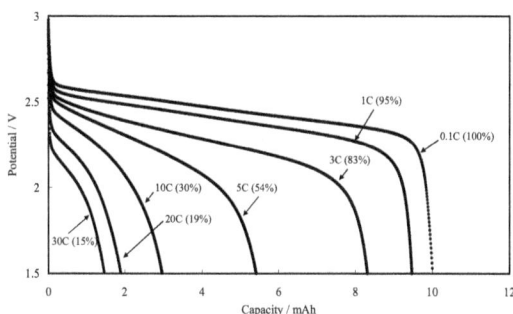

Figure 17. Discharge curves of lithium polymer cell at various current densities at 40 °C. The positive electrode = $LiMn_2O_4$; negative electrode = $Li_4Ti_5O_{12}$; electrolyte = 1.30 mol kg^{-1} of LiTFSI in DEME-TFSI and poly(DEMM-TFSI) composite. The concentration is defined in terms of molality = (mol solute/kg polymer-ionic liquid composite). The polymer concentration for the composite electrolyte was 5.4 wt%.

The discharge specific capacities and Columbic efficiency of the cell at the 1st was 11.8 mAh, and 82 %; at the second cycle, they were 11.4 mAh, and 92 %. After the 5th cycle, the Columbic efficiency of this cell remained at approximately 97-98 %.

The battery in Fig. 17 that combined the $Li_4Ti_5O_{12}$ anode with the $LiMn_2O_4$ cathode showed an excellent rate discharge character as for a lithium polymer battery. Evidently, this battery retains 83 % or more of the capacity maintenance rate at a 3C high power discharge. Thus, it was possible to create a new, leak-free battery with a vapor-free, practical discharge performance, and a prismatic cell design by selecting an electrode material with a high speed charge/discharge reaction when combined with an *LILP* system. The rate performances of the $Li_4Ti_5O_{12}$ negative cell and the hard carbon negative cells differ greatly, even though the electrolyte in each is the same *LILP* system. The interfacial compatibility of the active material and the electrolyte seem to have a bigger influence on the rate performance than the bulk ionic conductivity of the electrolyte itself. The cycling behavior of the cell, plotted in Fig. 18, indicated a good cycle durability, almost equivalent to that of a conventional lithium ion cell.

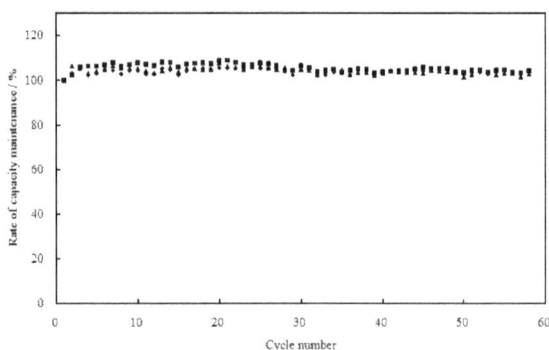

Figure 18. Cycle life of lithium polymer cell including $Li_4Ti_5O_{12}$ as negative active material. The charge- discharge process was performed at 0.1C at 40 °C. The cut off voltages were 3.0V and 1.5V. Three cells were tested.

As in the case of the hard carbon electrode, we expected that electrochemical degradation of the DEME-TFSI or poly(DEMM-TFSI) probably occurred. In our previous study, we suggested that some kinds of organic solvents, such as VC and ethylene carbonate, are effective as *SEI*-forming additives on the graphite used as an active material in the anode of a lithium ion cell with a binary Li-*IL* electrolyte. In this study, we did use a small amount of VC as a dilution solvent; however, because almost all the VC evaporates from the battery in the process of establishing the composite, it appears the small amounts of remained VC did not have large contribution to the SEI formation.

On the other hand, however, we should point out that the use of $Li_4Ti_5O_{12}$ as the negative electrode for the lithium-ion *LILP* cell avoids cathodic decomposition of the *LILP* composite during the charge/discharge cycling.

6. Conclusion

This chapter has reported the synthesis and physicochemcal properties of a quaternary ammonium polymerizable ionic liquid and of the corresponding polymer. In conclusion we find;

1. The polymerizable ionic liquid, DEMM-TFSI having a methacryl functional group as part of the ionic liquid's cation species has a remarkably low glass transition temperature (T_g = -68 °C), a relatively high ionic conductivity (0.6 mS/cm at 25 °C) and relatively wide potential windows.

2. The DEMM-TFSI monomer has a high reactivity for radical polymerization. We could prepare poly(DEMM-TFSI) with an ultra high molecular weight, with an M_w of over one million. However, in many cases, control of the molecular weight has been difficult because of the highly reactive nature of these monomers.

3. We succeeded in achieving molecular weight control by the radical polymerization of the ionic liquid monomer via three methods; atom transfer radical polymerization (ATRP), chain transfer radical polymerization (CTRP) and reversible addition-fragmentation chain transfer polymerization (RAFT). The control of the molecular weight occurred extremely easily in CTRP because of the addition of 3-mercapto-1-hexanol as a chain transfer reagent. In RAFT polymerization, molecular weight controlled poly(IL) with a low polydispersity index can be prepared.

4. A poly-cation type material such as poly(DEMM-TFSI) has bulky tetra-alkyl ammonium functional groups in each unit and has a relatively low T_g value due to it having a large free volume due to the effect of electrostatic repulsions of cationic groups. A polymer with a molecular weight that exceeds 100,000 has roughly a T_g of 50 °C while, on the other hand, a polymer with M_n about 10,000 had a 20 degrees lower T_g of 30 °C, presumably due to the chain ends effect.

5. The bulk poly(DEMM-TFSI) has a relatively low ionic conductivity that is not practical for use in electrochemical devices at room temperature. However, the temperature dependence of the ionic conductivity showed an interesting property, unlike that of a nonionic polymer electrolyte, such as polyethylene oxide. Ionic conduction also appeared below the glass transition temperature in the polymer matrix. The anion seemed to show ionic conduction, independent of the segmental motion of the polymer chain. A new ionic conduction mechanism in the poly-cation matrix might be discovered by a further research.

6. We measured the self-diffusion coefficient value D of the poly-cation, anion and solvent molecule in a high concentration solution of poly(DEMM-TFSI) by PGSE-NMR. It was clarified that the anion had a D value that was comparatively close to that of the solvent molecule. We think that the anion has a high D value by escaping from a strong interaction with the poly-cation.

7. We obtained a polymer electrolyte by combining a small amount of the poly(ILs) and a binary Li-IL. The system of a lithium salt dissolved in an ionic liquid polymer and ionic liquid composite (*LILP*) could potentially comprise an electrolyte having zero vapor pressure. The discharge performance of a cell with an *LILP* system might be expected to show a poor discharge performance, because the solidified ionic liquid decreases the mobility of the ionic species. However, we have conceived a method of making a polymer battery with a practicable performance by combining electrodes that can offer a

high-speed charge/discharge reaction. The experimental battery that combined a $Li_4Ti_5O_{12}$ anode with a $LiMn_2O_4$ cathode and included an LILP electrolyte showed an excellent rate discharge character for a lithium polymer battery; at a 3C current rate, it retained 83% of its discharge capacity, and relatively good cycle performance. A lithium ion cell with a LILP system performed, in terms of cell performance and cycle durability, at a level of practical utility. This novel lithium polymer cell, non-flammable and leak-free, is a promising candidate as a safe, large size lithium secondary battery.

Acknowledgements

The authors thank Ms. Shoko Marukane and Mrs. Saika Honma at Tsuruoka National College of Technology for various assistance in the lithium polymer cell preparation and measurements.

Author details

Takaya Sato, Takashi Morinaga and Takeo Ishizuka

*Address all correspondence to: takayasa@tsuruoka-nct.ac.jp

Tsuruoka National College of Technology, Japan

References

[1] Rogers, R. D., & Seddon, K. R. (2001). Ionic Liquids: Industrial Applications for Green Chemistry (Acs Symposium Series). *American Chemical Society*, 0-84123-789-1.

[2] Fung, Y. S. Room temperature molten salt as medium for lithium battery and alloy electrodeposition- fundamental and application. Trends Inorg Chem. (1998).

[3] Fung, Y. S., & Zhou, R. Q. (1999). Room temperature molten salt as medium for lithium battery. *J. Power Sources*, 81.

[4] Caja, T. D. J., Dunstan, D. M., & Ryan, V. (2000). P.C. Trulove (Ed.). Molten Salts XII. *Pennington, NJ: Electrochem. Soc.*, 150.

[5] Fung, Y. S., & Zhu, D. R. (2002). Electrodeposited tin coating as negative electrode material for lithium-ion battery in room temperature molten salt. *J. Electrochem. Soc.*, 149, A319.

[6] Nakagawa, H., Izuchi, S., Kuwana, K., & Aihara, Y. (2003). Liquid and polymer gel electrolytes for lithium batteries composed of room-temperature molten salt doped by lithium salt. *J. Electrochem. Soc.*, 150, A695.

[7] Sakaebe, H., & Matsumoto, H. (2003). N-Methyl-N-propylpiperidinium bis(trifluoro-methanesulfonyl)imide (PP13-TFSI)-novel electrolyte base for Li battery. *Electrochem. Commun.*, 5, 594.

[8] Shin, J.H., Henderson, W.A., Henderson, W. A., & Passerini, S. (2003). Ionic liquids to the rescue? *Overcoming the ionic conductivity limitations of polymer electrolytes. Electrochem. Commun.*, 5.

[9] Sato, T., Maruo, T., Marukane, S., & Takagi, K. (2004). Ionic liquids containing carbonate solvent as electrolytes for lithium ion cells. *J. Power Sources*, 138.

[10] Mc Ewen, A. B., Mc Devitt, S. F., & Koch, V. R. (1997). Nonaqueous electrolytes for electrochemical capacitors: imidazolium cations and inorganic fluorides with organic carbonates. *J. Electrochem. Soc.*, 144, L84.

[11] Mc Ewen, A. B., Ngo, H. L., Le Compte, K., & Goldman, J. L. (1999). Electrochemical properties of imidazolium salt electrolytes for electrochemical capacitor applications. *J. Electrochem. Soc.*, 146.

[12] Ue, M., & Takeda, M. (2002). Application of ionic liquids based on 1-ethyl-3-methylimidazolium cation and fluoroanions to double-layer capacitors. *J. Korean Electrochem. Soc.*, 5.

[13] Ue, M., Takeda, M., Toriumi, A., Kominato, A., Hagiwara, R., & Ito, Y. (2003). Application of low-viscosity ionic liquid to the electrolyte of double-layer capacitors. *J. Electrochem. Soc.*, 150, A499.

[14] Sato, T., Masuda, G., & Takagi, K. (2004). Electrochemical properties of novel ionic liquids for electric double layer capacitor applications. *Electrochim. Acta*, 49(2004), 3603-3611.

[15] Ohno, H., Yoshizawa, M., & Ogiwara, W. (2004). Development of new class of ion conductive polymers based on ionic liquid. *Electrochim. Acta*, 50(2004), 255-261.

[16] Nakajima, H., & Ohno, H. (2005). Preparation of thermally stable polymer electrolytes from imidazolium-type ionic liquid derivatives. *Polymer*, 46(2005), 11499-11504.

[17] Ogiwara, W., Washiro, S., Nakajima, H., & Ohno, H. (2006). Effect of cation structure on the electrochemical and thermal properties of ion conductive polymers obtained from polymerizable ionic liquids. *Electrochim. Acta*, 51(2006), 2614-2619.

[18] Lu, J., Yan, F., & Texter, J. (2009). Advanced applications of ionic liquids in polymer science. *J. Prog. Polym. Sci.*, 34(2009), 431-448.

[19] Mecerreyes, D. (2011). Polymeric ionic liquids: Broadening the properties and applications of polyelectrolytes. *J. Prog. Polym. Sci.*, 36-1629.

[20] Yuan, J., & Antonietti, M. (2011). Poly(ionic liquid)s: Polymers expanding classical property profiles. *polymer*, 52-1469.

[21] Benton, M. G., & Brazel, C. S. (2004). An Investigation of the Degree and Rate of Polymerization of Poly (methyl methacrylate) in the Ionic Liquid 1-Butyl-3- Methylimidazolium Hexafluorophosphate. *Polym. Int.*, 53.

[22] Sato, T., Marukane, S., Narutomi, T., & Akao, T. High rate performance of a lithium polymer battery using a novel ionic liquid polymer composite. *J. Power Sources*, 164(20042007), 390-396.

[23] Yu-H, Chang., Pei-Y, Lin., Ming-S, Wu., & King-F, Lin. (2012). Extraordinary aspects of bromo-functionalized multi-walled carbon nanotubes as initiator for polymerization of ionic liquid monomers. *Polymer*, 53(2012), 2008-2014.

[24] Sato, T., Morinaga, T., Marukane, S., Narutomi, T., Igarashi, T., Kawano, Y., Ohno, K., Fukuda, T., & Tsujii, Y. (2011). Novel solid-state polymer electrolyte of colloidal crystal decorated with ionic-liquid polymer brush. *Adv. Mater.*, 23(2011), 4868-4872.

[25] Appetecchi, G. B., Kim, G.T., montanino, M., Carewska, M., Marcilla, R., Mecerreyes, D., & De Meatza, I. (2010). Ternary polymer electrolytes containing pyrrolidinium-based polymeric ionic liquids for lithium batteries. *J. Power Sources*, 195(2010), 3668-3675.

[26] Kubisa, P. (2004). Application of ionic liquids as solvents for polymerization processes. *Prog. Polym. Sci.*, 29(2004), 3-12.

[27] Mori, H., Yahagi, M., & Endo, T. (2009). RAFT Polymerization of N-vinylimidazolium salt and synthesis of thermoresponsive ionic liquid block copolymers. *Macromolecules*, 42-8082.

[28] Santangelo, P. G., & Roland, C. M. (1998). Molecular weight dependence of fragility in polystyrene. *Macromolecules*, 31(1998), 4581-4585.

[29] Claudy, P., Letoffe, J. M., Camberlain, Y., & Pascault, J. P. (1983). Glass transition of polystyrene versus molecular weight. *Polym. Bull.*, 9(1983), 208-215.

[30] Teran, A. A., Tang, M. H., Mullin, S. A., & Balsara, N. P. (2011). Effect of molecular weight on conductivity of polymer electrolytes. *Solid State Ionics*, 203(2011), 18-21.

[31] Hayamizu, K. (2001). *Nihon-Denshi News*, 33(1), 6-10, http://www.jeol.co.jp/publication/ nihondenshi/j_backnumber/33/j33_all.pdf, (accessed 9 Jun 2012).

[32] Stejskal, E. O., & Tanner, J. E. (1965). Spin diffusion measurements: spin echoes in the presence of a time-dependent field gradient. *J. Chem. Phys.*, 42-288.

[33] Hayamizu, K. On Accurate Measurements of Diffusion Coefficients by PGSE NMR Methods -room- themperature ionic liquids- http://www.jeolusa.com/DesktopModules/Bring2mind/DMX/Download.aspx?EntryId=713&Command=Core_Download&PortalId=2&TabId=337 (accessed 9 Jun) (2012).

[34] Fuller, J., Breda, A. C., & Richard, R. T. (1998). Ionic liquid-polymer gel electrolytes from hydrophilic and hydrophobic ionic liquids. *J. Electroanal. Chem.*, 459.

[35] Abu-Bin, M., Hasan, S. T., Kaneko, T., Noda, A., & Watanabe, M. (2005). Ion Gels Prepared by in Situ Radical Polymerization of Vinyl Monomers in an Ionic Liquid and Their Characterization as Polymer Electrolytes. *J. Am. Chem. Soc.*, 127, 4876.

[36] Amine, K., Liu, J., Belharouak, I., & Park, S. H. (2006). Proceedings of advanced technology development review meeting at Sandia National Laboratory. *Carlsbad, NM, USA.*

[37] Holzapfel, M., Jost, C., & Nov'ak, P. (2004). Stable cycling of graphite in an ionic liquid based electrolyte. *Chem. Commun.*

Permissions

The contributors of this book come from diverse backgrounds, making this book a truly international effort. This book will bring forth new frontiers with its revolutionizing research information and detailed analysis of the nascent developments around the world.

We would like to thank Prof. Dr. Jun-ichi Kadokawa, for lending his expertise to make the book truly unique. He has played a crucial role in the development of this book. Without his invaluable contribution this book wouldn't have been possible. He has made vital efforts to compile up to date information on the varied aspects of this subject to make this book a valuable addition to the collection of many professionals and students.

This book was conceptualized with the vision of imparting up-to-date information and advanced data in this field. To ensure the same, a matchless editorial board was set up. Every individual on the board went through rigorous rounds of assessment to prove their worth. After which they invested a large part of their time researching and compiling the most relevant data for our readers. Conferences and sessions were held from time to time between the editorial board and the contributing authors to present the data in the most comprehensible form. The editorial team has worked tirelessly to provide valuable and valid information to help people across the globe.

Every chapter published in this book has been scrutinized by our experts. Their significance has been extensively debated. The topics covered herein carry significant findings which will fuel the growth of the discipline. They may even be implemented as practical applications or may be referred to as a beginning point for another development. Chapters in this book were first published by InTech; hereby published with permission under the Creative Commons Attribution License or equivalent.

The editorial board has been involved in producing this book since its inception. They have spent rigorous hours researching and exploring the diverse topics which have resulted in the successful publishing of this book. They have passed on their knowledge of decades through this book. To expedite this challenging task, the publisher supported the team at every step. A small team of assistant editors was also appointed to further simplify the editing procedure and attain best results for the readers.

Our editorial team has been hand-picked from every corner of the world. Their multi-ethnicity adds dynamic inputs to the discussions which result in innovative outcomes. These outcomes are then further discussed with the researchers and contributors who give their valuable feedback and opinion regarding the same. The feedback is then collaborated with the researches and they are edited in a comprehensive manner to aid the understanding of the subject.

Apart from the editorial board, the designing team has also invested a significant amount of their time in understanding the subject and creating the most relevant covers. They scrutinized every image to scout for the most suitable representation of the subject and create an appropriate cover for the book.

The publishing team has been involved in this book since its early stages. They were actively engaged in every process, be it collecting the data, connecting with the contributors or procuring relevant information. The team has been an ardent support to the editorial, designing and production team. Their endless efforts to recruit the best for this project, has resulted in the accomplishment of this book. They are a veteran in the field of academics and their pool of knowledge is as vast as their experience in printing. Their expertise and guidance has proved useful at every step. Their uncompromising quality standards have made this book an exceptional effort. Their encouragement from time to time has been an inspiration for everyone.

The publisher and the editorial board hope that this book will prove to be a valuable piece of knowledge for researchers, students, practitioners and scholars across the globe.

List of Contributors

I. Cota and F. Medina
Departament d'Enginyeria Química, Escola Tècnica Superior d'Enginyeria Química, Universitat Rovira I Virgili, Avinguda Països Catalans 26, Campus Sescelades, 43007 Tarragona, Spain

R. Gonzalez-Olmos
Laboratory of Chemical and Environmental Engineering (LEQUiA), Institute of the Environment, University of Girona, Campus Montilivi s/n, Faculty of Sciences, E-17071 Girona, Spain

M. Iglesias
Departamento de Engenharia Química, Escola Politécnica, Universidade Federal da Bahia, 40210-630 Salvador-Bahia, Brazil

Liangfang Zhu and Changwei Hu
Key Laboratory of Green Chemistry and Technology, Ministry of Education, College of Chemistry, Sichuan University, Chengdu, P.R. China

María N. Kneeteman
Área de Química Orgánica-Departamento de Química-Facultad de Ingeniería Química-Universidad Nacional del Litoral (UNL), Santa Fe, Argentina
Consejo Nacional de Investigaciones Científicas y Técnicas (CONICET), de la República Argentina

Luis R. Domingo
Departamento de Química Orgánica, Facultad de Química, Universidad de Valencia, España

Pedro M. E. Mancini, Carla M. Ormachea and Claudia D. Della Rosa
Área de Química Orgánica-Departamento de Química-Facultad de Ingeniería Química-Universidad Nacional del Litoral (UNL), Santa Fe, Argentina

Ahmed Al Otaibi and Adam McCluskey
Chemistry, School of Environmental & Life Sciences, The University of Newcastle, University Drive, Callaghan NSW, Australia

Marcos A. P. Martins, Izabelle M. Gindri, Aniele Z. Tier, Dayse N. Moreira, Jefferson Trindade Filho, Guilherme S. Caleffi, Lilian Buriol and Clarissa P. Frizzo
Department of Chemistry, NUQUIMHE, Federal University of Santa Maria, Brazil

Rohitkumar G. Gore and Nicholas Gathergood
School of Chemical Sciences and National Institute for Cellular Biotechnology, Dublin City University, Glasnevin, Dublin, Ireland

Ana P.M. Tavares, Oscar Rodríguez and Eugénia A. Macedo
LSRE - Laboratory of Separation and Reaction Engineering - Associate Laboratory LSRE/ LCM, Faculdade de Engenharia, Universidade do Porto, Porto, Portugal

Hidetaka Noritomi
Tokyo Metropolitan University Japan

Jun-ichi Kadokawa
Graduate School of Science and Engineering, Kagoshima University, Japan

Elaheh Kowsari
Department of Chemistry, Amirkabir University of Technology, Tehran, Iran

Susumu Kuwabata
Department of Applied Chemistry, Graduate School of Engineering, Osaka University, Japan
Japan Science and Technology Agency, CREST, Japan

Tsukasa Torimoto
Japan Science and Technology Agency, CREST, Japan
Department of Crystalline Material Sciences, Graduate School of Engineering, Nagoya University, Japan

Akihito Imanishi
Japan Science and Technology Agency, CREST, Japan
Department of Chemistry, Graduate School of Engineering Science, Osaka University, Japan Frontier Research Base for Global Young Researchers, Graduate School of Engineering, Japan

Tetsuya Tsuda
Department of Applied Chemistry, Graduate School of Engineering, Osaka University, Japan
Department of Chemistry, Graduate School of Engineering Science, Osaka University, Japan Frontier Research Base for Global Young Researchers, Graduate School of Engineering, Japan

Toshiro Kaneko, Shohei Takahashi and Rikizo Hatakeyama
Department of Electronic Engineering, Tohoku University, Sendai, Japan

Takaya Sato, Takashi Morinaga and Takeo Ishizuka
Tsuruoka National College of Technology, Japan

www.ingramcontent.com/pod-product-compliance
Lightning Source LLC
Chambersburg PA
CBHW070719190326
41458CB00004B/1030